AF806549

Advances in Biochemistry in Health and Disease

Volume 18

Series Editor

Naranjan S. Dhalla, Institute of Cardiovascular Sciences, St. Boniface Hospital, Winnipeg, MB, Canada

Editorial Board

Roberto Bolli, Medicine & Cardiology Department, University of Louisville, Louisville, KY, USA

Ramesh Goyal, Delhi Pharmaceutical Sciences & Research, New Delhi, India

Chandrasekharan Kartha, Cardiovascular Diseases and Diabetes Bio, Kerala Institute of Medical Sciences, Thiruvananthapuram, Kerala, India

Lorrie Kirshenbaum, St. Boniface General Hospital, Winnipeg, MB, Canada

Naoki Makino, Kyushu University, Fukuoka, Japan

Jawahar L. L. Mehta, Division of Cardiology, University of Arkansas for Medical Sciences, Little Rock, AR, USA

Bohuslav Ostadal, Institute of Physiology, Czech Academy of Sciences, Prague, Czech Republic

Grant N. Pierce, St. Boniface General Hospital, Winnipeg, MB, Canada

Jan Slezak, Institute for Heart Research, Slovak Academy of Sciences, Karlova Ves, Slovakia

Andras Varro, Pharmacology & Pharmacotherapy, University of Szeged, Szeged, Hungary

Karl Werdan, Martin Luther University Halle-Wittenber, Halle (Saale), Sachsen-Anhalt, Germany

William B. Weglicki, School of Medicine and Health Sciences, George Washington University, Washington, USA

Advances in Biochemistry in Health and Disease focus on the latest developments in biochemical research with implications for health and disease. This book series consists of original edited volumes and monographs, presented by leading experts in the field and provides an up to date and unique source of information for all those interested in the fundamental, biochemical processes of the latest and emerging topics and techniques.

Covering a wide variety of topics, this book series is a valuable source of information from those at the lab bench through to the Health Care workers.

More information about this series at https://link.springer.com/bookseries/7064

Lorrie A. Kirshenbaum
Editor

Biochemistry of Apoptosis and Autophagy

Springer

Editor
Lorrie A. Kirshenbaum
University of Manitoba
Winnipeg, MB, Canada

ISSN 2512-2142 ISSN 2512-2150 (electronic)
Advances in Biochemistry in Health and Disease
ISBN 978-3-030-78798-1 ISBN 978-3-030-78799-8 (eBook)
https://doi.org/10.1007/978-3-030-78799-8

© Springer Nature Switzerland AG 2022
This work is subject to copyright. All rights are reserved by the Publisher, whether the whole or part of the material is concerned, specifically the rights of translation, reprinting, reuse of illustrations, recitation, broadcasting, reproduction on microfilms or in any other physical way, and transmission or information storage and retrieval, electronic adaptation, computer software, or by similar or dissimilar methodology now known or hereafter developed.
The use of general descriptive names, registered names, trademarks, service marks, etc. in this publication does not imply, even in the absence of a specific statement, that such names are exempt from the relevant protective laws and regulations and therefore free for general use.
The publisher, the authors and the editors are safe to assume that the advice and information in this book are believed to be true and accurate at the date of publication. Neither the publisher nor the authors or the editors give a warranty, expressed or implied, with respect to the material contained herein or for any errors or omissions that may have been made. The publisher remains neutral with regard to jurisdictional claims in published maps and institutional affiliations.

This Springer imprint is published by the registered company Springer Nature Switzerland AG
The registered company address is: Gewerbestrasse 11, 6330 Cham, Switzerland

Dedication

This book is dedicated to the late Dr. Arnold H. Greenberg (1941–2001) for his exceptional leadership and ground-breaking discoveries in the areas of programmed cell death, immunology and cancer and unwavering steadfast mentorship. I considered Arnold to be not only an outstanding academician in the truest sense but also a role model for others who genuinely cared about helping people personally and professionally. Arnold's passion for science was truly infectious.

Arnold completed his Doctors of Medicine degree at the University of Manitoba in Winnipeg, Canada, followed by his Residency and Fellowships in Pediatrics at Johns Hopkins Hospital School of Medicine in Baltimore, Maryland. He later completed a Ph.D. in Immunology at the University of London in London, England. Upon returning to Winnipeg to join the Manitoba Institute of Cell Biology where he served as Director from 1988 to 1999. Arnold was a brilliant scientist. He was intelligent with a research acumen that was laser sharp, he remained always focused, and knew exactly what questions were important to ask to solve a given problem. His research achievements were truly remarkable. His pioneering work on programmed cell death contributed in a major way to advancing this emerging field. His many discoveries including Natural Killer T-cells, revealed for the first time a connection between the immune system and cancer. His work was consistently published in prestigious high impact journals such as *Science.* His landmark contributions were recognized at the highest level internationally by different scientific bodies through his many honors and awards. His other major discoveries included defining a role for Granzyme B and Porin as mediators of apoptotic cell death in cancer, which was soon to be followed by his later interest in a gene called Bnip3 for which both he and I shared a passion. As Director of the Manitoba Institute of Cell Biology, Arnold built a world class cancer research institute which is known worldwide and respected for the high caliber of its scientists and students it trained. On a personal note, I can attest that Arnold had a true gentleman, he had delightful soft-spoken personality and wit about him that was truly enviable. Despite being a complex man in many aspects of his life, Arnold loved simplicity, he exuded intelligence and wisdom without arrogance.

He excelled in everything he undertook; and was in constant pursuit of knowledge, he was an avid reader and a highly competitive sportsman, he loves playing tennis. Arnold was always generous with his time, and exhibited humility even throughout his battle with cancer. My friendship with Dr. Arnold Greenberg began more than 30 years ago, when I was a post-doctoral fellow at Baylor College of Medicine. My research interests were to explore mechanisms that underlie cell death in cardiac myocytes during heart failure and following myocardial ischemic injury. I recall contacting Arnold about my recruitment to the Institute of Cardiovascular Sciences at the University of Manitoba in the mid-1990s, expressing an interest to collaborate with him in this newly minted field. Arnold was elated my interest to explore cell death research in the heart and extended an opportunity to work and collaborate with his group. From that point onward, we had developed a great personal and professional relationship that often included lunches and dinners and many debates about the latest and greatest cell death publications. I fondly remember, my disappointment with what I believed at the time to be a failed experiment, only later to be enlightened by Arnold's sage explanation for the outcome. By analogy, he was Obi-Wan-Kenobi and I was Luke Skywalker. It was at this time, our passion for the study of programmed cell death further overlapped with the independent cloning of the Bnip3 gene by our laboratories. We worked together exchanging ideas and research directions on Bnip3 even during his illness. Arnold focused his attention on work and his family until the day before he died. He loved his wife Faye and his family Juliet, David, Marni, Rachel, and Kathy, and had deep admiration for his grandchildren. Arnold Greenberg was someone who I deeply admired; he was simply a wonderful human being.

Preface

Myocardial infarction is the leading cause of morbidity and mortality worldwide. In the context of the adult myocardium which exhibits a limited capacity for *de novo* myocyte regeneration after injury, the functional loss of cardiac cells leads to acute contractile dysfunction and as importantly, sets into motion a cascade of events that leads to adverse cardiac remodeling that ultimately manifest as heart failure. Cell death is a major component of this remodeling process. Perhaps one of the most compelling observations in the scientific literature over the past 40 years is the discovery that cell death is a highly regulated and predictable event. Indeed, the astute observations by Andrew Wylie and his team recognized the appearance of dysmorphic looking T-cells which exhibited features that were unlike other dying cells. By electron microscopy, the T-cells were pycnotic, the DNA was fragmented and marginalized to the periphery of the nucleus and unlike classical necrosis, the cell membrane was intact. These features have been utilized to discriminate between what is now referred to as programmed cell death of "apoptosis" which is derived from the Greek *apóptōsis*, or "falling off". Apoptosis is an important aspect of normal embryonic development as exemplified by the selective death of the interdigital spaces that give rise to the figures and toes as well as the involution of breast ductal tissue following pregnancy. There are many other examples where apoptosis plays a crucial physiological role which became more apparent after the discovery of critical cell death regulator genes in the worm *c. elegans*. Indeed, c.elegans provided a useful tool that led to the not only the discovery of certain genes that promoted cell survival but also genes that promoted cell death. Mammalian orthologues of cell death regulatory genes were soon discovered which resulted in an explosion in medicine which had implicated defects in the regulatory processes that govern cell death to be involved in some if not all aspects in the pathogenesis of human diseases such cancer, neurodegenerative diseases such as Parkinson's Disease, Alzheimer's Disease, Huntington Disease, and cardiovascular diseases. While the majority of cell death reported prior to the discovery of apoptosis advanced the concept that cell death was not only programmed but a highly regulated by the interplay between survival and death genes respectively, such as the B-cell Lymphoma (BCL) family. Indeed, discovery of Bcl-2 in more than 25 years ago as a translocation break-point mutation between chromosomes 8 and 4 in B-lymphoma was determined to conferred resistance of cancer cells to chemotherapeutic agents. This major break-through led to the subsequent discovery of the Bcl-2 gene family which can promote or prevent cell death in different cell types under different pathophysiological conditions by impinging on different cellular organelles such as mitochondria, endoplasmic reticulum, and nucleus to promote or prevent cell death. The discovery that apoptosis is a regulated event and potentially amenable to therapeutic interventions generated considerable excitement in the field, because it meant that disease entities resulting from either too little or too much apoptosis could be potentially cured with new therapies that modulate apoptotic cell death. The concept that cell death could be targeted therapeutically provided the impetus for developing new classes of agents that modulate cell death for treating human diseases.

In contrast, necrosis has been traditionally considered as an unregulated passive response to injury. While there is little doubt that necrosis induced by massive cellular trauma is likely an unregulated event, several lines of investigation including new exciting data from our laboratory have challenged this dogma and classical text-book definition that necrotic cell death is merely accidental and unregulated. This emerging and contemporary view is a paradigm shift in our thinking about how cell fate is regulated. In fact, the concept that necrotic death is regulated has tremendous implications for understanding the pathogenesis of diseases that were previously unexplored as well as developing novel therapies for conditions where necrosis is known to play a significant role.

In addition to apoptosis and necrosis, autophagy is another cellular process that has received considerable attention over the past decade for its remarkable ability to promote cell survival and cell death. Macro autophagy is a catabolic process that involves the selective and targeted removal of oxidized proteins, macromolecular structures, and organelles through an elaborate cellular process involving a lysosome mediated pathway. Defects in autophagy have been linked to a variety of diseases including Danon disease, Ischemia-reperfusion injury, cancer, and remodeling after myocardial infarction. A specialized form of autophagy that uniquely involves mito-chondrial clearance through a process called mitophagy plays an essential role in clearing damaged or superfluous mitochondria for maintaining mitochondrial quality control. Other forms of autophagy involving adapter proteins commonly referred to as chaperone mediated autophagy is a poorly understood process and involves the selective removal of cellular cargo by the ubiquitin-proteasome pathway.

The relationship between apoptosis, necrosis, and autophagy is profound and coupled to the mitochondria which serves as signaling platform for integrating biological signals such as hypoxia, nutrient stress, inflammation, and other cytotoxic stimuli. Hence there has been tremendous excitement surrounding the underlying mechanisms that regulate cell death and how these pathways can be manipulated therapeutically to mitigate human disease. Other regulated forms of cell death have been recently identified and include pyroptosis, ferroptosis, necroptosis, autosis, as well as others, many of these share overlapping features with apoptosis and necrosis respectfully. Understanding the biochemical signaling mechanisms that couple apop-tosis, necrosis, and autophagy are of great scientific and clinical importance toward the design of new therapies that will modulate the pathogenesis of human disease where excess cell death is the primary underlying defect. This book will provide new important insight into the biology of the mechanisms that regulate cell death and autophagy in the pathogenesis of human disease with a particular focus on cardiovascular system.

At this time, I would like to take the opportunity and express sincerest gratitude to all the authors and co-authors for their important contributors to this book which is greatly appreciated. I would also like to express my profound appreciation to Dr. Naranjan S. Dhalla for his continual enthusiastic support of this initiative as well as the staff at Springer Nature, in the preparation of this book. I would also like to acknowledge Ms. Shweta Sharma and Kairee Ryplanski for their diligence and administrative assistance in the preparation of this book.

Winnipeg, Canada Lorrie A. Kirshenbaum

Contents

Contributors

Kevin M. Alexander Stanford Cardiovascular Institute, Stanford University School of Medicine, Stanford, CA, USA

Allen M. Andres Smidt Heart Institute, Cedars-Sinai Medical Center, Los Angeles, CA, USA

Monika Bartekova Institute for Heart Research, Centre of Experimental Medicine, Slovak Academy of Sciences, Bratislava, Slovakia

Sukhwinder K. Bhullar Department of Physiology and Pathophysiology, Institute of Cardiovascular Sciences, Max Rady College of Medicine, St. Boniface Hospital Albrechtsen Research Centre, University of Manitoba, Winnipeg, Canada

Sikta Chattopadhyaya Department of Physiology and Pathophysiology, Rady Faculty of Health Sciences, University of Manitoba, Winnipeg, Canada

Michael P. Czubryt Department of Physiology and Pathophysiology, Rady Faculty of Health Sciences, University of Manitoba, Winnipeg, Canada;
Institute of Cardiovascular Sciences, St. Boniface Hospital Albrechtsen Research Centre, Winnipeg, MB, Canada

Naranjan S. Dhalla Institute of Cardiovascular Sciences, Research Centre, Department of Physiology and Pathophysiology, Max Rady College of Medicine, St. Boniface Hospital Albrechtsen, University of Manitoba, Winnipeg, Canada

Rimpy Dhingra Department of Physiology and Pathophysiology, The Institute of Cardiovascular Sciences, St. Boniface Hospital Albrechtsen Research Centre, Winnipeg, MB, Canada

Sanjiv Dhingra Department of Physiology and Pathophysiology, Rady Faculty of Health Sciences, St. Boniface Hospital Albrechtsen Research Centre, Institute of Cardiovascular Sciences, Max Rady College of Medicine, University of Manitoba, Winnipeg, Canada

Loren J. Field The Krannert Institute of Cardiology and the Wells Center for Pediatric Research, Indiana University School of Medicine, Indianapolis, USA

Jonathon Gerstein Department of Physiology and Pathophysiology, The Institute of Cardiovascular Sciences, St. Boniface Hospital Albrechtsen Research Centre, Winnipeg, MB, Canada

Roberta A. Gottlieb Smidt Heart Institute, Cedars-Sinai Medical Center, Los Angeles, CA, USA

Matthew Guberman Department of Physiology and Pathophysiology, The Institute of Cardiovascular Sciences, St. Boniface Hospital Albrechtsen Research Centre, Winnipeg, MB, Canada

Åsa B. Gustafsson Skaggs School of Pharmacy and Pharmaceutical Sciences, University of California San Diego, La Jolla, California, USA

Elissavet Kardami Institute of Cardiovascular Sciences, St. Boniface Albrechtsen Research Centre, Winnipeg, MB, Canada;
Department of Human Anatomy & Cell Sciences, University of Manitoba, Winnipeg, Canada;
Department of Physiology & Pathophysiology, University of Manitoba, Winnipeg, Canada

Lorrie A. Kirshenbaum Department of Physiology and Pathophysiology, The Institute of Cardiovascular Sciences, St. Boniface Hospital Albrechtsen Research Centre, Winnipeg, MB, Canada;
Department of Pharmacology and Therapeutics, Rady College of Medicine, Max Rady Faculty of Health Sciences, University of Manitoba, Winnipeg, MB, Canada

Navid Koleini Institute of Cardiovascular Sciences, St. Boniface Albrechtsen Research Centre, Winnipeg, MB, Canada;
Department of Physiology & Pathophysiology, University of Manitoba, Winnipeg, Canada;
School of Medicine, Division of Cardiology, Johns Hopkins University, Baltimore, USA

Jessica Lee NIH BD2K Center of Excellence for Biomedical Computing At UCLA (HeartBD2K), Los Angeles, USA;
Department of Physiology, UCLA School of Medicine, MRL Building, Los Angeles, CA, USA

Ronglih Liao Stanford Cardiovascular Institute, Stanford University School of Medicine, Stanford, CA, USA

David A. Liem NIH BD2K Center of Excellence for Biomedical Computing At UCLA (HeartBD2K), Los Angeles, USA;
Department of Physiology, UCLA School of Medicine, MRL Building, Los Angeles, CA, USA

Matthew Love Department of Physiology and Pathophysiology, The Institute of Cardiovascular Sciences, St. Boniface Hospital Albrechtsen Research Centre, Winnipeg, MB, Canada

Jessica MacLean Department of Pharmacology, Dalhousie University, Halifax, NS, Canada

Lynn A. Megeney Ottawa Hospital Research Institute, Sprott Centre for Stem Cell Research, Regenerative Medicine Program, Ottawa Hospital, Ottawa, ON, Canada;
Department of Cellular and Molecular Medicine, University of Ottawa, Ottawa, ON, Canada;
Department of Medicine, University of Ottawa, Ottawa, ON, Canada

Isabel Morgado Stanford Cardiovascular Institute, Stanford University School of Medicine, Stanford, CA, USA

Kishore B. S. Pasumarthi Department of Pharmacology, Dalhousie University, Halifax, NS, Canada

Peipei Ping NIH BD2K Center of Excellence for Biomedical Computing At UCLA (HeartBD2K), Los Angeles, USA;
Department of Physiology, UCLA School of Medicine, MRL Building, Los Angeles, CA, USA;
Department of Computer Science, Scalable Analytics Institute, UCLA School of Engineering, Los Angeles, CA, USA

Jennifer S. Polson NIH BD2K Center of Excellence for Biomedical Computing At UCLA (HeartBD2K), Los Angeles, USA;
Department of Physiology, UCLA School of Medicine, MRL Building, Los Angeles, CA, USA

Somayeh Pourpirali Smidt Heart Institute, Cedars-Sinai Medical Center, Los Angeles, CA, USA

Charis Putinski Ottawa Hospital Research Institute, Sprott Centre for Stem Cell Research, Regenerative Medicine Program, Ottawa Hospital, Ottawa, ON, Canada;
Department of Cellular and Molecular Medicine, University of Ottawa, Ottawa, ON, Canada

Inna Rabinovich-Nikitin Department of Physiology and Pathophysiology, The Institute of Cardiovascular Sciences, St. Boniface Hospital Albrechtsen Research Centre, Winnipeg, MB, Canada

Niketa Sareen Department of Physiology and Pathophysiology, Rady Faculty of Health Sciences, St. Boniface Hospital Albrechtsen Research Centre, Institute of Cardiovascular Sciences, Max Rady College of Medicine, University of Manitoba, Winnipeg, Canada

Anureet K. Shah Department of Kinesiology, Nutrition and Food Science, California State University, Los Angeles, CA, USA

Ding Wang NIH BD2K Center of Excellence for Biomedical Computing At UCLA (HeartBD2K), Los Angeles, USA;
Department of Physiology, UCLA School of Medicine, MRL Building, Los Angeles, CA, USA

Tania Zaglia Department of Cardiac, Thoracic, Vascular Sciences and Public Health, University of Padova, Padova, Italy;
Department of Biomedical Sciences, University of Padova, Padova, Italy;
Veneto Institute of Molecular Medicine, Padova, Italy

Chapter 1
A Protein-Centric Perspective of Autophagy and Apoptosis Signaling and Crosstalk in Health and Disease

Ding Wang, Jessica Lee, Jennifer S. Polson, David A. Liem, and Peipei Ping

Abstract Autophagy and apoptosis are pivotal pro-survival mechanisms of multicellular organisms activated in response to a variety of external and internal cues. Specifically, autophagy entails the degradation of excessive or aberrant cellular components to restore cellular homeostasis and ensure cell survival, whereas apoptosis initiates programmed cell death that sacrifices a subgroup of cells to preserve the viability of the whole organism. While seemingly opposing forces, these two distinct yet interconnected mechanisms serve to constrain stress-induced damage, eliminate potential triggers of pathogenesis, and recycle biological building blocks to support new biosynthesis in a living cell or organism. Accordingly, autophagy and apoptosis are tuned by delicate and complex signaling cascades, and inappropriate activation or disruption of these pathways have been associated with disease phenotypes. This book chapter focuses on delineating critical promoters of these two distinct biological processes as well as their interdependent molecular crosstalk from a protein-centric perspective. To better appreciate their clinical relevance, we also highlight the role of autophagy, apoptosis, and their crosstalk in diseases spanning four distinct organ types: heart, liver, central nervous system, and skeletal muscle. While the biological impact and signaling pathways of autophagy and apoptosis have been well-established, present studies have only touched the tip of the iceberg regarding the complex interplay between these two processes. Therefore, further exploration of the autophagy-apoptosis crosstalk opens the door to fully understanding the precise molecular mechanisms in many diseases and developing more effective therapeutic approaches.

D. Wang (✉) · J. Lee · J. S. Polson · D. A. Liem · P. Ping
NIH BD2K Center of Excellence for Biomedical Computing At UCLA (HeartBD2K), Los Angeles, USA
e-mail: dingwang@g.ucla.edu

Department of Physiology, UCLA School of Medicine, MRL Building, 675 Charles E Young Drive S Suite1609, Los Angeles, CA 90095, USA

P. Ping
Department of Computer Science, Scalable Analytics Institute, UCLA School of Engineering, Los Angeles, CA 90095, USA

© Springer Nature Switzerland AG 2022

L. A. Kirshenbaum (ed.), *Biochemistry of Apoptosis and Autophagy*,
Advances in Biochemistry in Health and Disease 18,
https://doi.org/10.1007/978-3-030-78799-8_1

Keywords Autophagy · Apoptosis · Signaling pathway · Crosstalk · Clinical relevance

Introduction and Rationale

During its lifespan, a living organism encounters innumerous internal or external, physical or chemical challenges that disrupt the organism's cellular homeostasis. When elevated to extreme levels, these stress conditions may endanger the organism's life. To survive, many preventative and protective mechanisms have been refined over millions of years of evolution. Among such pro-survival mechanisms, autophagy and apoptosis employ a fundamental strategy of sacrificing a fraction of cells or subcellular components to preserve the primary functions and the wellbeing of the organism as a whole. When triggered by internal or environmental signals, these two pivotal biological processes initiate the decomposition of subcellular components (e.g., aggregated proteins or damaged organelles), as well as cell death, to restrain local damages and simultaneously recycle metabolic elements. Perturbation of these processes can cause pathophysiological manifestations that affect diverse biological systems within the organism and across species, as demonstrated in several groundbreaking studies [1–4].

In this book chapter, we will cover the well-accepted pathways of autophagy and apoptosis, focusing on their individual signal transduction cascades, interdependent crosstalk, and biological effects. Furthermore, the clinical manifestations of autophagy and apoptosis in different organ systems will be summarized.

Signal Transduction Pathways

Several major signaling transduction pathways have been summarized to emphasize the key components of autophagy and apoptosis.

Major Pathways of Autophagy

In contrast to other degradative processes (e.g., degradation by mitochondrial proteases, ubiquitin-proteasomal system, or endopeptidase), canonical autophagy has unique characteristics involving membrane-aided isolation of degradation targets (i.e., cargo selection) and subsequent recruitment of lysosomes as sources of degradative enzymes [5]. Canonical autophagy can be categorized as macroautophagy, microautophagy, and chaperone-mediated autophagy according to differences in cargo selection and lysosomal lumen formation. The details regarding these subtypes of autophagy have been extensively reviewed [6–8]; thus, we will only focus on describing the major type, macroautophagy, and refer to it as "autophagy" for the rest of the book chapter.

Canonical autophagic formation of autophagosomes, which consists of four stages, is depicted in Fig. 1.1 [9–11]. During the initiation stage, AMPK phosphorylates UNC51-like kinase 1 (ULK1) to form the ULK1 complex [9].

The initiation stage of autophagy is modulated in response to nutrient supply (e.g., ATP, amino acids) via mTORC1 and AMPK [12]. At basal conditions, mTORC1 is tonically active and phosphorylates ULK1, which precludes AMPK from associating with the ULK1 complex. A low ATP/ADP ratio activates AMPK, providing ULK1 with activating phosphorylations and inhibiting mTORC1 binding [13]. The activated ULK1 forms the ULK1 complex, which phosphorylates the Atg protein Beclin 1 to form the PI3K complex. In the nucleation stage, PI3K complex activity produces the lipid phosphatidylinositol-3-phosphate (PI3P), which is essential to autophagosome formation from precursor membranes (e.g., omegasomes derived from the ER membrane). The third stage, featuring the elongation, cargo loading, and closure of

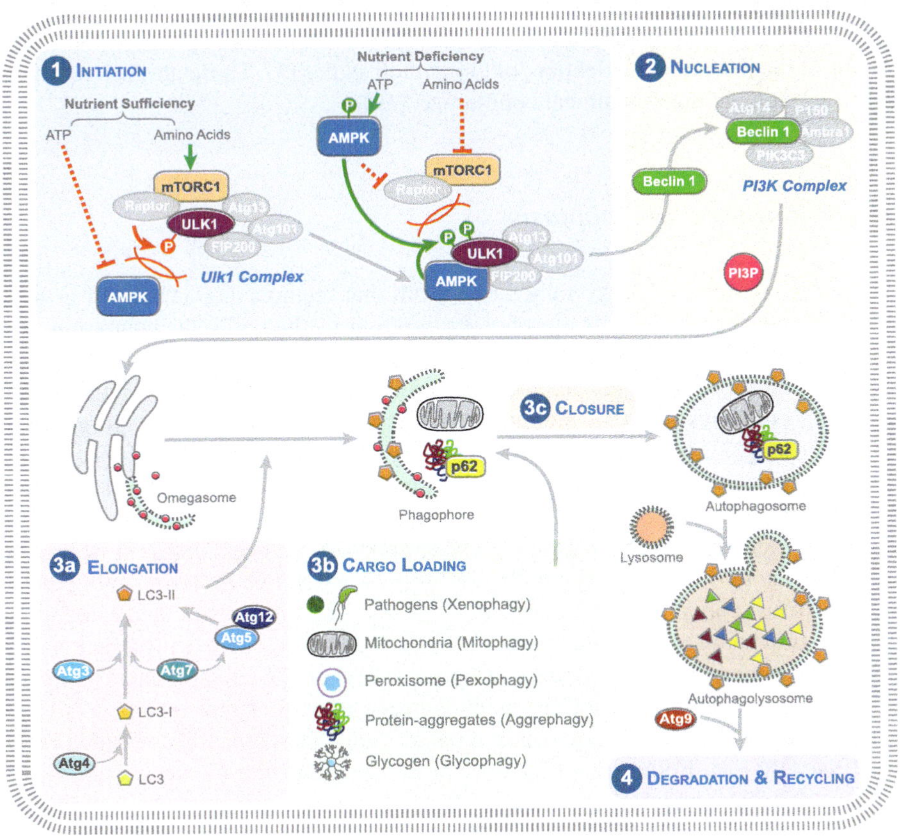

Fig. 1.1 During the initiation stage, AMPK phosphorylates

the nucleated membrane, termed the phagophore, is facilitated by two ubiquitin-like conjugation systems: the Atg12-Atg5 system and the microtubule-associated protein 1A/1B-light chain 3 (LC3) system. The Atg12-Atg5 system results in an Atg12-Atg5 conjugate that associates with the phagophore and enhances lipidation of LC3 with phosphatidylethanolamine (PE) through the LC3 system. The LC3-PE conjugate, termed LC3-II, and Atg12-Atg5 complex are both required for phagophore elongation [14]. During elongation, the p62 complex helps to load the cellular debris and other cargo by recognizing tags and trafficking it to the phagophore [15]. Along with aggregated proteins, polysaccharides, and pathogens, damaged mitochondria are common sources of autophagic cargo. Due to the biological significance of the mitochondria, mitophagy in particular has been extensively studied. Indeed, mitophagy has unique regulatory pathways, having both conserved portions and components unique to a certain species or organ type [16, 17]. After closure, the newly formed autophagosome can fuse with the lysosome to form the autophagolysosome. It is at that point that the contained cargo can be broken down into biomolecules. The degradation and recycling stage, which is dependent on Atg9, ensures that the digested units can be used by the cell for other biological processes [18].

Other forms of autophagy, classified under non-canonical autophagy, differ in the details of the initiation, nucleation, or elongation stages [5]. These alternative routes may complement their canonical counterpart [19].

Major Pathways of Apoptosis

Apoptosis is a form of programmed cell death that features degradation of genetic material and changes in cell morphology, leading to the cell's disintegration into apoptotic bodies [20]. A broad spectrum of stressors, including DNA damage and oxidative stress, can trigger apoptotic signaling cascades [21, 22]. Within these signaling pathways are also biological checkpoints that regulate the progression of apoptosis [22]. In particular, the autophagy checkpoint, where autophagic and apoptotic processes interface, will be explored in more detail in Sect. 1.3 of this chapter.

While each type of stressor may activate apoptosis through different triggering mechanisms, these signals eventually converge at one of two major apoptotic pathways: extrinsic apoptosis and intrinsic apoptosis. The progression and integration of these two pathways are illustrated in Fig. 1.2 [20, 22]. The extrinsic pathway is initiated by ligand-binding of death receptors (DRs) on the cell surface. The ligand tumor necrosis factor alpha (TNF-α) binds tumor necrosis factor receptor 1 (TNF-R1), whereas TNF-related apoptosis-inducing ligand (TRAIL) and Fas ligand (FasL) bind the DR Fas. Both types of DRs can interact with the Fas-associated death domain (FADD), which is responsible for binding and activating pro-caspase 8. Active caspase 8 subsequently stimulates the effector caspases 3, 6, and 7 to perform caspase-dependent apoptosis [23].

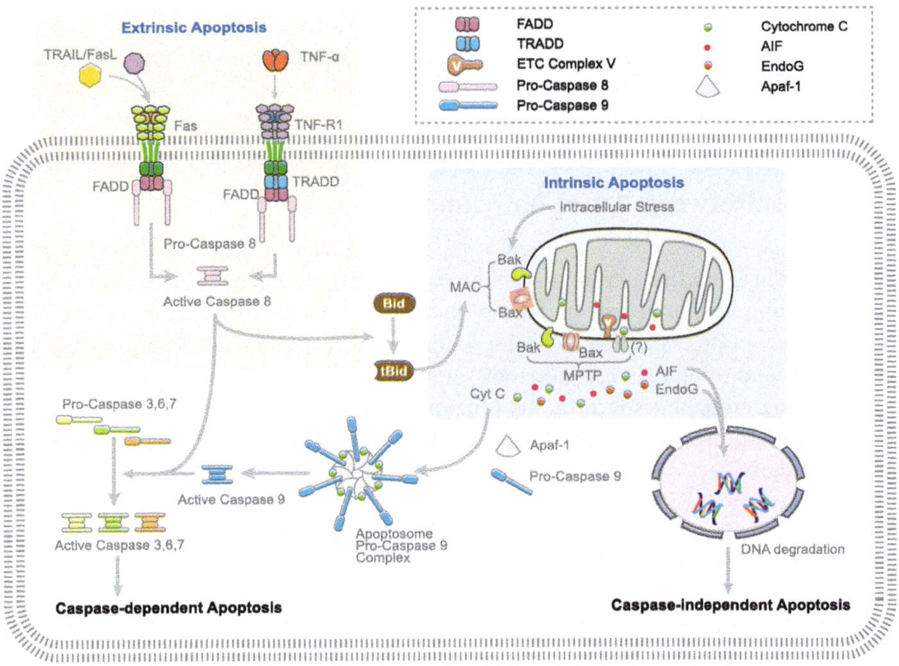

Fig. 1.2 Both types of DRs can interact

Intrinsic apoptosis, which occurs inside the cell, could be initiated by permeabilization of the mitochondrial outer membrane (OMM)—termed mitochondrial outer membrane permeabilization (MOMP)—and/or of the mitochondrial inner membrane (IMM), termed mitochondrial permeability transition (MPT). MPT requires the establishment of MPT pores (mPTPs), whereas MOMP can be initiated by either mPTPs or the formation of mitochondrial apoptosis-induced channels (MACs), composed of the pro-apoptotic Bcl-2 homology 3 (BH3)-only proteins BAK and BAX [22, 24]. Though the precise composition of mPTPs remains under intense debate, there is recent evidence supporting BAX and BAK as the OMM components and the F_1F_0-ATP synthase (ETC complex V) as the IMM component [22, 25]. Formation of mPTPs results in swelling of the mitochondrial matrix and consequent rupture of the OMM, releasing factors such as cytochrome C (Cyt C), apoptosis inducing factor (AIF), and endonuclease G (EndoG) into the cytosol [26]. Cyt C complexes with apoptotic protease activating factor 1 (Apaf-1) and pro-caspase 9 to form the apoptosome, which activates pro-caspase 9. Similar to caspase 8, active caspase 9 proceeds to stimulate the caspase-dependent apoptotic cascade [27]. In parallel, AIF and EndoG translocate to the nucleus, where they instigate the degradation of nuclear DNA and the signaling cascade for caspase-independent apoptosis [28].

Notably, the extrinsic and intrinsic pathways may intersect through a BH3-only protein, BH3-interacting domain death agonist (Bid). Instead of activating the caspase signaling cascade directly, DR-activated caspase 8 from the extrinsic pathway cleaves Bid into its truncated, active form (tBid), which promotes the MOMP essential to intrinsic apoptosis [29].

Crosstalk Between Autophagy and Apoptosis

Efforts by the scientific community in the past four to five decades have advanced the understanding of autophagic and apoptotic signal transduction pathways, enabling in-depth recognition of their complex interplay. The molecular crosstalk between autophagy and apoptosis are inherently linked by shared metabolic intermediates as triggering signals or by proteins (enzymes) acting at intersections of signaling cascades, enabling the two biological processes to inhibit or enhance the other. Many of these metabolites and proteins coalesce at the mitochondria, which has been long recognized as the central hub hosting metabolic processes and regulatory networks [30]. In this section, several protein families substantially participating in the autophagy-apoptosis crosstalk will be discussed.

Atg Proteins [31]

Autophagy-related proteins (Atgs) were named after their pivotal roles in the initiation of autophagy. As we mentioned before, autophagy is a subcellular process to remove aggregated proteins and damaged mitochondria, which requires the differentiation of degradable targets and ones to preserve. Such sensitive selection is conducted via an ubiquitin-like tagging mechanism supported by Atg proteins (see Sect. 2.1). Generally, damaged mitochondria release Cyt C that, when beyond the cell-tolerable threshold, activates Apaf-1 and subsequently triggers caspase-dependent apoptosis (Fig. 1.2) [32]. Autophagy promoted by Atgs can remove ROS-damaged proteins as well as damaged mitochondria (Fig. 1.1), thus removing a source of apoptotic inducers and increasing the chance of cell survival [33]. For instance, Atg4, a cysteine protease involved in the lipidation of LC3 during the autophagy initiation phase, serves as a sensor for low ROS levels and is upregulated during autophagy and downregulated during apoptosis [34]. In addition to Atg4, Atg5 and Atg12 are two proteins whose dual impacts on both biological processes have been gradually recognized. Atg5 recruits FADD, a major factor that triggers extrinsic apoptosis, and competes for the association between FADD and the death receptors, thus reducing the risk for apoptosis initiation [35]. Furthermore, the calpain-truncated form of Atg5 may associate with Bcl-2 and translocate to mitochondria, which contributes to the anti-apoptotic impact of Bcl-2 by competing with other regulatory partners (e.g., NOXA and PUMA) [36].

Caspases

The caspase-dependent signaling cascade is a major apoptotic pathway involving sequential activation of caspases 3, 6, 7, 8, and 9 (Fig. 1.2). Several key members of the Atg family, including Atg3, Atg4, and Beclin 1, can be degraded by these caspases, thus inhibiting autophagy [33]. This autophagy-centric subcellular apparatus is also indispensible for apoptosis. Using the autophagosomal membrane as the assembly matrix, caspase 8 associates with Atg5 and colocalizes with LC3-II and p62, which is critical for autophagy-dependent caspase 8 activation [35].

BH3-Only Proteins

According to sequence homology, BH3-only proteins are members of the Bcl-2 protein family that share a BH3 region. While Bcl-2 family members are known to play pivotal roles in apoptosis, a few BH3-only proteins appear to trigger both apoptosis and autophagy [37]. One such protein, Bnip3, is known primarily for stimulating cell death via mitochondrial dysfunction, but has also been shown to work with the LC3 conjugate to stimulate autophagy [38]. Another BH3-only protein, Bid, is involved in activating intrinsic apoptosis as well as the nucleation phase in autophagy by obstructing the inhibitory effects of Bcl-2 on these processes. Meanwhile, Bcl-2-interacting mediator of cell death (Bim), similarly binds Bcl-2 to activate apoptosis but associates with Beclin 1 to inhibit autophagy [20].

P53 Protein

The tumor suppressor protein p53 is best known for its pro-apoptotic role in preventing cancer, but it can also promote autophagy. Under non-stress conditions, p53 resides in the cytosol, where it binds the ULK1 complex component FIP200 to block initiation of autophagy. However, when the cell suffers DNA damage, p53 translocates to the nucleus, where it induces the expression of pro-apoptotic and autophagic genes such as BAX and AMPK. p53 may also translocate to the mitochondrial matrix to effect MPT, though the severity of the resulting mitochondrial dysfunction determines whether p53 prompts mitophagy or cell death [20].

Clinical Relevance of Autophagy and Apoptosis

Although autophagy and apoptosis serve central purposes for survival, insufficient or excessive activity can contribute to both cell death and disease. Since the discovery of these processes, there have been numerous studies documenting their roles in disease, including cancer, infectious diseases, aging, (drug) intoxications, cardiovascular diseases, neurological illnesses, liver diseases, and skeletal muscle diseases. To highlight their clinical relevance, we have itemized the most common diseases linked to autophagy (Table 1.1) and apoptosis (Table 1.2) among four distinct organ types, namely the heart, liver, central nervous system (CNS), and skeletal muscle.

Autophagy in the Pathogenesis of Four Organ Types

Autophagic processes can result in distinct clinical manifestations. In the heart, autophagy contributes to ischemia–reperfusion (I/R) injury and oxidative stress, cardiac remodeling and cardiomyopathies, heart failure, hypertension, atherosclerosis, and lysosomal storage diseases such as Pompe's disease. Both degradation and recycling are important in organs with high metabolism such as the liver and skeletal muscle. Correspondingly, autophagy is linked to viral hepatitis infections, alcoholic liver disease, non-alcoholic fatty liver disease (NAFLD) and metabolic syndrome, hepatocellular carcinoma (HCC), starvation, and I/R injury in liver cells. Perturbations in autophagy have been implicated in CNS diseases such as ataxias, epilepsy, and prion disease, with the majority in neurodegenerative diseases such as Alzheimer's disease (AD), Parkinson's disease (PD), amyotrophic lateral sclerosis (ALS) and Huntington's disease. Lastly, in skeletal muscle, autophagy has been described in exercise, insulin-signaling, diabetes, muscle atrophy, myopathies, and Duchennes muscular dystrophy. Overall, autophagy appears to be a key player in a broad range of pathological mechanisms and organ types.

Apoptosis in the Pathogenesis of Four Organ Types

While apoptosis is an integral part of organism survival, it has the propensity to lead to disease phenotypes. Numerous cardiac studies conclude that apoptosis extends myocardial infarction and heart failure, and is a key factor in the development of cardiac fibrosis. In highly metabolic organs such as the liver, apoptotic caspases are major players in the pathology of alcoholic- and non-alcoholic liver diseases, drug induced hepatotoxicity, cholestasis, and viral hepatitis. In the CNS, apoptosis contributes to cell death in ischemic strokes and neurodegenerative diseases such as AD, PD, HD, and MS. Molecular signaling of apoptosis underlies muscle atrophy and sarcopenia in skeletal muscle. Similar to autophagy, apoptosis plays a fundamental role in the pathology of multiple organ types.

Table 1.1 Clinical manifestations of autophagy in four distinct organ types

Heart		Liver		Central nervous system		Skeletal muscle	
Condition	Mechanism	Condition	Mechanism	Condition	Mechanism	Condition	Mechanism
I/R injury Myocardial infarction	• Hypoxia triggers autophagosome formation adjacent to fragmented mitochondria [39] • Autophagy has protective effects in I/R by neutralizing ROS, pro-apoptotic factors, and swollen mitochondria [40] • Glycogen synthase-3β, AMPK, and Beclin 1 regulate autophagy at different time stages in I/R [41]	**Virus infection** – HBV – HCV	• Autophagy in HBV infection is increased • Autophagic vacuoles may act as a membranous web of complex viral proteins and serve as transient scaffolds to enhance efficient translation of viral RNA • Inhibition of autophagy may be a potential therapeutic target of HCV and HBV [42, 43]	**Parkinson's disease (PD)**	• Disruption of autophagy at several levels of the formation and transport of autophagolysosomes leads to the degeneration of dopaminergic axons and dendrites • Autophagy is important in the degradation of α-syn-containing Lewy bodies in the substantia niga • Accumulation of α-syn and Lewy bodies due to lack of autophagy is an important hallmark of Alzheimer's disease and leads to damaged mitochondria	**Exercise**	• Regulatory mechanisms of exercise-induced autophagy has been linked to skeletal muscle oxidative function • Exercise may stimulate the degradation of cytoplasmic components in skeletal muscle through the initiation of autophagy (exercise-induced autophagy), as shown by the conversion of LC3-I to LC3-II and reduced content of P62. This process may be coordinated by skeletal muscle PGC-1α [23]

(continued)

Table 1.1 (continued)

Heart		Liver		Central nervous system		Skeletal muscle	
Condition	Mechanism	Condition	Mechanism	Condition	Mechanism	Condition	Mechanism
Cardiac remodeling Cardio-myopathy (CM)	• DCM may be attributed to upregulated Akt-mTOR signaling, suppressed mitophagy and mitofusin-2, and impaired autophagy • Elevated mTOR signaling impairs autophagy, while inhibition leads to cell survival through restoration of autophagic flux and cardiac function • Genetic ablation of Beclin 1 accelerates CM • Upregulation of p62 protects cardiomyocytes from protein misfolding [16, 44]	**Alcoholic liver disease** – Hepatitis – Fatty Liver – Cirrhosis	• Autophagy by alcohol mostly targets the degradation of lipid droplets (lipophagy) and mitochondria (mitophagy), which are the two organelles mostly affected by alcohol • Chronic alcohol consumption has been reported to reduce autophagy secondary to inhibition of AMPK, or to the increase of hepatic lipids both as a result of lipogenesis [43, 45]	**Alzheimer's disease (AD)**	• AD is a progressive degenerative disorder of the brain, characterized by the accumulation of misfolded proteins, inflammatory changes, and oxidative damage with a global dysfunction of the proteolytic capacity of lysosomes • Restoring lysosomal function attenuated the progression of AD • Autophagy is inhibited by mutated presinillin-1 in a familial form of AD [46–48]	**Insulin-signaling diabetes**	• Increased skeletal muscle autophagy is important in improving glucose tolerance • Insulin signaling has been reported to suppress autophagy in skeletal muscle, thereby inhibiting degradation of cellular components • Conversely, both glucagon and insulin resistance activate autophagy [49]

(continued)

Table 1.1 (continued)

Heart		Liver		Central nervous system		Skeletal muscle	
Condition	Mechanism	Condition	Mechanism	Condition	Mechanism	Condition	Mechanism
Heart failure	• Autophagy may be an adaptive response to HF, as increased expression of Atg7 in mice ameliorates cardiac function • Pressure overload mediates angiotensin-II and ROS release, thereby enhancing a maladaptive autophagic response [16, 44] • Mechanical unloading by left ventricualr assist devices (LVAD) attenuates autophagy [50]	**Non-alcoholic fatty liver disease (NAFLD) metabolic syndrome**	• Upregulated autophagy supports a protective role in NAFLD while insulin may suppress autophagy • Both glucagon and insulin-resistance in metabolic syndrome may augment autophagy • The role of autophagy in the mobilization of hepatic lipids supports a protective role against NAFLD and lipotoxicity [43, 45]	**Amyotrophic lateral sclerosis (ALS)**	• Post-mortem ALS biopsies show impaired autophagy • Autosomal forms of ALS are associated with mutant forms of valosin-containing proteins essential for the formation of autophagosomes • The cause of mitochondrial malfunction in ALS is linked to accumulation of SOD1 and TDP-43, both of which are degraded through autophagy [46–48]	**Muscle atrophy myopathy muscle weakness**	• Studies in preclinical animal models with ablation of the muscle-specific autophagy gene Atg7 led to sarcomere disorganization, myo-fiber degeneration, oxidative stress, abnormal mitochondria, and accumulation of unfolded proteins • The defective autophagy is accompanied by persistent activation of the Akt-mTOR axis [23]

(continued)

Table 1.1 (continued)

Heart		Liver		Central nervous system		Skeletal muscle	
Condition	Mechanism	Condition	Mechanism	Condition	Mechanism	Condition	Mechanism
Hypertension	• Autophagy evokes pulmonary hypertension by increased expression of LC3B-II and reduction of P62, leading to proliferation of smooth muscle cells [51]	**Hepato-cellular carcinoma (HCC)**	• In HCC, most studies report a tumor-suppressor role for autophagy • Human liver biopsies of HCC show an increase of P62 and induction of Nrf2, while experimental blockage of P62 reduces cell growth [52]	**Huntington's disease**	• Autophagosome formation, trafficking, and fusion are dysfunctional, leading to a buildup of toxic materials in the cell, such as Htt43Q [46–48]	**Duchenne muscular dystrophy (DMD)**	• Deficient autophagy leads to damaged mitochondria, aggregation of proteins, and distention of sarcoplasmic reticula, which are cellular markers of DMD • Muscle biopsies from DMD patients display severely impaired autophagy through an Akt-mTOR axis and reduction of the lipidated form of LC3, which is a common marker for autophagy induction [53]
Athero-sclerosis	• Autophagy favors plaque stabilization and promotes cell survival through the regulation of lipid metabolism [54]	**Starvation** – Hepatic – Ketogenesis	• Starvation induces liver autophagy, which regulates lipid droplet formation in hepatocytes for energy homeostasis and ketogenesis [45]	**Prion diseases:** – Creutzfeldt-Jacob (CJD)	• Brain tissues obtained from CJD patients show increased autophagosomes • Synaptic autophagy contributes to the overall synaptic loss in brains affected in prion diseases [55]		

(continued)

Table 1.1 (continued)

Heart		Liver		Central nervous system		Skeletal muscle	
Condition	Mechanism	Condition	Mechanism	Condition	Mechanism	Condition	Mechanism
Lysosomal storage disease-pompe's disease	• Disrupted lysosomal storage is ameliorated by ATG7 deficiency and inhibited autophagosome formation [56]	**Liver I/R injury**	• Autophagy has protective effects during liver I/R injury by removing the damage caused by oxidative stress, ROS, and swollen mitochondria [42]	**Inherited Ataxias**	• Spinocerebellar ataxia type 3 patients show active P62 in brain areas of neurodegeneration [46, 47, 48]	**Sarcopenia**	• Most studies indicate a functional defect in autophagy-dependent signaling during sarcopenia • Dysregulated autophagy contributes to the age-related accumulation of defective mitochondria and ROS release [23]
				Epilepsy	• Brain biopsies from patients with tuberous sclerosis complex indicate impaired autophagy [46–48]		

Table 1.2 Clinical manifestations of apoptosis in four distinct organ types

Heart		Liver		Central nervous system		Skeletal muscle	
Condition	Mechanism	Condition	Mechanism	Condition	Mechanism	Condition	Mechanism
Ischemic heart disease (I/R injury)	• Decreased availability of oxygen may stimulate apoptosis • Mitochondrial ROS release activates caspases • Pharmacological intervention of apoptosis during I/R may reduce oxidative damage • Nucleosomal ladders of DNA were detected in ischemic-reperfused rabbit hearts but not in ischemic only. This finding was further referred to as "reperfusion injury" • DNA nick end-labeling-positive human heart cells in areas of acute infarction were shown in contrast to normal heart cells adjacent from the infarct • Cell death by apoptosis is greater than cell death by necrosis after coronary occlusion in rats [57, 58]	**Viral hepatitis** HCV	• Both HBV and HCV mediate adaptive and immune responses in the liver, which induces target cell apoptosis • HCV upregulates pro-apoptotic factors CD95, FADD, TRADD, Bax, Bad, and caspases, and sensitizes TRAIL-induced apoptosis • HBV enhances hepatocyte apoptosis through TNF-α, TNFR1, and Fas expression [59]	**Alzheimer's disease (AD)**	• Loss of neurons in AD is apoptotic signaling intrinsic apoptotic signaling pathway with mitochondrial release of cytochrome C, caspase-9, and caspase-3 • GSK3 promotes intrinsic apoptotic signaling in AD, leading to neuronal loss • Ceramides may initiate a cascade of released ROS and cytochrome C, Bcl-2 depletion, and caspase-3 activation [60]	**Skeletal muscle atrophy**	• Mitochondrial function and its alterations are considered major key players in muscle atrophy • Damaged mitochondria in muscle atrophy display increased amounts of ROS generation and interfere with cellular quality control mechanisms, both leading to a greater propensity to apoptosis • Administration of mitochondria-targeted antioxidants has been shown to attenuate ROS production and myofibril atrophy in hind-limb muscles • Excessive mitochondrial ROS generation during muscle inactivity and atrophy may be due to calcium dyshomeostasis and/or de-arrangements in bioenergetics pathways and decreased protein synthesis, followed by apoptosis • During muscle atrophy, entire DNA units can be removed via myonuclear apoptosis, leading to atrophy or fiber removal • During muscle disuse, mitochondrial dynamics shift towards fission and over-activation of autophagy. ROS generation by mitochondria, together with increased fission, will trigger apoptosis [61, 62]

(continued)

Table 1.2 (continued)

Heart		Liver		Central nervous system		Skeletal muscle	
Condition	Mechanism	Condition	Mechanism	Condition	Mechanism	Condition	Mechanism
		Cholestasis (Toxic bile acids)	• Cholestatic liver injury initiates necrosis in the acute phase and apoptosis in the chronic phase, thereby exhibiting co-existence at a certain stage of the liver injury • Cholestatic liver injury initiates Fas signaling and mitochondrial dysfunction and downstream caspases [59]	**Ischemic stroke**	• In acute cerebral ischemia, matrix metalloproteinases (MMP)-2 and MMP-9 in nuclei are activated together with DNA damage • MMPs facilitate oxidative DNA damage and inflammatory cytokines and cleave ARP-1 and XRCC1, thereby promoting DNA fragmentation and neuronal apoptosis [63]		

(continued)

Table 1.2 (continued)

Heart		Liver		Central nervous system		Skeletal muscle	
Condition	Mechanism	Condition	Mechanism	Condition	Mechanism	Condition	Mechanism
Cardiomyopathy Heart failure	• Heart biopsies from end-stage DCM patients show DNA fragmentation [64] • Caspase activation in mice induces apoptosis and severe DCM and HF [65] • Hypoxic conditions could stimulate cardiomyocyte apoptosis during ischemia, leading to ischemic CM [58, 64, 66] • Cardiomyocyte apoptosis may be responsible for the development of HF following pressure overload • Preventing apoptosis, either pharmacologically or genetically, may prevent cardiac remodeling and failure [58, 64, 66]	**Alcoholic liver disease (ALD),** (Alcoholic Hepatitis, Fatty Liver, Cirrhosis)	• Alcohol metabolism produces ROS, which then induce lipid peroxidation and GSH depletion, leading to apoptosis • Ethanol stimulates both the intrinsic and extrinsic pathways of apoptosis and upregulates TNF-α and CD95 [65]	**Parkinson's disease (PD)**	• Post-mortem PD brain biopsies identified apoptotic neurons using in situ nick labeling • Post-mortem brain biopsies demonstrated higher expressions of p53, CD95, active caspase-3, and TNF-α [63]		
		Non-alcoholic fatty liver disease (NAFLD)	• NAFLD exhibits mitochondrial dysfunction and oxidative stress, leading to activation of pro-apoptotic caspases [65] • Release of ROS may damage mitochondrial proteins and DNA, lipids, and cardiolipin, inducing apoptosis in NAFLD [67]	**Huntington's disease (HD)**	• Brains from patients with HD exhibit mitochondrial dysfunction • Lymphoblasts derived from HD patients show increased stress-induced apoptotic cell death associated with caspase-3 activation • Lymphoblasts from HD patients demonstrate an increased mitochondrial depolarization [68]	**Sarcopenia**	• Alterations in mitochondrial function, followed by increased ROS release, are considered major factors underlying sarcopenia [61] • Gastrocnemius muscle in senile rats exhibits increased Apaf-1 and cleaved caspase-9 in conjunction with DNA fragmentation Interestingly, Cyt-c and procaspase-9 were not changed [62] • Conversely, two other components of apoptosis, pro-caspase-9 and cytochrome C, were not increased in gastrocnemius muscle from senile rats • Senile rat muscle shows increases in HSP27, HSP60, and HSP70. In parallel, it was also shown that Bax was increased, Bcl-2 was decreased, and Bak showed no change [61, 62]

(continued)

Table 1.2 (continued)

Heart		Liver		Central nervous system		Skeletal muscle	
Condition	Mechanism	Condition	Mechanism	Condition	Mechanism	Condition	Mechanism
Myocardial fibrosis	• Cardiac fibrosis provokes pathophysiological alterations, leading to hypertrophy and apoptosis and culminating in cardiac dysfunction and HF • Collagen-secreting fibroblasts undergo apoptosis and provoke pathological changes in the heart, ultimately leading to remodeling and HF [69]	**Acetaminophen-Induced hepatotoxicity**	• Hepatotoxic injury by acetaminophen induces cell death by necrosis and apoptosis • Cytochrome P450 metabolizes Acetaminophen, thereby triggering mitochondrial oxidative stress • Acetaminophen exposure in mice and humans show TUNEL labeling and DNA fragmentation [59]	**Multiple sclerosis (MS)**	• PBMCs from MS patients have shown in the relapsing phase activation of T cell receptor signaling, upregulation of cytokines, and suppression of cell death by apoptosis • Early lesions in MS brain biopsies are mainly areas of apoptotic oligodendrocytes • Mice expressing caspase-9, ROS, and TNF-α induce demyelination similar to MS [70]		

Crosstalk of Autophagy and Apoptosis in the Pathogenesis of Ischemia–reperfusion Injury

I/R injury is a pathological condition that occurs in all four aforementioned organ types. For instance, ischemic heart disease, which is precipitated by an acute thrombotic coronary artery occlusion at the site of an atherosclerotic plaque, is the leading cause of death and disability in the world. Reperfusion therapy induces an excessive release of ROS, causing an additional bout of myocardial damage and cell death via necrosis and apoptosis. Comparably, I/R injury is also a common pathological condition in the liver during transplantation surgery, the CNS during ischemic strokes, and in skeletal muscle during exertion-induced claudication of the extremities or during acute muscle infarction. Conversely, autophagy is generally believed to be protective during I/R injury (Table 1.1). During prolonged ischemia as well as I/R, activation of autophagy inhibits apoptosis, indicating a physiological interplay between these two phenomena during this pathological condition. Moreover, emerging evidence pinpointed fundamental proteins in autophagy and apoptosis that interact with each other. For instance, when Beclin 1 expression is reduced, autophagy is inhibited, followed by increased I/R injury. Furthermore, Mst1, which is a fundamental protein in promoting apoptosis, is able to phosphorylate the Thr108 residue in the BH3 domain of Beclin 1, thus enhancing the interaction between Beclin 1 and several Bcl-2 protein family members that regulate apoptosis. Taken together, the extent of I/R injury may be affected by a functional balance between autophagy and apoptosis.

Conclusions and Future Perspectives

It is well known that autophagy is a biological process related to the orderly degradation of unnecessary or abnormal proteins, organelles, and other cytoplasmic constituents. Accordingly, autophagy is fundamental to maintaining a balanced physiology and cellular homeostasis under normal conditions. With respect to disease, autophagy can be considered an adaptive response to stress in order to preserve cell survival; in certain pathological conditions, it appears to promote morbidity and cell death among different organs. In conjunction with autophagy, apoptosis is a highly regulated and controlled process of cell death that occurs in multicellular organisms. It is characterized by a cascade of biochemical events that culminate as chromatin condensation, cell and nuclear shrinkage, and fragmentation of nuclei and chromosomal DNA. Similar to autophagy, numerous studies have documented the pathophysiological role of apoptosis over a wide range of diseases. The biological function and molecular understanding of apoptosis and autophagy have been extensively characterized for four decades; however, recent findings have made it clear that we have only witnessed the tip of the iceberg in both processes, particularly regarding their "crosstalk" signals. Most interestingly, emerging evidence emphasizes that both processes may work in synergy rather than in isolation. The complex interplay between autophagy and apoptosis in pathological conditions presents a potential avenue for therapeutic interventions of diseases and a major challenge for future investigations. In this respect, direct protein–protein interactions between autophagic and apoptotic proteins, as well as their post-translational modifications, remain to be further explored. Moreover, it is still a matter of investigation whether the common proteins involved in both autophagy and apoptosis modulate each process separately or simultaneously in true crosstalk.

This book chapter has paid special attention to the translational and clinical relevance of both autophagy and apoptosis. It is remarkable that both processes are well documented over a wide range of diseases in four distinct organ types, including cardiovascular diseases, neurodegeneration of the CNS, and diseases in highly metabolic organs such as the liver and skeletal muscle. Lastly, while both autophagy and apoptosis are extensively studied in preclinical animal models as well as in humans, the paucity of autophagy-apoptosis crosstalk studies among different species still stands as a major challenge for clinical translation and insights. In the future, crosstalk between autophagy and apoptosis may be a novel avenue for gaining mechanistic insights and developing disease therapies.

References

1. Kung G, Dai P, Deng L et al (2014) A novel role for the apoptosis inhibitor ARC in suppressing TNFalpha-induced regulated necrosis. Cell Death Differ 21(4):634–644
2. Chen Y, Dorn GW 2nd (2013) PINK1-phosphorylated mitofusin 2 is a Parkin receptor for culling damaged mitochondria. Science 340(6131):471–475
3. Shirakabe A, Zhai P, Ikeda Y et al (2016) Drp1-dependent mitochondrial autophagy plays a protective role against pressure overload-induced mitochondrial dysfunction and heart failure. Circulation 133(13):1249–1263
4. Kubli DA, Cortez MQ, Moyzis AG, et al. (2015) PINK1 Is Dispensable for Mitochondrial Recruitment of Parkin and Activation of Mitophagy in Cardiac Myocytes. PloS one 10(6):e0130707.
5. Codogno P, Mehrpour M, Proikas-Cezanne T (2012) Canonical and non-canonical autophagy: variations on a common theme of self-eating? Nat Rev Mol Cell Biol 13(1):7–12
6. Kundu M, Thompson CB (2005) Macroautophagy versus mitochondrial autophagy: a question of fate? Cell Death Differ 12(Suppl 2):1484–1489
7. Jia G, Sowers JR (1852) (2015) Autophagy: a housekeeper in cardiorenal metabolic health and disease. Biochim Biophys Acta 2:219–224
8. Li WW, Li J, Bao JK (2012) Microautophagy: lesser-known self-eating. Cell Mol Life Sci 69(7):1125–1136
9. Kim J, Kundu M, Viollet B et al (2011) AMPK and mTOR regulate autophagy through direct phosphorylation of Ulk1. Nat Cell Biol 13(2):132–141
10. Carling D, Thornton C, Woods A et al (2012) AMP-activated protein kinase: new regulation, new roles? Biochem J 445(1):11–27
11. Sinha S, Levine B (2008) The autophagy effector Beclin 1: a novel BH3-only protein. Oncogene 27(Suppl 1):S137–S148
12. Russell RC, Yuan HX, Guan KL (2014) Autophagy regulation by nutrient signaling. Cell Res 24(1):42–57
13. Galluzzi L, Pietrocola F, Levine B et al (2014) Metabolic control of autophagy. Cell 159(6):1263–1276

14. Rogov V, Dotsch V, Johansen T et al (2014) Interactions between autophagy receptors and ubiquitin-like proteins form the molecular basis for selective autophagy. Mol Cell 53(2):167–178
15. Rusten TE, Stenmark H (2010) p62, an autophagy hero or culprit? Nat Cell Biol 12(3):207–209
16. Saito T, Sadoshima J (2015) Molecular mechanisms of mitochondrial autophagy/mitophagy in the heart. Circ Res 116(8):1477–1490
17. Lavandero S, Troncoso R, Rothermel BA et al (2013) Cardiovascular autophagy: concepts, controversies, and perspectives. Autophagy 9(10):1455–1466
18. Reggiori F, Tooze SA (2012) Autophagy regulation through Atg9 traffic. J Cell Biol 198(2):151–153
19. Wang DW, Peng ZJ, Ren GF et al (2015) The different roles of selective autophagic protein degradation in mammalian cells. Oncotarget 6(35):37098–37116
20. Marino G, Niso-Santano M, Baehrecke EH et al (2014) Self-consumption: the interplay of autophagy and apoptosis. Nat Rev Mol Cell Biol 15(2):81–94
21. Li M, Gao P, Zhang J (2016) Crosstalk between autophagy and apoptosis: potential and emerging therapeutic targets for cardiac diseases. Int J Mol Sci 17(3):332
22. Green DR, Galluzzi L, Kroemer G (2014) Cell biology. Metab Control Cell Death Sci 345(6203):1250256
23. Fritzen AM, Frosig C, Jeppesen J et al (2016) Role of AMPK in regulation of LC3 lipidation as a marker of autophagy in skeletal muscle. Cell Signal 28(6):663–674
24. Madamba SM, Damri KN, Dejean LM et al (2015) Mitochondrial Ion channels in cancer transformation. Front Oncol 5:120
25. Karch J, Kwong JQ, Burr AR et al (2013) Bax and Bak function as the outer membrane component of the mitochondrial permeability pore in regulating necrotic cell death in mice. Elife 2:e00772
26. Gouspillou G, Sgarioto N, Kapchinsky S et al (2014) Increased sensitivity to mitochondrial permeability transition and myonuclear translocation of endonuclease G in atrophied muscle of physically active older humans. FASEB J: Off Publ Fed Am Soc Exp Biol 28(4):1621–1633
27. Zhou M, Li Y, Hu Q et al (2015) Atomic structure of the apoptosome: mechanism of cytochrome c- and dATP-mediated activation of Apaf-1. Genes Dev 29(22):2349–2361
28. Guo W, Zhang Y, Ling Z et al (2015) Caspase-3 feedback loop enhances Bid-induced AIF/endoG and Bak activation in Bax and p53-independent manner. Cell Death Dis 6:e1919
29. Jacob SF, Wurstle ML, Delgado ME et al (2016) An analysis of the truncated Bid- and ROS-dependent spatial propagation of mitochondrial permeabilization waves during apoptosis. J Biol Chem 291(9):4603–4613
30. Dorn GW 2nd, Kitsis RN (2015) The mitochondrial dynamism-mitophagy-cell death interactome: multiple roles performed by members of a mitochondrial molecular ensemble. Circ Res 116(1):167–182
31. Wesselborg S, Stork B (2015) Autophagy signal transduction by ATG proteins: from hierarchies to networks. Cell Mol Life Sci 72(24):4721–4757
32. Li M, Tan J, Miao Y et al (2015) The dual role of autophagy under hypoxia-involvement of interaction between autophagy and apoptosis. Apoptosis 20(6):769–777
33. Kaminskyy VO, Zhivotovsky B (2014) Free radicals in cross talk between autophagy and apoptosis. Antioxid Redox Signal 21(1):86–102
34. Young MM, Takahashi Y, Khan O et al (2012) Autophagosomal membrane serves as platform for intracellular death-inducing signaling complex (iDISC)-mediated caspase-8 activation and apoptosis. J Biol Chem 287(15):12455–12468
35. Jaganathan S, Malek E, Vallabhapurapu S et al (2014) Bortezomib induces AMPK-dependent autophagosome formation uncoupled from apoptosis in drug resistant cells. Oncotarget 5(23):12358–12370
36. Shamas-Din A, Brahmbhatt H, Leber B et al (1813) (2011) BH3-only proteins: orchestrators of apoptosis. Biochim Biophys Acta 4:508–520
37. Hanna RA, Quinsay MN, Orogo AM et al (2012) Microtubule-associated protein 1 light chain 3 (LC3) interacts with Bnip3 protein to selectively remove endoplasmic reticulum and mitochondria via autophagy. J Biol Chem 287(23):19094–19104

38. Kroemer G, Galluzzi L, Brenner C (2007) Mitochondrial membrane permeabilization in cell death. Physiol Rev 87(1):99–163
39. Mei Y, Thompson MD, Cohen RA et al (2015) Autophagy and oxidative stress in cardiovascular diseases. Biochim Biophys Acta 2:243–251
40. Decker RS, Poole AR, Crie JS et al (1980) Lysosomal alterations in hypoxic and reoxygenated hearts. II. Immunohistochemical and biochemical changes in cathepsin D. Am J Pathol 98(2):445–56
41. Maejima Y, Isobe M, Sadoshima J (2015) Regulation of autophagy by Beclin 1 in the heart. J Mol Cell Cardiol 95:19–25
42. Schneider JL, Cuervo AM (2014) Liver autophagy: much more than just taking out the trash. Nat Rev Gastroenterol Hepatol 11(3):187–200
43. Madrigal-Matute J, Cuervo AM (2016) Regulation of liver metabolism by autophagy. Gastroenterology 150(2):328–339
44. Nishida K, Otsu K (2015) Autophagy during cardiac remodeling. J Mol Cell Cardiol 95:11–18
45. Takagi A, Kume S, Kondo M et al (2016) Mammalian autophagy is essential for hepatic and renal ketogenesis during starvation. Sci Rep 6:18944
46. Lipinski MM, Zheng B, Lu T et al (2010) Genome-wide analysis reveals mechanisms modulating autophagy in normal brain aging and in Alzheimer's disease. Proc Natl Acad Sci U S A 107(32):14164–14169
47. Vidal RL, Matus S, Bargsted L et al (2014) Targeting autophagy in neurodegenerative diseases. Trends Pharmacol Sci 35(11):583–591
48. Yang Y, Coleman M, Zhang L et al (2013) Autophagy in axonal and dendritic degeneration. Trends Neurosci 36(7):418–428
49. Barlow AD, Thomas DC (2015) Autophagy in diabetes: beta-cell dysfunction, insulin resistance, and complications. DNA Cell Biol 34(4):252–260
50. Wellnitz K, Taegtmeyer H (2010) Mechanical unloading of the failing heart exposes the dynamic nature of autophagy. Autophagy 6(1):155–156
51. Wang ZV, Rothermel BA, Hill JA (2010) Autophagy in hypertensive heart disease. J Biol Chem 285(12):8509–8514
52. Dash S, Chava S, Chandra PK et al (2016) Autophagy in hepatocellular carcinomas: from pathophysiology to therapeutic response. Hepat Med 8:9–20
53. De Palma C, Perrotta C, Pellegrino P et al (2014) Skeletal muscle homeostasis in duchenne muscular dystrophy: modulating autophagy as a promising therapeutic strategy. Front Aging Neurosci 6:188
54. Martinet W, De Meyer GR (2009) Autophagy in atherosclerosis: a cell survival and death phenomenon with therapeutic potential. Circ Res 104(3):304–317
55. Sikorska B, Liberski PP, Giraud P et al (2004) Autophagy is a part of ultrastructural synaptic pathology in Creutzfeldt-Jakob disease: a brain biopsy study. Int J Biochem Cell Biol 36(12):2563–2573
56. Raval KK, Tao R, White BE et al (2015) Pompe disease results in a Golgi-based glycosylation deficit in human induced pluripotent stem cell-derived cardiomyocytes. J Biol Chem 290(5):3121–3136
57. Gottlieb RA, Burleson KO, Kloner RA et al (1994) Reperfusion injury induces apoptosis in rabbit cardiomyocytes. J Clin Invest 94(4):1621–1628
58. Jose Corbalan J, Vatner DE, Vatner SF (2016) Myocardial apoptosis in heart disease: does the emperor have clothes? Basic Res Cardiol 111(3):31
59. Wang K (2014) Molecular mechanisms of liver injury: apoptosis or necrosis. Exp Toxicol Pathol 66(8):351–356
60. Jazvinscak Jembrek M, Hof PR, Simic G (2015) Ceramides in Alzheimer's disease: key mediators of neuronal apoptosis induced by oxidative stress and Abeta accumulation. Oxid Med Cell Longev 2015:346783
61. Calvani R, Joseph AM, Adhihetty PJ et al (2013) Mitochondrial pathways in sarcopenia of aging and disuse muscle atrophy. Biol Chem 394(3):393–414

62. Marzetti E, Leeuwenburgh C (2006) Skeletal muscle apoptosis, sarcopenia and frailty at old age. Exp Gerontol 41(12):1234–1238
63. Kimura-Ohba S, Yang Y (2016) Oxidative DNA damage mediated by intranuclear MMP activity is associated with neuronal apoptosis in ischemic stroke. Oxid Med Cell Longev 2016:6927328
64. Narula J, Haider N, Virmani R et al (1996) Apoptosis in myocytes in end-stage heart failure. N Engl J Med 335(16):1182–1189
65. Williams JA, Manley S, Ding WX (2014) New advances in molecular mechanisms and emerging therapeutic targets in alcoholic liver diseases. World J Gastroenterol 20(36):12908–12933
66. Wencker D, Chandra M, Nguyen K et al (2003) A mechanistic role for cardiac myocyte apoptosis in heart failure. J Clin Invest 111(10):1497–1504
67. Paradies G, Paradies V, Ruggiero FM et al (2014) Oxidative stress, cardiolipin and mitochondrial dysfunction in nonalcoholic fatty liver disease. World J Gastroenterol 20(39):14205–14218
68. Sawa A, Wiegand GW, Cooper J et al (1999) Increased apoptosis of Huntington disease lymphoblasts associated with repeat length-dependent mitochondrial depolarization. Nat Med 5(10):1194–1198
69. Travers JG, Kamal FA, Robbins J et al (2016) Cardiac fibrosis: the fibroblast awakens. Circ Res 118(6):1021–1040
70. Macchi B, Marino-Merlo F, Nocentini U et al (2015) Role of inflammation and apoptosis in multiple sclerosis: comparative analysis between the periphery and the central nervous system. J Neuroimmunol 287:80–87
71. Lev N, Melamed E, Offen D (2003) Apoptosis and Parkinson's disease. Prog Neuropsychopharmacol Biol Psychiatry 27(2):245–250

Chapter 2
Adrenergic Receptor Signaling Pathways in the Regulation of Apoptosis and Autophagy in the Heart

Jessica MacLean and Kishore B. S. Pasumarthi

Abstract Despite the discovery of several new therapeutic interventions, heart disease remains the number one cause of death worldwide. Many forms of heart disease are associated with loss of functional heart muscle cells, cardiac hypertrophy, remodeling and chamber dilation as well as deterioration in cardiac function. Significant changes in the levels of mediators regulating apoptosis and autophagy have been reported in both patients and experimental models of heart disease. Endogenous catecholamines, norepinephrine and epinephrine have been shown to play a critical role in cell death and pro-survival pathways in the heart by binding to β1-, β2- and α1- adrenergic receptors (ARs). While there is scant information on the mechanisms regulating cell death and pro-survival pathways, a deeper understanding of these processes and their interplay is essential for development of new therapies for patients suffering from heart disease. Here, we have briefly reviewed various signal transduction mechanisms linking adrenergic receptors with apoptosis and autophagy in the heart. Notably, elevated plasma levels of norepinephrine or β1-AR autoantibodies can induce apoptosis via stimulation of the β1-AR/cAMP/protein kinase A dependent and independent pathways and activation of β2- or α1-ARs can have pro-survival functions in cardiomyocytes. Whereas both AR agonists and antagonists can either increase or decrease autophagy depending on the cardiovascular cell type and experimental model system used. In addition, we have provided a comprehensive survey of the literature relevant to the effects of various adrenergic drugs on the signaling mechanisms regulating apoptosis and autophagy and discussed the relevance of these events in heart disease.

Keywords Cell death and survival pathways · Heart disease · Adrenergic receptors · Catecholamines · Adrenergic receptor agonists and α- and β-adrenergic receptor blockers

---

J. MacLean · K. B. S. Pasumarthi (✉)
Department of Pharmacology, Dalhousie University, Sir Charles Tupper Building, 5850 College Street, P.O. Box 15000, Halifax, NS B3H 4R2, Canada
e-mail: kpasumar@dal.ca

© Springer Nature Switzerland AG 2022

L. A. Kirshenbaum (ed.), *Biochemistry of Apoptosis and Autophagy*, Advances in Biochemistry in Health and Disease 18,
https://doi.org/10.1007/978-3-030-78799-8_2

Introduction

The adrenergic system consists of the endogenous catecholamines norepinephrine and epinephrine and the adrenergic receptors. Norepinephrine from nerve endings and epinephrine from the adrenal gland are secreted when the sympathetic nervous system is activated [1]. The systemic effects of the sympathetic nervous system, particularly on the heart and circulatory system, enable the human body's fight or flight response. To this end, sympathetic activation elicits an increased heart rate, increased contractility and relaxation of the heart, and it causes vasoconstriction with a subsequent increase in blood pressure. These effects are caused by the actions of epinephrine and norepinephrine on the adrenergic receptors [2]. Given that norepinephrine is the catecholamine released at the cardiac sympathetic nerve terminals, it can modulate the heart's activity more directly than epinephrine which is released by the adrenal glands in a systemic manner [2, 3]. In addition, some of the norepinephrine released from the postganglionic fibers innervating the heart can also reach the general circulation when it is produced in large amounts in the myocardium [2, 4]. Primarily epinephrine, as well as a small amount of norepinephrine, are produced by chromaffin cells located in the medulla of the adrenal glands [3, 4]. Under normal physiological circumstances 80% of catecholamine production will consist of epinephrine with the remaining 20% being norepinephrine [2].

Heart failure results from the heart's inability to function effectively as a pump due to diseases such as a myocardial infarction or ischemia. Chronic activation of adrenergic signaling is an important feature of many forms of heart disease. Elevated plasma norepinephrine levels in individuals with heart failure are strongly correlated with decreased survival and increased risk of death from chronic heart failure [5]. In heart failure, overactivation of the sympathetic system occurs with the primary goal of maintaining cardiac output in the failing heart [1, 4]. A study by Chidsey et al. [6] was one of the first to highlight the association between heart failure and increased functioning of sympathetic nervous system, as illustrated by higher norepinephrine secretion [6]. Decreased cardiac output in heart failure patients can also impact on the clearance of catecholamines from the plasma, and this, along with increased activation of the adrenal gland further contributes to increased catecholamines in the plasma of heart failure patients [2]. In untreated heart failure patients, the levels of circulating norepinephrine can reach approximately 50 times when compared to those in normal individuals who are exercising to their maximum ability [2, 7]. Plasma levels of norepinephrine in patients with congestive heart failure are significantly higher than those reported for angina patients with no clinical features of heart failure [8, 9]. Elevated epinephrine and norepinephrine levels have also been found in patients with hypertension [10]. In addition to the overactivation of adrenergic system in heart failure, decreased cardiac output increases production of angiotensin II and endothelin, which collectively promote systemic vasoconstriction and increase the afterload. These systemic changes further reduce ejection fraction and cardiac output, and the cycle repeats. The downward spiral is continued until a new steady state is reached in which cardiac output is lower and afterload is higher than optimal for normal activity.

Despite the discovery of several therapeutic interventions, heart failure due to ischemia remains the number one cause of death worldwide [11]. In adults, cardiomyocytes (CMs) that die in response to aging and pathological insults are replaced by scar tissue instead of new muscle cells [12, 13]. While recent reports suggest an intrinsic capacity for the mammalian myocardium to regenerate via endogenous stem/progenitor cells, the magnitude of such response appears to be minimal and not yet realized in CV patients [14, 15]. Current therapies can increase the life expectancy of end-stage heart failure patients by 2–3 years [16]. Considering these issues, a deeper understanding of molecular mechanisms regulating cell death and survival is essential to combat the increasing burden of heart failure. Studies from multiple laboratories revealed that ischemia and cytotoxic levels of catecholamines can induce autophagy and programmed cell death or apoptosis in many forms of heart disease [17–19]. These two events can account in part for the loss of function in the diseased heart. In this chapter, we have briefly reviewed various signal transduction mechanisms linking adrenergic receptors with apoptosis and autophagy in the heart. In addition, we have provided a brief survey of the literature related to the effects of adrenergic receptor blockers on apoptosis and autophagy in the heart. Subsequently, we have identified some important research questions for future studies.

Adrenergic Receptors in the Heart

Norepinephrine and epinephrine elicit their effects by binding to α- and β-adrenergic receptors (AR). The β-adrenergic receptors (β-ARs) that are located in the mammalian heart are of three subtypes β1, β2, and β3 [20]. Under normal physiological circumstances only β1 and β2 receptors are active [1, 21]. Additionally, α1-ARs of the subtypes 1A, 1B, and 1D are also present in the heart, although their levels of expression are lower than the β-ARs. α1-AR subtypes 1A and 1B are found on cardiomyocytes whereas 1D receptors are found on smooth muscle cells in the coronary arteries. The role of α1-ARs in the heart is not well defined, but studies have shown that they have a protective effect against apoptosis and necrosis and these receptors are also reported to increase the strength of cardiac contractions. Through interactions with α1-ARs in blood vessels epinephrine and norepinephrine can elicit vasoconstriction [1, 22].

Both β1- and β2-ARs are coupled to adenylyl cyclase (AC) through the stimulatory G protein, G_s. Activation of AC leads to an increase in the intracellular levels of the second messenger, cyclic adenosine monophosphate (cAMP) which can subsequently activate protein kinase A (PKA). β2-ARs can also interact with the inhibitory G protein, G_i. Under normal conditions, cardiomyocytes predominantly express the β1 subtype, whereas cardiac fibroblasts predominantly express the β2 subtype [23–27]. The β3-ARs are believed to interact with the G_i protein and subsequently inhibit the activation of adenylyl cyclase, ultimately resulting in myocardial relaxation. These receptors are thought to counteract the effects that the β1- and β2-ARs have on heart rate, conduction speed and inotropy. However, the β3-AR mediated actions are more relevant in disease states as these receptors become activated in response to higher concentrations norepinephrine compared with β1- or β2-ARs [28].

In the failing heart, the adrenergic signalling system changes in response to chronically increased levels of catecholamines. During heart failure, the ratio of β1 to β2 receptors changes from 3:1 to 1:1. Additionally, β1-AR expression at the plasma membrane is decreased and both β1- and β2-ARs become uncoupled from their associated G proteins and this process is termed as β-AR desensitization [2, 29]. The expression level of α1-ARs has been found to either remain constant or increase during heart failure [22].

Role of Adrenergic Receptor Signaling in the Regulation of Apoptosis and Autophagy in the Heart

Apoptosis is a controlled form of cell death. Unlike necrotic cell death where the contents of a dying cell leak out, apoptotic cells are not detrimental to surrounding cells nor do they initiate an inflammatory response [30]. Apoptosis is initiated by a cascade of caspase enzymes ultimately leading to the cleavage of critical proteins required for cell survival and the release of caspase activated DNAse enzymes. Subsequently, cells undergoing apoptosis shrink, lose their cytoskeleton and the cellular DNA is fragmented to the nucleosomal level. Apoptotic cells are then phagocytosed by surrounding cells or macrophages which recognize specific proteins on the surface of dying cells [30]. Autophagy is a cellular process activated in response to cellular stress and is also active in a form of non-apoptotic programmed cell death known as autophagic cell death [23]. Levine and Kroemer [31] suggested that autophagy is an adaptive mechanism used by cells to deal with injurious cellular components such as injured organelles or aggregates of protein. The pathway by which autophagy occurs can be different according to the cell type/environment [31]. The adrenergic signaling pathways associated with cardiovascular apoptosis and autophagy are discussed in subsequent sections.

Chronic catecholamine stimulation can lead to apoptosis in the heart [26, 32, 33]. When cardiomyocytes die, cardiac fibroblasts proliferate to maintain the architecture of the heart and this results in pathological remodeling and contributes to the development of heart failure [34]. Norepinephrine-induced apoptosis occurs via stimulation of the β1-AR and subsequent activation of cAMP-PKA signaling as well as via PKA independent pathways (Fig. 2.1). Conversely, activation of the β2-AR opposes apoptosis. The anti-apoptotic actions of the β2-AR stimulation occur via activation of the G_i protein and subsequent phosphoinositide-3-kinase (PI3K) and Akt activity [2, 32, 33, 35, 36] (Fig. 2.1). The p38 MAP kinase, a known downstream target of PKA, has been shown to play both pro- and anti-apoptotic roles in β-AR

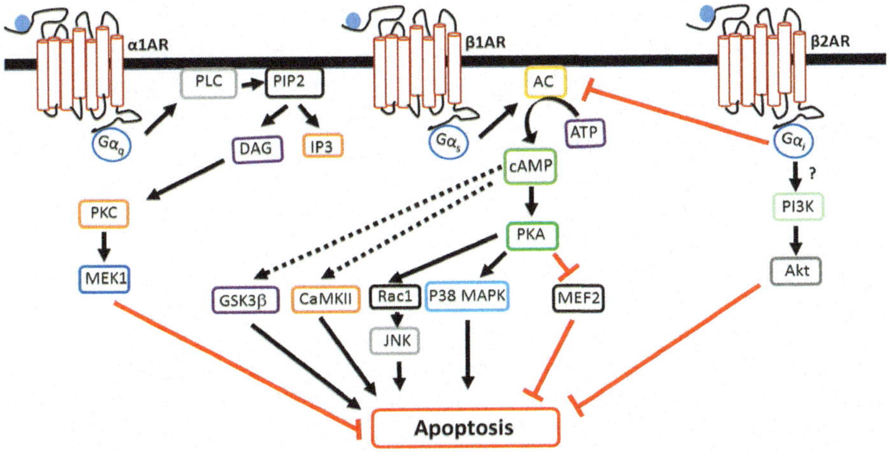

Fig. 2.1 Schematic diagram depicting various mediators involved in the adrenergic receptor (AR) agonist mediated effects on apoptosis in the heart. PLC: phospholipase C; PIP2: phosphotidyl inositol 4,5 biphosphate; IP3: inositol triphosphate; DAG: diacyl glycerol; PKC: protein kinase C; MEK1: mitogen activated protein kinase kinase; AC: adenylyl cyclase; ATP: adenosine triphosphate; cAMP: cyclic adenosine monophosphate; PKA: protein kinase A, p38 MAPK: p38 mitogen activated protein kinase, MEF2: myocyte enhancer factor 2; CaMKII: calcium dependent calmodulin kinase II; GSK3β: glycogen synthase kinase 3β; JNK: c-Jun N-terminal kinase; PI3K: phosphotidyl inositol-3 kinase and Akt: protein kinase B. Unknown signaling connections are indicated by question marks. Solid black arrows indicate the stimulatory events, dashed lines with arrowheads indicate stimulatory events that occur independent of PKA and solid red lines indicate inhibitory events

mediated apoptosis in cardiomyocytes [37, 38]. Notably, chronic stimulation of β1-ARs can also promote cardiomyocyte apoptosis independent of PKA via activation of Ca^{2+}/calmodulin kinase II (CaMKII) [39]. Other studies found that stimulation of β1-ARs can increase cardiomyocyte apoptosis via activation of either Rac1/c-jun N-terminal kinase pathway [40] or GSK-3β [41]. Stimulation of the β1-ARs in neonatal cardiomyocyte cultures by isoproterenol (ISO) or 8-Br-cAMP induced apoptosis [35]. Notably, ISO induced apoptosis could be blocked by co-treatment with a PKA inhibitor and 8-Br-cAMP induced apoptosis was blocked by co-treatment with an α1-AR agonist, phenylephrine via the $G_{\alpha q}$/PKC/MEK-1 pathway (Fig. 2.1) [35].

In addition to the increased levels of catecholamines, autoantibodies against the β1-ARs are frequently generated in patients with dilated cardiomyopathy [42] and experimental models of heart disease [43]. Recent studies indicated that myocardial damage is sufficient to stimulate the generation of β1-AR specific autoantibodies in rats subjected for aortic banding or adriamycin treatment [43] and the long-term presence of these antibodies can significantly decrease cardiac function [44]. While these autoantibodies were shown to activate canonical β1-AR associated cAMP and PKA signaling pathway, sustained stimulation of β1-ARs with these antibodies was shown to increase levels of activated caspase-3 and cardiomyocytes apoptosis [45, 46]. Notably, β1-AR autoantibodies were also shown to increase the cardiomyocyte endoplasmic reticulum stress and apoptosis via β1-AR coupled CaMKII, p38 MAPK and ATF6 pathway [47]. Long-term exposure to β1-AR autoantibodies was also shown to decrease the myocardial autophagy response and the cardiac dysfunction associated with autoantibodies could be reversed by upregulation of autophagy using the mTOR inhibitor, rapamycin [48].

Consistent with a role for adrenergic receptor signaling in the modulation of cardiomyocyte autophagy, stimulation of β-ARs using a non-selective agonist, isoproterenol was shown to activate Epac1 and increase autophagy in mouse hearts as well as primary rat cardiomyocytes [49]. Notably, pharmacological inhibition or germline deletion of Epac1 prevented the induction of β-AR mediated remodeling and cardiomyocyte autophagy in response to isoproterenol treatment. Specifically, it was shown that agonist actions on β-ARs can activate Epac1 via cAMP, leading to activation of Rap2B and phospholipase C which subsequently activate the Ca^{2+}/CaMKKβ/AMPK pathway (Fig. 2.2). This pathway is proposed to inhibit mTORC1 and thereby stimulate autophagy [49]. In this scenario, autophagy is proposed to be an adaptive response of cardiomyocytes to oppose the effects of Epac1 induced pathological cardiac remodeling [49]. Activation of the β2-ARs in adult rat cardiac fibroblasts by both isoproterenol and salbutamol (a β2-AR selective agonist) stimulated autophagy (Fig. 2.2), which augmented type I collagen degradation [23]. This study proposed that autophagy in response to high levels of catecholamines

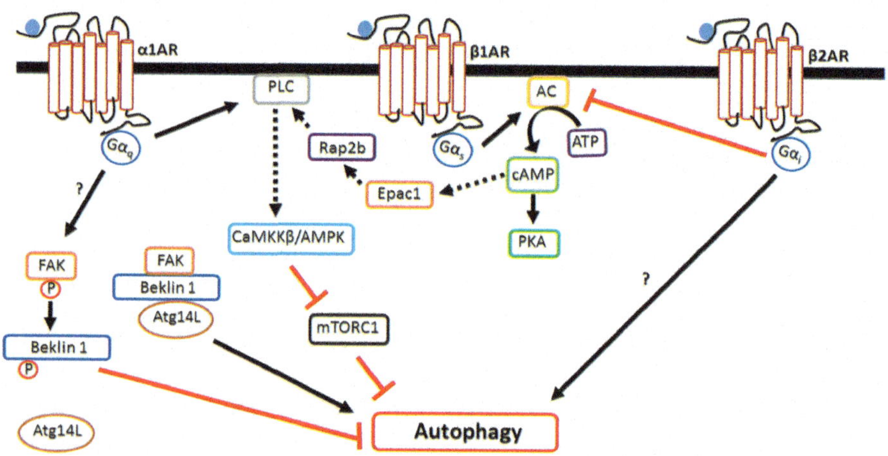

Fig. 2.2 Schematic diagram depicting various mediators involved in the adrenergic receptor (AR) agonist mediated effects on autophagy in the heart. FAK: focal adhesion kinase; Atg14L: autophagy related gene 14-like; AC: adenylyl cyclase; ATP: adenosine triphosphate; cAMP: cyclic adenosine monophosphate; PKA: protein kinase A; Epac1: exchange protein directly activated by cAMP 1; PLC: phospholipase C; CaMKKβ: calcium dependent calmodulin kinase kinase β; AMPK: 5' adenosine monophosphate-activated protein kinase; mTORC1: mammalian target of rapamycin 1. Unknown signaling connections are indicated by question marks. Solid black arrows indicate the stimulatory events, dashed lines with arrowheads indicate stimulatory events that occur independent of PKA and solid red lines indicate inhibitory events

may be an attempt to decrease the fibrosis by balancing the changes in extracellular matrix composition resulting from increased adrenergic stimulation [23]. A recent study demonstrated that stimulation of α1-ARs with phenylephrine in cardiomyocyte cultures or in vivo can suppress autophagy and promote hypertrophy (Fig. 2.2) via focal adhesion kinase mediated phosphorylation of Beclin1 and subsequent disruption of Beclin1-Atg14L interaction required for autophagosome formation (Fig. 2.2) [50]. While these observations underscore the importance of α1-AR signaling mediated suppression of cardiomyocyte autophagy in compensatory hypertrophy, it is not known whether this pathway plays any regulatory role in pathological hypertrophy and heart failure.

β-AR Blockers and Heart Disease

Three generations of β-AR blocking drugs have been developed over the past several decades. The first generation of β-blockers are non-selective, meaning that they block both the β1- and β2-AR subtypes. These drugs include propranolol, nadolol and timolol. The second generation β-blockers bind preferentially to the β1-AR subtype and include metoprolol, atenolol and bisoprolol. The third-generation blockers are either selective for the β1 subtype (e.g. celiprolol and nebivolol) or non-selective (e.g. carvedilol). Importantly, the third-generation blockers can also cause vasodilation in the periphery. Carvedilol does so by blocking the α1-AR, while nebivolol elicits this effect by interacting with the endothelium and increasing production of nitric oxide (NO) [28, 51]. Additionally, carvedilol also has antioxidant properties due to its ability to "scavenge free radicals" [52, 53]. β-blockers are effective in treating chronic heart failure as can be seen by their ability to mitigate adverse left ventricular remodeling, decreasing heart rate which then decreases the amount of oxygen required by the heart and the workload of the heart. These outcomes are accomplished by decreasing the toxic actions of the elevated levels of norepinephrine, upregulating the number of β-ARs, among others [1, 2].

The difference in β-AR selectivity between the various β-blockers must be considered when selecting a drug for clinical use. The majority of the "therapeutic actions" of these drugs are due to interactions with β1-ARs, whereas many "adverse effects" are caused by interactions with the β2-AR such as blockade of this receptor in lungs can lead to bronchoconstriction [52]. This interaction has important implications for selecting an appropriate β-blocker for patients suffering from both heart disease and asthma [54, 55]. Additionally, it is important to note that while some drugs are cardioselective by preferentially binding to the β1-AR, such selectivity can be overcome at higher doses as in the case with metoprolol. Nebivolol was found to be more cardioselective compared to other β-blockers such as metoprolol or bisoprolol [52, 56]. The ability of third generation β-blockers to cause vasodilation may be of use in certain circumstances, as discussed below.

A meta-analysis conducted on the use of atenolol to treat uncomplicated hypertension found that it had no impact on the occurrence of myocardial infarction or mortality compared to placebo or no treatment, despite an ability to decrease blood pressure [57]. It was found that treatment of hypertension with third generation β-blockers had lower incidence of mortality compared with atenolol [57, 58]. Okamoto et al. investigated the actions of nebivolol versus metoprolol on hypertension in patients with autonomic failure to distinguish the β-AR effects from the vasodilatory effects through increasing NO. Nebivolol was able to significantly decrease the systolic blood pressure compared to control whereas metoprolol did not, and thus the authors concluded that nebivolol elicited these antihypertensive effects independent of its actions at the β1-AR [59]. Although another study found that nebivolol decreased the incidence of cardiovascular events more effectively than did metoprolol in patients with acute myocardial infarction (MI) and left ventricular dysfunction, there was no large difference between treatment with metoprolol and treatment with carvedilol with respect to cardiovascular events, nor was there a difference between the carvedilol and nebivolol groups [60]. The authors of this study suggest that carvedilol may be less effective in MI patients since blocking of β2-ARs can impair the pro-survival pathways [33]. Whereas, β2-AR blocking, and antioxidant properties of carvedilol are thought to be beneficial for the treatment of heart failure patients since β2-AR levels and oxidative stress are elevated in those patients [60]. Based on recent reports, it may be beneficial to keep both β2- and α1-AR signaling pathways intact in acute MI patients, as stimulation of β2-AR can increase autophagy and collagen degradation in cardiac fibroblasts [23] and α1-AR stimulation can promote hypertrophy by decreasing cardiomyocyte autophagy [50]. It is possible that blocking both β2- and α1-AR pathways in acute MI patients using carvedilol may lead to precocious fibrosis and impaired compensatory hypertrophic responses in addition to loss of pro-survival signals (Figs. 2.1 and 2.2), subsequently increasing adverse cardiovascular events. Thus, more research should be conducted to confirm the effects of these new generation β-blockers on pathways linking autophagy and apoptosis in addition to a clear focus on clinical outcomes.

New Mechanistic Insights for the Actions of β-AR Blockers in the Regulation of Apoptosis and Autophagy

It is clear from both experimental and clinical literature that β-AR blockers are highly effective in decreasing cardiac workload, oxidative stress, apoptosis, remodeling and arrhythmias in animal models or patients suffering from heart disease. Although β-AR blockers were initially not preferred for the treatment of heart failure due to their negative inotropic effects, positive results from clinical trials combined with new findings related to the novel actions of these drugs have increased a great deal of interest in the development of new generation drugs for the treatment of several forms of heart disease. Notably, the third-generation β-AR blockers nebivolol and carvedilol are also shown to act as antioxidants and increase the activity of endothelial nitric oxide synthase (NOS). These additional features are helpful since heart disease is often accompanied by oxidative stress [61, 62].

Carvedilol has been implicated in modulating levels of microRNAs (miR), which are known to repress the expression of certain genes following transcription [61]. A study by Xu et al. [63] found that carvedilol upregulated miR-133 in the infarcted rat heart as well as in hydrogen peroxide treated neonatal rat cardiomyocyte cultures [63]. Further, co-treatment of cardiomyocyte cultures with hydrogen peroxide and carvedilol resulted in suppression of caspase 9 expression and associated apoptosis which were normally seen with peroxide treatment alone [63]. However, the mechanistic link between carvedilol treatment and miR-133 upregulation is yet to be established. Activation of β1-ARs in neonatal cardiomyocyte cultures using ISO was shown to induced apoptosis via cAMP/PKA mediated suppression of myocyte enhancer factor 2 (MEF2) transcriptional activity and expression of pro-survival genes such as KLF6 [24]. Whereas, inhibition of β-ARs using atenolol enhanced the MEF2 transcriptional activity and promoted cardiomyocyte survival (Fig. 2..1) [34]. Although clinical use of atenolol over other β-AR blockers in the treatment of hypertension is debated [57, 58], new findings related to atenolol mediated pro-survival pathways could be of use shortly after a myocardial infarction as this could prevent cardiomyocyte apoptosis and the subsequent pathological remodeling [34].

Ahmet and colleagues investigated the long-term effects of β1-AR antagonism in combination with β2-AR agonism using a rat model of chronic heart failure over a period of one year [64]. Rats treated in this manner had a 34% increase in survival when compared to rats that were not treated. Left ventricular remodeling and function were improved with the combination treatment. Additionally, the size of the infarct did not increase, and cardiomyocyte apoptosis was significantly decreased in the combination treatment group [64]. In a rat model of coronary artery ligation and cardioplegia-induced cardiac arrest, carvedilol treatment was found to decrease the amount of cardiomyocyte apoptosis that occurred following the cardioplegia and subsequent reoxygenation via activation of PI3K and MEK (mitogen activated protein kinase kinase) [65]. These observations underscore the potential for carvedilol to be used prior to cardiac surgery for prevention of cardiomyocyte apoptosis. Using a rat model of acute MI, another study found that carvedilol treatment can mitigate the upregulation of toll like receptor 4 (TLR-4) that is commonly seen following an infarction. TLR4 activates inflammatory pathways and has been implicated in pro-apoptotic signaling in the infarcted heart, suggesting this may be the mechanism by which carvedilol decreases apoptosis [66].

Compared to the effects of β-AR blockers on apoptosis, there is scant information on the effects of these drugs on autophagy in the heart. In a rat cardiac resuscitation model, treatment with a short acting β1-AR blocker, esmolol, was effective in reducing the levels of epinephrine induced cardiomyocyte apoptosis, and this protective effect was also correlated with a significant reduction in cardiomyocyte autophagy, as evidenced by reductions in Beclin-1 and parkin expression levels [67]. In a mouse model of pressure-overload induced cardiac dysfunction, treatment with metoprolol (β1-AR blocker) in combination with quiliqiangxin (QL; a Chinese herbal medication) over a 4-week period significantly improved cardiac function, decreased remodeling and apoptosis compared to QL treatment alone [68]. Although QL+ metoprolol treatment was effective in reducing the levels of an autophagy marker protein (LC3-II) in the diseased heart when compared to no drug treatment group, this combination was not as effective as other combination treatments such as QL+ an angiotensin II receptor blocker or an angiotensin converting enzyme inhibitor [68]. In contrast to the suppressive effects of esmolol and metoprolol on autophagy, an earlier study showed that acute treatment of healthy rats with propranolol (a non-selective β-AR blocker) can significantly increase the volume fraction and density of autophagic vacuoles (AVs) in the left ventricular myocardium compared to the control hearts within 2–4 h after treatment. This study suggested that autophagy is an early cardiac adaptation response to reduced workload since similar effects on AVs were also observed after acute treatment with a calcium channel blocking drug verapamil [69]. Similar to the effects of propranolol on AVs, treatment of rats subjected to acute MI with carvedilol for 2 weeks also revealed a significant increase in the number of AVs, autophagy- and anti-apoptosis related proteins (e.g. Beclin-1 and Bcl-XL) in the infarct region and the region bordering infarction compared to those regions in untreated MI hearts [70]. Notably, treatment of non-cardiac HepG2 cells with propranolol in the absence or presence of a β2-AR agonist clenbuterol significantly increased the LC3-II protein levels [71]. However, propranolol treated HepG2 cells also revealed a significant increase in SQSTM1/p62 protein levels as well as a significant decrease in the number of autolysosomes suggesting that autophagosome flux was blocked at late stages such as lysosomal fusion, acidification or protease digestion [71]. Future studies on autophagosome flux with regards to the status of SQSTM1/p62 levels and autolysosome formation may provide additional insights into the mechanisms underlying protective effects of β-AR blockers in different cardiovascular cell types.

Conclusion

Here, we provide evidence from the literature that adrenergic signaling plays a critical role in the regulation of apoptosis and autophagy in cardiovascular system and both adrenergic receptor agonists and antagonists can be used to modulate the levels of apoptosis and autophagy in the heart. Although common signaling mediators can be readily identified for both pathways, more work needs to be done to define the relationship between apoptosis and autophagy and their relative significance in the context of cardiovascular disease. Some studies suggest that upregulation of autophagy can improve cardiac function [72] while other studies suggest that downregulation of autophagy may be beneficial during disease states [73, 74]. It has also been suggested that inadequate autophagy responses may contribute to cardiac dysfunction [75]. Some in vitro studies suggest that the degree of ischemic injury may dictate the manifestations of autophagy, apoptosis and necrosis in the heart. For example, mild ischemic injury may result in autophagy and apoptosis whereas moderate or severe ischemia may cause apoptosis and necrosis [76]. Thus, it is generally thought that autophagy can be an early adaptive response to disease and increased autophagy can delay the onset of both apoptotic and necrotic cell death. However, increased autophagy is not always associated with a concomitant decrease in apoptosis as seen with the in vitro effects of different adrenergic receptor agonists and antagonists on these processes. A better understanding of molecular events involved in autophagy and apoptosis and the interplay between these pathways in heart disease as well as additional mechanistic information related to the effects of adrenergic drugs specifically on these processes will enable us to develop efficacious treatments.

Acknowledgements We thank the Heart and Stroke Foundation of Canada for the grant support (G-15-0009233 and G-18-0022140).

References

1. Triposkiadis F, Karayannis G, Giamouzis G et al (2009) The sympathetic nervous system in heart failure physiology, pathophysiology, and clinical implications. J Am Coll Cardiol 54:1747–1762
2. Lymperopoulos A, Rengo G, Koch WJ (2013) Adrenergic nervous system in heart failure: pathophysiology and therapy. Circ Res 113:739–753
3. Lymperopoulos A, Rengo G, Koch WJ (2007) Adrenal adrenoceptors in heart failure: fine-tuning cardiac stimulation. Trends Mol Med 13:503–511
4. Pepper GS, Lee RW (1999) Sympathetic activation in heart failure and its treatment with beta-blockade. Arch Intern Med 159:225–234
5. Cohn JN, Levine TB, Olivari MT et al (1984) Plasma norepinephrine as a guide to prognosis in patients with chronic congestive heart failure. N Engl J Med 311:819–823
6. Chidsey CA, Braunwald E, Morrow AG (1965) Catecholamine excretion and cardiac stores of norepinephrine in congestive heart failure. Am J Med 39:442–451
7. Morris MJ, Cox HS, Lambert GW et al (1997) Region-specific neuropeptide Y overflows at rest and during sympathetic activation in humans. Hypertension 29:137–143
8. Levine TB, Francis GS, Goldsmith SR et al (1982) Activity of the sympathetic nervous system and renin-angiotensin system assessed by plasma hormone levels and their relation to hemodynamic abnormalities in congestive heart failure. Am J Cardiol 49:1659–1666
9. Viquerat CE, Daly P, Swedberg K et al (1985) Endogenous catecholamine levels in chronic heart failure. Relation to the severity of hemodynamic abnormalities. Am J Med 78:455–460
10. de Champlain J, Farley L, Cousineau D et al (1976) Circulating catecholamine levels in human and experimental hypertension. Circ Res 38:109–114
11. Mathers C, Stevens G, Mahanani WR et al (2017) WHO methods and data sources for country-level causes of death 2000–2015. Glob Health Esitmates Tech Pap Ser 1–85
12. Pasumarthi KB, Nakajima H, Nakajima HO et al (2005) Targeted expression of cyclin D2 results in cardiomyocyte DNA synthesis and infarct regression in transgenic mice. Circ Res 96:110–118
13. Gaspard GJ, Pasumarthi KB (2008) Quantification of cardiac fibrosis by colour-subtractive computer-assisted image analysis. Clin Exp Pharmacol Physiol 35:679–686
14. McMullen NM, Pasumarthi KB (2007) Donor cell transplantation for myocardial disease: does it complement current pharmacological therapies? Can J Physiol Pharmacol 85:1–15
15. Murry CE, Field LJ, Menasche P (2005) Cell-based cardiac repair: reflections at the 10-year point. Circulation 112:3174–3183

16. Lloyd-Jones DM (2001) The risk of congestive heart failure: sobering lessons from the Framingham heart study. Curr Cardiol Rep 3:184–190
17. Lavandero S, Chiong M, Rothermel BA et al (2015) Autophagy in cardiovascular biology. J Clin Invest 125:55–64
18. Mughal W, Dhingra R, Kirshenbaum LA (2012) Striking a balance: autophagy, apoptosis, and necrosis in a normal and failing heart. Curr Hypertens Rep 14:540–547
19. Mughal W, Kirshenbaum LA (2011) Cell death signalling mechanisms in heart failure. Exp Clin Cardiol 16:102–108
20. Lohse MJ, Engelhardt S, Eschenhagen T (2003) What is the role of beta-adrenergic signaling in heart failure? Circ Res 93:896–906
21. Skeberdis VA, Gendviliene V, Zablockaite D et al (2008) beta3-adrenergic receptor activation increases human atrial tissue contractility and stimulates the L-type Ca^{2+} current. J Clin Invest 118:3219–3227
22. Jensen BC, O'Connell TD, Simpson PC (2014) Alpha-1-adrenergic receptors in heart failure: the adaptive arm of the cardiac response to chronic catecholamine stimulation. J Cardiovasc Pharmacol 63:291–301
23. Aranguiz-Urroz P, Canales J, Copaja M et al (2011) Beta(2)-adrenergic receptor regulates cardiac fibroblast autophagy and collagen degradation. Biochim Biophys Acta 1812:23–31
24. Gustafsson AB, Brunton LL (2000) beta-adrenergic stimulation of rat cardiac fibroblasts enhances induction of nitric-oxide synthase by interleukin-1beta via message stabilization. Mol Pharmacol 58:1470–1478
25. Lau YH, Robinson RB, Rosen MR et al (1980) Subclassification of beta-adrenergic receptors in cultured rat cardiac myoblasts and fibroblasts. Circ Res 47:41–48
26. Lefkowitz RJ, Rockman HA, Koch WJ (2000) Catecholamines, cardiac beta-adrenergic receptors, and heart failure. Circulation 101:1634–1637
27. Leicht M, Greipel N, Zimmer H (2000) Comitogenic effect of catecholamines on rat cardiac fibroblasts in culture. Cardiovasc Res 48:274–284
28. Rehsia NS, Dhalla NS (2010) Mechanisms of the beneficial effects of beta-adrenoceptor antagonists in congestive heart failure. Exp Clin Cardiol 15:e86-95
29. Bristow MR, Ginsburg R, Umans V et al (1986) Beta 1- and beta 2-adrenergic-receptor subpopulations in nonfailing and failing human ventricular myocardium: coupling of both receptor subtypes to muscle contraction and selective beta 1-receptor down-regulation in heart failure. Circ Res 59:297–309
30. Fink SL, Cookson BT (2005) Apoptosis, pyroptosis, and necrosis: mechanistic description of dead and dying eukaryotic cells. Infect Immun 73:1907–1916
31. Levine B, Kroemer G (2009) Autophagy in aging, disease and death: the true identity of a cell death impostor. Cell Death Differ 16:1–2
32. Chesley A, Lundberg MS, Asai T et al (2000) The beta(2)-adrenergic receptor delivers an anti-apoptotic signal to cardiac myocytes through G(i)-dependent coupling to phosphatidylinositol 3'-kinase. Circ Res 87:1172–1179
33. Communal C, Singh K, Sawyer DB et al (1999) Opposing effects of beta(1)- and beta(2)-adrenergic receptors on cardiac myocyte apoptosis: role of a pertussis toxin-sensitive G protein. Circulation 100:2210–2212
34. Hashemi S, Salma J, Wales S et al (2015) Pro-survival function of MEF2 in cardiomyocytes is enhanced by beta-blockers. Cell Death Discov 1:15019
35. Iwai-Kanai E, Hasegawa K, Araki M et al (1999) alpha- and beta-adrenergic pathways differentially regulate cell type-specific apoptosis in rat cardiac myocytes. Circulation 100:305–311
36. Zhu WZ, Zheng M, Koch WJ et al (2001) Dual modulation of cell survival and cell death by beta(2)-adrenergic signaling in adult mouse cardiac myocytes. Proc Natl Acad Sci U S A 98:1607–1612
37. Communal C, Colucci WS, Singh K (2000) p38 mitogen-activated protein kinase pathway protects adult rat ventricular myocytes against beta-adrenergic receptor-stimulated apoptosis. Evidence for Gi-dependent activation. J Biol Chem 275:19395–19400

38. Peter PS, Brady JE, Yan L et al (2007) Inhibition of p38 alpha MAPK rescues cardiomyopathy induced by overexpressed beta 2-adrenergic receptor, but not beta 1-adrenergic receptor. J Clin Invest 117:1335–1343
39. Zhu WZ, Wang SQ, Chakir K et al (2003) Linkage of beta1-adrenergic stimulation to apoptotic heart cell death through protein kinase A-independent activation of Ca^{2+}/calmodulin kinase II. J Clin Invest 111:617–625
40. Ito M, Adachi T, Pimentel DR et al (2004) Statins inhibit beta-adrenergic receptor-stimulated apoptosis in adult rat ventricular myocytes via a Rac1-dependent mechanism. Circulation 110:412–418
41. Menon B, Johnson JN, Ross RS et al (2007) Glycogen synthase kinase-3beta plays a pro-apoptotic role in beta-adrenergic receptor-stimulated apoptosis in adult rat ventricular myocytes: role of beta1 integrins. J Mol Cell Cardiol 42:653–661
42. Magnusson Y, Wallukat G, Waagstein F et al (1994) Autoimmunity in idiopathic dilated cardiomyopathy. Characterization of antibodies against the beta 1-adrenoceptor with positive chronotropic effect. Circulation 89:2760–2767
43. Liu HR, Zhao RR, Jiao XY et al (2002) Relationship of myocardial remodeling to the genesis of serum autoantibodies to cardiac beta(1)-adrenoceptors and muscarinic type 2 acetylcholine receptors in rats. J Am Coll Cardiol 39:1866–1873
44. Zuo L, Bao H, Tian J et al (2011) Long-term active immunization with a synthetic peptide corresponding to the second extracellular loop of beta1-adrenoceptor induces both morphological and functional cardiomyopathic changes in rats. Int J Cardiol 149:89–94
45. Jane-wit D, Altuntas CZ, Johnson JM et al (2007) Beta 1-adrenergic receptor autoantibodies mediate dilated cardiomyopathy by agonistically inducing cardiomyocyte apoptosis. Circulation 116:399–410
46. Gao Y, Liu HR, Zhao RR et al (2006) Autoantibody against cardiac beta1-adrenoceptor induces apoptosis in cultured neonatal rat cardiomyocytes. Acta Biochim Biophys Sin (Shanghai) 38:443–449
47. Mao W, Fukuoka S, Iwai C et al (2007) Cardiomyocyte apoptosis in autoimmune cardiomyopathy: mediated via endoplasmic reticulum stress and exaggerated by norepinephrine. Am J Physiol Heart Circ Physiol 293:H1636-1645
48. Wang L, Hao H, Wang J et al (2015) Decreased autophagy: a major factor for cardiomyocyte death induced by beta1-adrenoceptor autoantibodies. Cell Death Dis 6:e1862
49. Laurent AC, Bisserier M, Lucas A et al (2015) Exchange protein directly activated by cAMP 1 promotes autophagy during cardiomyocyte hypertrophy. Cardiovasc Res 105:55–64
50. Cheng Z, Zhu Q, Dee R et al (2017) Focal adhesion Kinase-mediated phosphorylation of beclin1 protein suppresses cardiomyocyte autophagy and initiates hypertrophic growth. J Biol Chem 292:2065–2079
51. Maffei A, Di Pardo A, Carangi R et al (2007) Nebivolol induces nitric oxide release in the heart through inducible nitric oxide synthase activation. Hypertension 50:652–656
52. Mason RP, Giles TD, Sowers JR (2009) Evolving mechanisms of action of beta blockers: focus on nebivolol. J Cardiovasc Pharmacol 54:123–128
53. Yue TL, Cheng HY, Lysko PG et al (1992) Carvedilol, a new vasodilator and beta adrenoceptor antagonist, is an antioxidant and free radical scavenger. J Pharmacol Exp Ther 263:92–98
54. Baker JG, Gardiner SM, Woolard J et al (2017) Novel selective beta1-adrenoceptor antagonists for concomitant cardiovascular and respiratory disease. FASEB J 31:3150–3166
55. Matthys H, Doshan HD, Ruhle KH et al (1985) The bronchosparing effect of celiprolol, a new beta 1-alpha 2-receptor antagonist on pulmonary function of propranolol-sensitive asthmatics. J Clin Pharmacol 25:354–359
56. Gupta S, Wright HM (2008) Nebivolol: a highly selective beta1-adrenergic receptor blocker that causes vasodilation by increasing nitric oxide. Cardiovasc Ther 26:189–202
57. Carlberg B, Samuelsson O, Lindholm LH (2004) Atenolol in hypertension: is it a wise choice? Lancet 364:1684–1689
58. Pedersen ME, Cockcroft JR (2009) What is the role, if any, for beta-blockers as initial therapy for uncomplicated hypertension? Curr Opin Cardiol 24:325–332

59. Okamoto LE, Gamboa A, Shibao CA et al (2014) Nebivolol, but not metoprolol, lowers blood pressure in nitric oxide-sensitive human hypertension. Hypertension 64:1241–1247
60. Ozaydin M, Yucel H, Kocyigit S et al (2016) Nebivolol versus carvedilol or metoprolol in patients presenting with acute myocardial infarction complicated by left ventricular dysfunction. Med Princ Pract 25:316–322
61. Kurdi M, Booz GW (2014) Carvedilol protects the infarcted heart by upregulating miR-133: first evidence that disease state affects beta-adrenergic arrestin-biased signaling? J Mol Cell Cardiol 76:12–14
62. Vanhoutte PM, Gao Y (2013) Beta blockers, nitric oxide, and cardiovascular disease. Curr Opin Pharmacol 13:265–273
63. Xu C, Hu Y, Hou L et al (2014) beta-Blocker carvedilol protects cardiomyocytes against oxidative stress-induced apoptosis by up-regulating miR-133 expression. J Mol Cell Cardiol 75:111–121
64. Ahmet I, Krawczyk M, Zhu W et al (2008) Cardioprotective and survival benefits of long-term combined therapy with beta2 adrenoreceptor (AR) agonist and beta1 AR blocker in dilated cardiomyopathy postmyocardial infarction. J Pharmacol Exp Ther 325:491–499
65. Yeh CH, Chen TP, Wang YC et al (2013) Carvedilol treatment after myocardial infarct decreases cardiomyocytic apoptosis in the peri-infarct zone during cardioplegia-induced cardiac arrest. Shock 39:343–352
66. Liu Q, Zhang J, Xu Y et al (2013) Effect of carvedilol on cardiomyocyte apoptosis in a rat model of myocardial infarction: a role for toll-like receptor 4. Indian J Pharmacol 45:458–463
67. Lu J, Shen Y, Liu LJ et al (2015) Combining epinephrine and esmolol attenuates excessive autophagy and mitophagy in rat cardiomyocytes after cardiac arrest. J Cardiovasc Pharmacol 66:449–456
68. Ye Y, Gong H, Wang X et al (2015) Combination treatment with antihypertensive agents enhances the effect of qiliqiangxin on chronic pressure overload-induced cardiac hypertrophy and remodeling in male mice. J Cardiovasc Pharmacol 65:628–639
69. Bahro M, Pfeifer U (1987) Short-term stimulation by propranolol and verapamil of cardiac cellular autophagy. J Mol Cell Cardiol 19:1169–1178
70. Zhang JL, Lu JK, Chen D et al (2009) Myocardial autophagy variation during acute myocardial infarction in rats: the effects of carvedilol. Chin Med J (Engl) 122:2372–2379
71. Farah BL, Sinha RA, Wu Y et al (2014) beta-Adrenergic agonist and antagonist regulation of autophagy in HepG2 cells, primary mouse hepatocytes, and mouse liver. PLoS One 9:e98155
72. Bhuiyan MS, Pattison JS, Osinska H et al (2013) Enhanced autophagy ameliorates cardiac proteinopathy. J Clin Invest 123:5284–5297
73. Sun A, Cheng Y, Zhang Y et al (2014) Aldehyde dehydrogenase 2 ameliorates doxorubicin-induced myocardial dysfunction through detoxification of 4-HNE and suppression of autophagy. J Mol Cell Cardiol 71:92–104
74. Zhu H, Tannous P, Johnstone JL et al (2007) Cardiac autophagy is a maladaptive response to hemodynamic stress. J Clin Invest 117:1782–1793
75. Nakai A, Yamaguchi O, Takeda T et al (2007) The role of autophagy in cardiomyocytes in the basal state and in response to hemodynamic stress. Nat Med 13:619–624
76. Loos B, Genade S, Ellis B et al (2011) At the core of survival: autophagy delays the onset of both apoptotic and necrotic cell death in a model of ischemic cell injury. Exp Cell Res 317:1437–1453

Chapter 3
Apoptosis in Ischemic Heart Disease

Monika Bartekova, Anureet K. Shah, and Naranjan S. Dhalla

Abstract Ischemic heart disease (IHD) due to reduced coronary blood flow over a prolonged period as well as reperfusion of the ischemic myocardium are associated with irreversible cellular damage and contractile failure. Although several major mechanisms including inflammation, oxidative stress and intracellular Ca^{2+}-overload have been identified to induce cardiac injury in IHD, the occurrence of cardiomyocyte cell death due to apoptosis has been considered to play a critical role in the development of heart dysfunction. The activation of several apoptotic and anti-apoptotic proteins belonging to both extrinsic and intrinsic pathways of apoptosis has been demonstrated in IHD. Some major proteins of the extrinsic apoptotic pathway include Fas/Fas ligand, TNF-α, TRAIL as well as Caspase-8 and Caspase-10. On the other hand, mediator proteins of the intrinsic apoptotic pathway are: Bcl-2, Bax, Cytochrome C and Caspase-9. Both extrinsic and intrinsic pathways converge on the terminal apoptotic pathway in which different proteins such as Caspase-3, Caspase-6 and Caspase-7 are activated leading to the degradation of cellular constituents in the ischemic myocardium. Although several agents targeting different apoptotic proteins have been shown to exert cardioprotective effects in experimental studies, the results in various clinical trials in IHD have not been encouraging.

Keywords Apoptosis · Apoptotic proteins · Anti-apoptotic proteins · Ischemic heart disease (IHD) · Ischemia–reperfusion (IR) injury

---

M. Bartekova (✉)
Institute for Heart Research, Centre of Experimental Medicine, Slovak Academy of Sciences, Bratislava, Slovakia
e-mail: monika.bartekova@savba.sk

A. K. Shah
Department of Kinesiology, Nutrition and Food Science, California State University, Los Angeles, USA

N. S. Dhalla
Institute of Cardiovascular Sciences, Research Centre, Department of Physiology and Pathophysiology, Max Rady College of Medicine, St. Boniface Hospital Albrechtsen, University of Manitoba, Winnipeg, Canada

© Springer Nature Switzerland AG 2022

L. A. Kirshenbaum (ed.), *Biochemistry of Apoptosis and Autophagy*,
Advances in Biochemistry in Health and Disease 18,
https://doi.org/10.1007/978-3-030-78799-8_3

Introduction

Ischemic heart disease (IHD), a major cause of death worldwide, is characterized by decreased blood flow in coronary arteries due to reduction of their lumen, resulting in an insufficient supply of oxygen and nutrients to the heart. IHD is mostly caused by the formation of atherosclerotic plaque in the vessel wall and the reduced blood flow in the ischemic heart is usually restored by reperfusion upon thrombolytic therapy, angioplasty and coronary by-pass surgery. There also occurs the formation of collateral vessels as an adaptive mechanism for reperfusion that may help to restore the supply of oxygen and nutrients to the ischemic heart. However, if the reperfusion does not occur or is not instituted within a critical time period, it may even exacerbate the damage to the heart and is called ischemia-reperfusion (I/R) injury [1, 2].

Numerous mechanisms have been documented to be involved in the pathophysiology of IHD; these include inflammation and endothelial dysfunction due to invasion of low-density lipoprotein (LDL) particles and macrophages, leading to development of atherosclerosis [2–5]. Furthermore, several mechanisms, including the stimulation of intracellular protein kinase cascades, oxidative stress leading to mitochondrial dysfunction, intracellular Ca^{2+}-overload resulting in proteolysis. Altered gene expression leading to malfunctioning proteins, as well as activation of cell death pathways such as apoptosis and necrosis, are activated in the heart leading to cardiac dysfunction and cardiomyocyte death due to I/R injury in IHD [1, 2, 5]. The present article summarizes the current knowledge on the role of apoptosis in the pathophysiology of cardiac dysfunction in the ischemic-reperfused myocardium. In addition, it is intended to discuss the potential of anti-apoptotic therapy in the treatment of IHD.

Development of Apoptosis

Apoptosis is a conserved and highly regulated process of cell death that takes place in several physiological and pathological conditions in the body. It helps in the exclusion of malfunctional cells from the organism as well as balance the proper ratio between cell proliferation and cell death in physiological conditions and in fact plays an important role in the prevention of cancer development in highly proliferative organs. On the other hand, it participates in the progression of various diseases including cardiovascular diseases (CVD). In addition, it contributes to the organ damage due to different pathologies such as myocardial damage due to I/R injury [6]. It is noteworthy that the adult myocardium is a non-proliferative tissue and apoptosis occurs at a low rate in the heart. The apoptotic rate increases in the heart due to pathological stimuli such as ischemia, inflammation, oxidative stress, and thus apoptosis is considered to contribute to the development of majority of CVD including IHD. [5, 6].

There is a large body of evidence to indicate that apoptosis is mediated via two main signaling pathways namely intrinsic and extrinsic, which converge on the terminal common pathway involving executive caspases [5–14]. The intrinsic pathway involves molecular mechanisms consisting of cascades of activation/deactivation of certain proteins, which can in general be divided into pro- and anti-apoptotic proteins. A typical example of a pro-apoptotic protein is Bax (Bcl-2-associated X) protein whereas a typical anti-apoptotic protein is Bcl-2 (B-cell lymphoma 2) protein. Both of these proteins, together with other pro- and anti-apoptotic proteins of similar structure belong to the Bcl-2 family of apoptotic proteins. Usually, the ratio between Bcl-2 and Bax levels in the cell determines whether the apoptosis process will start or the cell will remain alive. Anti-apoptotic proteins are capable of hetero-oligomerizing with pro-apoptotic proteins thereby preventing the formation of homo-oligomers of pro-apoptotic proteins. If the balance between anti- and pro-apoptotic proteins will shift in favor of apoptosis, homo-oligomers of pro-apoptotic proteins are formed in membranes of the mitochondria and allow the flow of mitochondrial proteins (e.g. Cytochrome c) into the cytoplasm. They then act as pro-apoptotic signals and stimulate the proteolytic cascade of caspases (specific proteases), namely caspase-9 and caspase-3. The terminal caspase is caspase-3, which proteolytically stimulates caspase-activated deoxyribonuclease and specifically fragments genomic DNA.

The extrinsic pathway, also termed as death receptor pathway, is triggered by the action of extracellular “death signals”, as in the case of a bacterial infection associated with the activation of immune system. It starts with the activation of death receptors due to tumor necrosis factor (TNF) family; this signaling pathway is activated by the binding of death ligands with TNF at appropriate receptors. When these ligands bind, the death receptor activates the so-called adaptor protein of this signaling pathway within the cell, for example FADD (Fas-associated protein with death domain). The adaptor protein binds to the initiating pro-caspases-8 and -10, and forms a complex called death-inducing signaling complex (DISC). The DISC activates pro-caspase-8 and pro-caspase-10 for the activation of executive caspases-3, -6, and -7 leading to cell death. The DISC is inhibited by c-FLIP protein which shows homology to caspase-8 but does not have the caspase activity. In some cell types, both intrinsic and extrinsic pathways of apoptosis may be cross-linked through cleavage of BID (BH_3-interacting death domain agonist) protein by caspase-8 to truncated BID (tBID), which is then translocated to the mitochondria and causes permeabilization of the mitochondrial membrane required for the activation of intrinsic pathway of apoptosis.

It is pointed out that the executive caspases -3, -6 and -7 in the terminal apoptotic pathway are common enzymes for both the extrinsic and intrinsic pathways of apoptosis and their activation leads to cleavage of various substrates, which in turn results in the destruction of cells by the apoptotic pathway [7]. The overview of apoptotic pathways is outlined in the Fig. 3.1. It may also be noted that dysregulation of apoptotic processes (either enhanced or insufficient) has been shown to be involved in many different diseases including cancer [8], neurological diseases [9, 10], as well as infectious and autoimmune diseases [11–13]. Finally, altered apoptosis plays a significant role in different types of CVD including IHD [17]. In fact, it has been well documented that apoptosis contributes to cardiomyocyte cell death in acute myocardial infarction (AMI) [15, 16] and LV remodeling after MI [17]. Particularly,

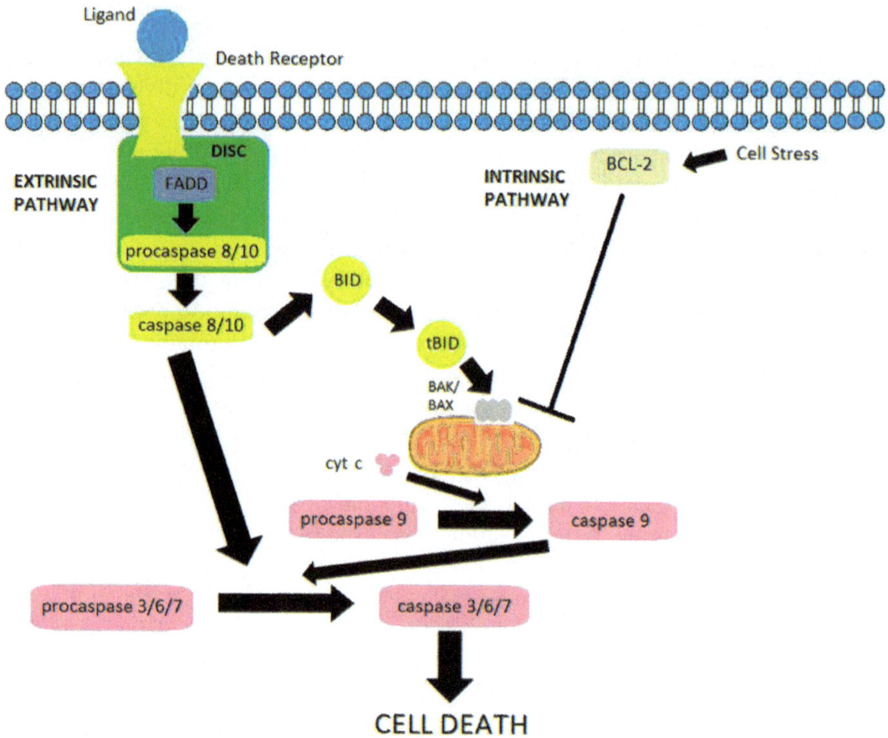

Fig. 3.1 Sketch for the apoptotic activation pathways. The intrinsic pathway consists of activation of pro- and anti-apoptotic proteins of the Bcl-2 family. If the equilibrium between them is shifted in favor of apoptosis, oligomers of pro-apoptotic proteins allow the flow of mitochondrial proteins (e.g. cytochrome c) into the cytoplasm. These then stimulate the proteolytic cascade of caspases, which in turn leads to cell death. On the other hand, the extrinsic pathway begins with the activation of death receptor, which stimulates an adaptor protein, such as FADD (Fas-associated protein with death domain). The adaptor protein together with the initiation caspase forms a death-inducing signaling complex (DISC) activating the initiation pro-caspases-8 and -10), which in turn activates the executive caspases-3, -6, and -7) which are common enzymes of both pathways. The whole process finally leads to cleavage of various substrates, and ultimately to cell destruction by the apoptotic pathway. In some cells, there occurs an interconnection of both pathways mediated through BID ("BH3-interacting death domain agonist"). This Figure is modified from the article by Zaman et al. [7]

apoptosis has been considered to determine the infarct size, as well as the extent of left ventricular remodeling, which contribute to the development of heart failure due to MI. Since the rate of apoptosis can be monitored by changes in various pro- and anti-apoptotic proteins in the heart tissue, the following discussion is planned to describe the participations of these mediator proteins for the extrinsic, intrinsic and common terminal apoptotic pathways in IHD.

Mediator Proteins of the Extrinsic Apoptotic Pathway

Fas/Fas ligand in IHD:

The Fas (First apoptosis signal) receptor, also known as Apo-1 or CD95, is a transmembrane protein of the TNF receptor family, which binds Fas ligand (FasL or CD95L or CD178). The interaction between Fas and FasL leads to the formation of the DISC apoptotic complex and activates the extrinsic apoptotic pathway. Fas is considered as a critical mediator of cardiomyocyte apoptosis in I/R injury [18, 19] and proposed as a potential target for the therapeutic strategies against chronic heart failure after AMI [20]. However, different results were obtained in different experimental models of cardiac I/R injury and human IHD patients. In an experimental AMI model in rats, mRNA and protein expressions of Fas/FasL in the heart tissue were significantly upregulated as compared to non-infarcted hearts, and were enhanced following the time extension of I/R [19]. On the other hand, in human AMI patients, the levels of soluble Fas receptor (sFas) observed in IHD patients differ from the results obtained by measuring the levels of FasL. Further, circulating levels of soluble FasL measured in patients 1 month after AMI were not associated with changes in LV monitored up to 1-year post MI [21]. Similar results were obtained in patients with ST-elevation myocardial infarction (STEMI) prior to percutaneous coronary intervention (PCI) and 24 h after the procedure, where levels of both sFas and sFasL did not correlate with infarct size or LV dysfunction [22]. In concordance, serum levels of sFas were not associated with prognosis of patients with acute coronary syndrome during 6-month follow-up period [23].

It was documented that genetic deletion of Fas/FasL ligands from Fas or FasL knock-out mice does not reduce infarct size after acute global I/R in isolated mouse hearts, suggesting that Fas/FasL is not the primary determinant of infarct size and ventricular dysfunction after AMI [24]. On the other hand, serum levels of sFasL were found elevated in patients with CAD as compared to patients with normal coronary arteries pointing to association of FasL levels with the severity of CAD and suggesting sFasL as a biochemical marker of coronary atherosclerosis [25]. Prognostic value of sFas was demonstrated also in patients with various forms of heart failure, where higher sFas concentrations were associated with higher risk of mortality or hospitalization for heart failure [26–28]. Finally, the inhibition of Fas-associated death domain-containing protein (FADD) protected mouse hearts against I/R injury in a mouse model of heart failure [29]. Taken together, the association between Fas/FasL-mediated apoptosis and cardiac I/R injury has been widely documented; however, the data are rather controversial and inconclusive to predict the potential of Fas/FasL as a therapeutic target or prognostic biomarker for IHD.

TNF-α in IHD:

TNF-α is a pro-inflammatory cytokine which has various biological functions, including important role in heart disease [5]. It plays an important role in mediating cardiomyocyte apoptosis due to I/R [30]. TNF-α has been also reported to be involved in Ca^{2+}-overload-induced cardiac apoptosis (Ca^{2+}-paradox) [31]. Depressed production of TNF-α by phosphodiesterase inhibitor pentoxifylline was associated with a dose-dependent cardioprotective effects in isolated rat hearts exposed to I/R manifested by markedly improved recovery of heart function upon reperfusion. This was associated with altered expression of NF-κB suggesting a role of inflammatory processes and/or apoptosis in the detrimental effects of TNF-α in cardiac I/R injury [32]. Similar findings have been obtained recently using 22-oxacalcitriol [33] or arginase-2 [34] to reduce the I/R injury in rats where both treatments suppressed the NF-κB/TNF-α pathway and reduced cardiomyocyte apoptosis, thus confirming the important role of TNF-α in apoptosis and inflammation in I/R injury.

Regarding molecular mechanisms underlying TNF-α mediated cardiomyocyte apoptosis, it is suggested that this includes the binding of TNF-α to TNF receptors or Fas thus triggering the classical extrinsic apoptotic pathway [30, 35]. Later it has been documented that TNF-α mediated cardiomyocyte apoptosis involves the activation of caspase-12 and calpain but not caspase-9 [36]. Recently, it has been found that stromal cell-derived factor-1 (SDF-1), the elevated levels of which are seen in a variety of clinical situations including AMI, induces TNF-mediated apoptosis in cardiac myocytes [37]. Furthermore, for the role of TNF-α in cardiac I/R injury, it should be mentioned that TNF-α may exert a dual effect in I/R depending on its concentration and the type of its receptor. It was documented that low doses of TNF-α administered during post-ischemic reperfusion to isolated rat hearts improved myocardial function, whereas high doses of TNF-α increased myocardial injury following I/R. When TNF-α was blocked with monoclonal anti-TNF-α antibody, the effects of TNF-α on myocardial contractility and relaxation were completely abolished [38]. It was demonstrated in an in vivo murine model of AMI that TNF-α by acting on TNF receptor 1 (TNFR1) exerts cardiotoxic effects but it is cardioprotective through TNF receptor 2 (TNFR2). In addition, TNFR2-knockout mice showed worst post-MI survival, greater ventricular dysfunction, as well as higher myocyte hypertrophy and fibrosis as compared to TNFR1-knockout mice, suggesting that selective blockade of TNFR1 may be a suitable therapeutic intervention for AMI [39]. However, clinical trials using compounds that antagonize TNF-α revealed diverse and largely disappointing results in CVD, notably in heart failure, most likely due to the ambivalent role of TNF-α and its receptors in cardiac function [40].

TRAIL in IHD:

TNF-related apoptosis stimulating ligand (TRAIL) is a protein that belongs to TNF superfamily and functions as a ligand inducing apoptosis by binding to death receptors. In addition to TNF-α, TRAIL has been widely documented in both animal models of cardiac I/R and in human patients with AMI to be involved in IHD. In experimental studies, TRAIL was found to be released from the post-ischemic hearts during early reperfusion period in isolated rat and mouse hearts [41]. The exact mechanism of TRAIL action in I/R has not been fully clarified yet; however, the role of TRAIL receptors in mediating pro-survival and proliferation signals has been pointed out [42]. In 2015, some investigators [43] used an in vivo mouse model to test the role of TRAIL in I/R, whereas cultured human microvascular endothelial cells were used to reveal molecular mechanisms for the vascular effects of TRAIL. It was demonstrated that TRAIL promotes angiogenesis in mice and that the ischemia-induced neovascularization is mediated via NADPH oxidase 4 (NOX4) and nitric oxide (NO)-dependent mechanisms. Accordingly, it was suggested that TRAIL may be a potential therapeutic target for improving the angiogenic response to ischemia and may increase the reperfusion recovery in patients with CVD [43]. In addition, an enhanced TRAIL expression has been reported in peripheral blood mononuclear cells after myocardial infarction, and the expression of TRAIL receptors 1 and 2 was documented immunohistochemically in human cardiomyocytes. These observations have suggested that TRAIL may be involved in the induction of cardiomyocyte apoptosis after AMI [44]. It has also been shown in macrophages that TRAIL facilitates cytokine expression and macrophage migration during hypoxia/reoxygenation via ER stress-dependent NF-κB pathway. It was suggested that TRAIL affects the functional activities of macrophages during I/R injury, and may represent a potential therapeutic target for the treatment of IHD [45]. Regarding potential therapeutic use of TRAIL in treating CVD, it has been shown that administration of soluble recombinant TRAIL reduced the development of cardiomyopathy in the ApoE(−/−) diabetic mice [46]. Systemic TRAIL delivery also induced the anti-atherosclerotic effect in diabetic mice [47]. Although these findings are not directly related to I/R injury, these data indicate certain therapeutic potential of TRAIL in cardioprotection.

Clinical trials were also found to support the relationship between TRAIL and IHD. It was reported that serum levels of TRAIL were significantly decreased in patients with AMI at baseline (within 24 h from admission) as compared to healthy controls and were inversely correlated with levels of negative prognostic markers, such as CK, CK-MB and BNP. Furthermore, the serum levels of TRAIL increased progressively at discharge and became normalized at 6–12 months after AMI. Notably, low serum TRAIL levels at the discharge time were associated with increased incidence of heart failure and cardiac death in the 12-month follow-up period. This study clearly indicated that the levels of circulating TRAIL might predict cardiovascular events independently of conventional cardiac risk markers [48]. This suggestion was further confirmed in a study by other investigators [49] where low levels of TRAIL were associated with increased risk of death during a 6-year period in older patients with CVD. Finally, low concentration of soluble TRAIL (sTRAIL) was found to be a strong predictor for poor prognosis in patients with acute coronary syndrome independently of age, ejection fraction, index peak troponin level, concentration of BNP or serum creatinine [23]. Soluble TRAIL concentration was found to be related to the composition of atheromatous plaques in patients with stable angina or positive for ischemia [50]. TRAIL concentrations were markedly reduced and associated with heart failure also in patients with established rheumatoid arthritis [51]. Recently it has been shown that TRAIL reached the lowest serum concentration after reperfusion in STEMI patients treated with primary PCI. In addition, low TRAIL levels were associated with worsened LVEF (LV ejection fraction) in acute STEMI and one month after STEMI. Thus, higher TRAIL level appears to be beneficial for making TRAIL as a potential protective mediator in post-AMI period [52]. In line with this suggestion, sTRAIL levels were found increased by ischemic post-conditioning and correlated with reduced infarct size and better LV function in patients with STEMI [53] pointing to the positive role of TRAIL in cardioprotective effects of ischemic conditioning.

Caspases 8 and 10 in IHD:

Caspase-8 is synthesized as inactive zymogen procaspase, and is activated by proteolytic cleavage through either autoactivation or trans-cleavage by other caspases. Activated caspase-8 propagates the apoptotic signal either directly by activating downstream caspases via their cleavage, or by cleaving the BH_3/Bcl-2 interacting protein, which evokes the release of cytochrome c from mitochondria [54]. Similarly, as other enzymes of the extrinsic apoptotic pathway, the activation of caspase-8 has been documented in cardiac I/R-induced apoptosis, and is consider to contribute in the myocardial damage in IHD. It was reported that simulated I/R in neonatal rat cardiomyocytes leads to marked conversion of inactive pro-caspase-8 to its cleaved active form; this conversion was also observed in cardiomyocytes of isolated hearts exposed to I/R [55, 56]. Increased levels of active caspase-8 were found in ischemic area of the mouse heart exposed to I/R by coronary occlusion in vivo [57]. In addition, sevoflurane anesthetic preconditioning blocked the increase in caspase-8 after I/R in isolated hearts, and this was associated with reduced cytochrome c release, improved functional recovery of hearts and decreased ischemic injury [58]. The cardioprotective effect of rosiglitazone against the I/R injury in rats in vivo was associated with the downregulation of active caspase-8 [59]. Interestingly, TNF-α induced caspase-8 activation leads to leaky ryanodine receptor channels which may also contribute to myocardial remodeling after I/R [60]. Finally, it was reported that transplantation of human mesenchymal cells with reduced caspase-8 activity (modified by adenovirus transfection of small hairpin RNA) significantly improved infracted heart function and attenuated cardiac fibrosis post MI [61].

It should also be pointed out that caspase-10 also exists as an inactive proenzyme that is cleaved to produce two subunits, large and small, which dimerize to form the active enzyme. Caspase-10 cleaves and activates caspases 3 and 7, and itself is processed by caspase 8. Regarding its role in IHD, there are no data documenting changes in the activity and/or expression of this enzyme in cardiac ischemia-related pathologies so far. The information regarding some of the mediator proteins involved in the extrinsic apoptosis pathway in IHD is summarized in Table 3.1.

Mediator Proteins of the Intrinsic (Apoptotic) Pathway

Bcl-2 family proteins in IHD:

Bcl-2 family consists of a number of proteins that share Bcl-2 homology domains. The Bcl-2 family members control apoptosis by affecting, permeability of the outer mitochondrial membrane, a key step in the intrinsic pathway of apoptosis. Most studied representatives of this family are Bcl-2 and Bax proteins; however,

Table 3.1 Some studies showing the role of various apoptotic mediators of the extrinsic pathway in myocardial I/R injury

Proteins	Injury/disease	Interventions	Observations	References
Fas/FasL	AMI in rats	None	↑mRNA/protein expression of Fas/FasL in heart tissue	Liu et al. [19]
	AMI in humans	None	↑sFas; unchanged FasL no association of Fas/FasL and IS, LV remodeling and patients prognosis	Fertin et al. [21], Nilsson et al. [22], Osmancik et al. [23]
	I/R in isolated mouse hearts	Knock-out of Fas/FasL	No reduction of IS due to deleted Fas/FasL	Tekin et al. [24]
	CAD patients	None	↑serum levels of sFasL	Sahinarslan et al. [25]
	Patients with HF	None	↑sFas associated with ↑risk of mortality or hospitalization for HF	Kawakami et al. [26], Tsutamoto et al. [27], Niessner et al. [28]
TNF-α	I/R in isolated rat hearts	Pentoxifilline treatment	↓TNF-α associated with improved heart function and ↓NF-κB	Zhang et al. [32]
	in vivo I/R in rat hearts	Treatment with 22-oxacalcitriol or arginase-2	↓I/R injury; ↓NF-κB/TNF-α ↓cardiomyocyte apoptosis	Zhou et al. [33], Huang et al. [34]
	I/R in isolated rat hearts	Administration of TNF-α	↓dose of TNF-α improved heart function ↑dose of TNF-α increased injury post-I/R	Asgeri et al. [38]
TRAIL	I/R in isolated rat/mouse hearts	None	↑release of TRAIL from hearts during early reperfusion	Jeremias et al. [41]

(continued)

other members such as Bak, Bcl-X_L, Bcl-w, Mcl-1 have also been shown to be involved in the apoptotic processes. The anti-apoptotic proteins including Bcl-2, Bcl-X_L, Bcl-w and Mcl-1 reside mainly in the mitochondria, protecting them against mitochondrial membrane permeabilization, while pro-apoptotic proteins like Bax (cytosolic) and Bak (associated with the outer mitochondrial membrane) are required for mitochondrial membrane permeabilization [62].

Table 3.1 (continued)

Proteins	Injury/disease	Interventions	Observations	References
	AMI in humans	None	↓serum levels of TRAIL in AMI patients TRAIL levels negatively correlated with AMI markers ↓serum levels of TRAIL associated with ↑incidence of HF and death	Secchiero et al. [48], Volpato et al. [49]
	STEMI patients	IPostC	↑levels of sTRAIL due to IPostC correlated with ↓IS and better LV function	Luz et al. [53]
Caspase-8	I/R in cultured CMs or isoleted hearts	None	Conversion of inactive pro-caspase-8 to active cleaved caspase-8	Stephanou et al. [55], Scarabelli et al. [56]
	in vivo I/R in mouse hearts	None	↑caspase-8 in ischemic area of hearts	Roubille et al. [57]
	I/R in isolated rat hearts	Sevoflurane preconditioning	↓caspase-8 associated with ↓cytochrome c release and improved functional recovery of hearts	Lu et al. [58]
	in vivo I/R in rat hearts	Rosiglitazone treatment	↓caspase-8 associated with cardioprotection	Palee et al. [59]

IS-Infarct size; I/R-Ischemia/reperfusion; AMI-Acute myocardial infarction; CAD-Coronary artery disease; HF-Heart failure; NF-κB-Nuclear factor kappa-B; IPostC-Ischemic post-conditioning; STEMI-Elevated ST segment myocardial infarction; LV-Left ventricle; ↑-Increase; ↓-Decrease

Regarding role of the intrinsic apoptotic pathway in IHD, it was demonstrated that cardiac overexpression of Bcl-2 in transgenic mice significantly reduced I/R injury, and this reduction correlated with the attenuation of cardiomyocyte apoptosis [63]. Expression of Bcl-2 and Bax was also examined in the post-mortem autopsied human hearts of patients who died of MI. It was documented that Bcl-2 was induced in salvaged myocytes at the acute stage of MI, whereas Bax was overexpressed at later stages of MI, suggesting that changes in Bcl-2/Bax expression may play an important pathophysiological role in acceleration of the apoptosis after I/R or protection of myocytes against I/R injury [64]. More recently, it was reported that several efficient cardioprotective interventions against I/R are accompanied with alterations in the intrinsic pathway of apoptosis in term of anti-apoptotic effects, suggesting that Bcl-2 family proteins may represent potential therapeutic target for protecting myocardium in IHD. For example, treatment of rats with a traditional Chinese medicine alkaloid, anisodamine, protected hearts from in vivo I/R injury, and this cardioprotection was associated with increased Bcl-2/Bax ratio in cardiomyocytes [65]. Association between altered Bcl-2/Bax expression and cardioprotection against I/R injury has also been reported in human patients undergoing valve replacement under cardiopulmonary bypass treated with L-carnitine. It was shown that L-carnitine reduced the surgery-induced myocardial apoptosis through enhanced expression of Bcl-2 and reduced expression of Bax, resulting in a protective effect of L-carnitine against cardiac I/R injury [66].

Several microRNAs (miRs) have been shown to be associated with Bcl-2/Bax-mediated apoptotic changes in I/R injury. For example, both miR-1 and miR-195 levels were downregulated and their common putative target gene Bcl-2 was upregulated after remote ischemic perconditioning (RIPerC) in patients with rheumatic valvular disease undergoing valve replacement surgery, suggesting the important role of gene expression in Bcl-2-mediated anti-apoptotic cardioprotective effects induced by RIPerC [67]. Recently, it was shown that hypoxia-induced mesenchymal stem cell-derived exosomes promoted cardiac repair in MI mice through miR-125b-mediated amelioration of apoptosis. In addition, exosomes with miR-125b knockdown lost the ability to suppress the expression of the pro-apoptotic BAK1 gene in cardiomyocytes [68]. The association between various miRs and Bcl-2 family proteins in cardiac I/R injury or cardioprotection has been widely documented in other studies using both in vitro and in vivo models of cardiac I/R injury [69–71]; these observations have opened a new avenue in this research field aimed to modulate apoptosis via exosome/miR-based therapies. Thus, there is no doubt that the intrinsic apoptotic pathway via its mediators from Bcl-2 protein family plays an important role in mediating the I/R-induced cardiac damage, and represents a potential target for the anti-ischemic therapies.

Cytochrome c in IHD:

Release of cytochrome c from mitochondria is one of the key processes involved in the intrinsic pathway of apoptosis induction. It may be noted that cytochrome c is an essential component of the mitochondrial electron transport chain which transfers electrons between Complexes III (Coenzyme Q-Cyt C reductase) and IV (Cyt C oxidase). Several studies have shown that the loss of cytochrome c from mitochondria is associated with cardiac I/R injury [72]; this point was not only documented in studies performed with isolated cardiac mitochondria [73, 74], but also in cardiac I/R injury [75, 76]. These observations were further supported by findings showing decreased cytochrome c release into the cytoplasm as a consequence of various cardioprotective interventions such as ischemic preconditioning [77] or treatment with natural antioxidants [78] in different models of cardiac I/R. In contrast to experimental studies, there are no clinical trials documenting cytochrome c release in IHD. However, in a human study in which some association of cytochrome c with IHD was found when plasma levels of cytochrome c were monitored in patients with chronic heart failure due to coronary heart disease as well as post-infarction cardiosclerosis after treatment with refracterin, a mixture containing cytochrome c and β-acetyldigoxin, NAD, oxyfedrine, and inosine [79]. In this study, increased levels of cytochrome c due to refracterin treatment were associated with increased activities of catalase and SOD, increased levels of NADH and NADPH, as well as increased total antioxidant activity. Notably, these alterations were also associated with improved cardiac function in these patients.

Caspase-9 in IHD:

Caspase-9 is an initiator caspase in the caspases-activating cascade, which results in apoptotic cell death after caspase-3 action in the terminal apoptotic pathway. Caspase-9 is activated by the release of cytochrome c from mitochondria and activation of apaf-1 (apoptosome) by the cleavage of pro-caspase-9. Similar to the above mentioned pro-apoptotic processes of the intrinsic pathway of apoptosis, increased activation of caspase-9 has been documented to be associated with cardiac I/R injury. In an in vitro model of hypoxia/reoxygenation of neonatal rat cardiomyocytes, 6-h of ischemia resulted in processing of pro-caspase-9 to active caspase-9 [55]. In an ex vivo model of I/R in isolated perfused rat hearts, pro-caspase-9 was processed to caspase-9 mainly in endothelial cells; addition of an inhibitor of caspase-9 to the perfusion solution prior to the ischemia prevented endothelial apoptosis in these hearts [56]. Cleaved caspase-9 was found upregulated also in in vivo infarcted rabbit hearts [80]. In addition, several cardioprotective interventions such as treatment with natural antioxidants [81, 82] or with opioid receptors agonists [83] were found to inhibit the cleavage of caspase-9 in I/R and improve heart function.

It is noteworthy that several experimental studies have indicated beneficial effects of different anti-apoptotic interventions for the intrinsic pathway. The data from some of the studies describing the role of several mediator proteins in the intrinsic apoptotic pathways are summarized in Table 3.2.

Mediator Proteins of the Terminal Apoptotic Pathway

Caspase-3 in IHD:

Caspase-3 is the most important enzyme of the terminal apoptotic pathway, as its activation results in degradation of cellular constituents followed by their fragmentation into apoptotic bodies, which are quickly removed by phagocytes. It was reported that increased cardiac specific caspase-3 was associated with increased infarct size and depressed cardiac function after in vivo I/R injury in transgenic mice overexpressing caspase-3 [84]. In addition, increased conversion from the inactive pro-caspase form to the active caspase-3 was found in the heart tissue after in vivo myocardial I/R injury in rabbits [85]. Caspase-3 was also up-regulated in the ischemic zone at

Table 3.2 Some studies showing the role of various apoptotic mediators of the intrinsic pathway in myocardial I/R injury

Proteins	Injury/disease	Interventions	Observations	References
Bcl-2/Bax	In vivo I/R in mouse hearts	Transgenic Bcl-2 overexpression	↑Bcl-2 reduced I/R injury and ↓cardiomyocyte apoptosis	Chen et al. [63]
	In vivo I/R in rat hearts	Anisodamine treatment	↑Bcl-2/Bax in cardiomyocytes associated with protection against I/R	Xing et al. [65]
	AMI in human patients	None	↑Bcl-2 in salvaged myocytes at the acute stage of MI ↑Bax in heart tissue of old MI	Misao et al [64]
	CPB patients valve replacement	L-carnitine treatment	↑Bcl-2 and ↓Bax reduced surgery-induced myocardial apoptosis	Li et al. [66]
	Rheumatic valvular disease patients	RIPerC	↑Bcl-2 associated with cardioprotection by RIPerC	Hu et al. [67]
Cytochrome c	In vivo I/R in rat hearts	IPC	↓cytochrome c release improved mitochondrial function	Lundberg and Szweda [77]
	I/R in isolated rat hearts	Anthocyanine treatment	↓cytochrome c release, ↓caspase activation improved mitochondrial respiration	Skemiene et al. [78]
Caspase-9	In vitro H/R in CMs	None	↑cleavage of pro-caspase-9 to active caspase-9	Stephanou et al. [55]
	I/R in isolated rat hearts	None	↑processing of pro-caspase-9 to caspase-9 in cardiac endothelial cells	Scarabelli et al. [56]

(continued)

Table 3.2 (continued)

Proteins	Injury/disease	Interventions	Observations	References
	In vivo MI in rabbit hearts	None	Upregulated cleaved caspase-9 in heart tissue	Qin et al. [80]

I/R-Ischemia/reperfusion; AMI-Acute myocardial infarction; MI-Myocardial infarction; CPB-Cardiopulmonary bypass; RI PerC-Remote ischemic preconditioning; IPC-Ischemic preconditioning; H/R-Hypoxia/reoxygenation; CMs-Cardiomyocytes; ↑-Increase; ↓-Decrease

5 h post I/R injury in pigs [86]. On the other hand, lentivirus-induced downregulation of caspase-3 was associated with decreased infarct size, reduced apoptosis in cardiomyocytes and improved heart function in an experimental model of AMI in rats [87]. Furthermore, downregulated expression of caspase-3 against I/R injury was documented to be the outcome of cardioprotective interventions such as ischemic conditioning [88, 89], flavonoid treatment [90, 91], anisodamine alkaloid treatment [65] and irisin (hormone, exercise-induced myokine) treatment [92]. Finally, altered caspase-3 expression has been associated with different miRs-mediated effects of I/R; up-regulated caspase-3 with miR-124-mediated detrimental effects [93], and down-regulated caspase-3 in cardioprotective effects mediated by miR-24-3p [94] or miR-322 [95].

Caspases 6 and 7 in IHD:

Caspase-6 is processed by caspases-7, -8 and -10, and can also undergo self-processing without other members of the caspase family. The precursor of caspase-7 is cleaved by caspase-3, -9, and -10. Both caspases-6 and -7 are involved in common terminal pathway of apoptosis and it was documented that gene and protein expression of these caspases were significantly down-regulated in mesenchymal stem cells preconditioned with hyperoxia. Since stem cell therapy may decrease infarct size and improve LV function post MI, these findings have been suggested to be of potential relevance in the treatment of AMI using stem cells [96]. In cardiac tissues of patients at the terminal stage of heart failure due to ischemic cardiomyopathy, no cleavage of caspase-7 has been observed [97]. On the other hand, a selective inhibitor of caspases-3/7 (MMPSI) was found to reduce myocardial injury due to I/R in isolated rabbit hearts [98]. It is pointed out that cardioprotection against in vivo cardiac I/R injury in rats by tilianin (flavonoid) [81], against oxygen–glucose deprivation/reperfusion in vitro injury in H9c2 cells by propofol [99], as well as in patients undergoing coronary artery bypass graft by N-acetylcysteine [100], was associated with a decrease in the expression of caspase-7. Notably, the downregulation of caspase-7 is, in fact exclusively, accompanied with the downregulation of caspase-3, thus pointing to hand-in hand action of these two caspases in IHD. The roles of these different proteins in the terminal apoptotic pathway are summarized in Table 3.3.

Table 3.3 Some studies showing the role of different apoptotic mediators of the terminal pathway in myocardial I/R injury

Proteins	Injury/disease	Interventions	Observations	References
caspase-3	In vivo I/R in mouse hearts	Transgenic caspase-3 overexpression	↑caspase-3 in heart tissue associated with ↑IS and ↓cardiac function	Condorelli et al. [84]
	In vivo I/R in rabbit hearts	None	↑conversion of pro-caspase-3 to active caspase-3	Holly et al. [85]
	In vivo I/R in pigs	None	↑caspase-3 in the ischemic zone at 5 h post I/R	Pavo et al. [86]
	AMI in rats	Transgenic caspase-3 downregulation	↓caspase-3 associated with ↓IS, ↓apoptosis and ↑heart function	Liu [87]
	In vivo I/R in rat hearts	RIC	↓caspase-3 associated with ↓IS and ↑heart function	You et al. [89]
	I/R in isolated rat hearts	Flavonoid treatment	↓caspase-3 associated with ↑heart function post I/R	Bartekova et al. [90]
caspase-7	In vivo I/R in rat hearts	Flavonoid treatment	↓caspase-7 associated with cardioprotection	Zeng et al. [81]
	CABG patients	Antioxidant treatment	↓caspase-7 associated with cardioprotection	Fischer et al. [100]

I/R-Ischemia/reperfusion; CABG-Coronary artery bypass graft; AMI-Acute myocardial infarction; RIC-Remote ischemic conditioning; IS-Infarct size; ↑-Increase; ↓-Decrease

Targeting Apoptosis for the Treatment of IHD

In view of the above mentioned observations, targeting apoptotic proteins can be seen to represent a promising therapeutic approach to prevent or treat myocardial I/R injury. In addition to non-specifically altered activity or expression of apoptotic proteins by diverse cardioprotective interventions including ischemic conditioning and antioxidant therapies, effects of several pharmacological interventions directly targeting concrete mediators of apoptotic pathways have been examined to prevent cardiac I/R injury. The targets of such interventions, primarily, TNF-α, and TNF-α inhibitors, were tested not only in experimental studies but also in clinical trials. Inhibition of TNF-α can be achieved by using monoclonal antibodies such as infliximab, adalimumab, certolizumab, or golimumab, etanercept, as well as simple molecules such as xanthine derivatives including pentoxifylline. In some studies, both groups of substances in combination were used to inhibit TNF-α in cardiac I/R injury.

It was demonstrated that inhibition of TNF-α with pentoxifylline improved cardiac recovery after I/R in isolated rat hearts [32]. Controversial results have been achieved in human studies in this regard. In the study of Aslanabadi et al. [101], pentoxifylline did not show away positive effect on peri-procedural myocardial injury in patients undergoing percutaneous coronary intervention whereas in the study of Mansourian et al. [102], it significantly improved the LV ejection fraction as well as decreased the intensive care unit stay and ventilation time in patients undergoing coronary artery bypass graft. It was also reported that the risks of AMI and heart failure are reduced in anti-TNF-treated patients with rheumatoid arthritis or psoriasis [103–107]. On the other hand, two large clinical trials namely RECOVER and RENAISSANCE, testing etanercept in patients with chronic heart failure were terminated prematurely since no clinical benefits were observed [108]. Etanercept provided no immediate benefit also in patients with AMI [109]. Finally, infliximab did not improve clinical condition of patients with moderate-to-severe heart failure in the ATTACH trial and in fact high doses of infliximab caused adverse effects in the patients [110].

Conclusions

It has become evident that apoptosis is one of the major players in IHD as a consequence of atherosclerotic plaque formation, as well as during the development of cardiac injury due to I/R. Extensive research has been performed to map the contributory roles of different proteins, members of both extrinsic and intrinsic apoptotic pathways, to the development of cardiac ischemic injury (Fig. 3.2) as well as to reveal how these proteins are influenced by various cardioprotective interventions aimed to prevent I/R injury. While experimental studies in laboratory animals have uniformly documented activation/upregulation of apoptotic proteins and depression of anti-apoptotic proteins due to ischemic insult, human studies and clinical trials have revealed controversial data regarding the role of apoptosis in cardiac I/R injury. Likewise, several experimental studies have shown promising data regarding the potential use of drugs targeting apoptosis in preventing myocardial I/R injury, whereas different clinical trials have revealed conflicting and largely disappointing results in preventing IHD using apoptosis-targeting substances, mainly TNF-α antibodies. Thus, further research is needed to fully uncover the particular contribution of apoptosis to myocardial I/R injury as well as to reveal the efficiency of anti-apoptotic substances in preventing cardiac ischemic damage. There is a real challenge to translate experimental knowledge to clinical use of apoptosis-targeting drugs for the benefit of patients in the treatment of IHD, and the hope is still there.

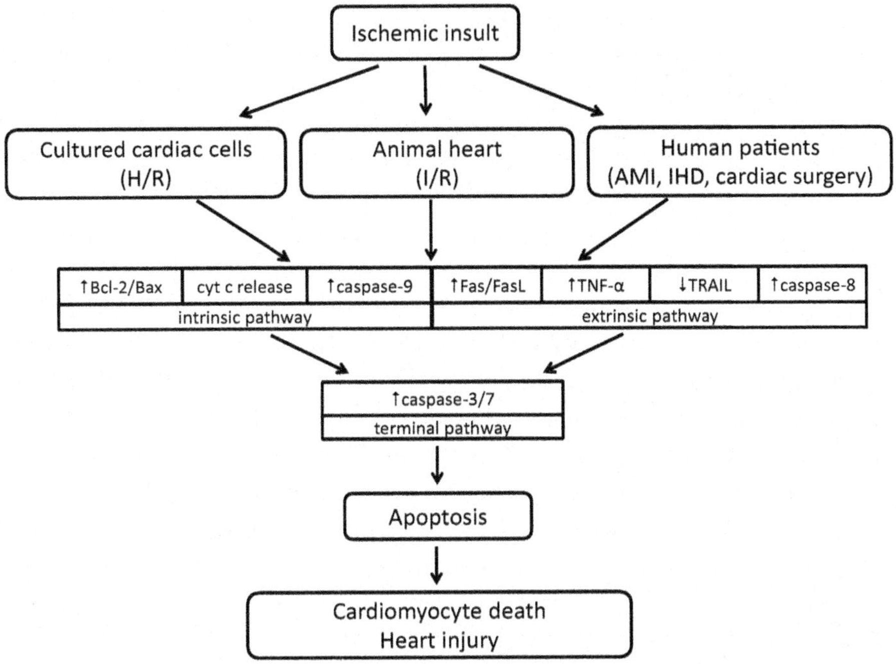

Fig. 3.2 The role of different mediator proteins from both intrinsic and extrinsic pathways in the development of apoptosis in cardiac I/R injury. Ischemic insult in cultured cardiac cells or animal heart, as well as any kind of ischemia in human heart, leads to the activation of both intrinsic and extrinsic apoptotic pathways via modulating their mediator proteins such as Bcl-2 family proteins, TNF-α or initiation caspases. These consequently activate common/terminal apoptotic pathway via activating executive caspases, thus leading to cardiomyocyte apoptosis, cell death, and finally to the heart injury due to I/R

Acknowledgements M.B. is supported by grant from the Scientific Grant Agency of the Ministry of Education, Science, Research and Sport of the Slovak Republic and the Slovak Academy of Sciences VEGA no. 2/0104/20. The infrastructure support for this project was provided by the St. Boniface Hospital Research Foundation, Winnipeg, Canada.

References

1. Bartekova M, Barancik M, Dhalla NS (2016) Role of oxidative stress in subcellular defects in ischemic heart disease. In: Gelpi RJ, Boveris A, Paderosa JJ (eds) Biochemistry of oxidative stress: physiopathology and clinical aspects; Adv Biochem Health Dis 16: 129–146
2. Bartekova M, Ferenczyova K, Jelemensky M, Dhalla NS (2019) Role of oxidative stress and cardiovascular risk factors in ischemic heart disease. In: Chakraborti S, Dhalla NS, Ganguly NK, Dikshit M (eds) Oxidative stress in heart diseases. Springer Nature Singapore, pp 375–394. ISBN 978-981-13-8272-7

3. Schaftenaar F, Frodermann V, Kuiper J, Lutgens E (2016) Atherosclerosis: the interplay between lipids and immune cells. Curr Opin Lipidol 27(3):209–215. https://doi.org/10.1097/MOL.0000000000000302
4. Geovanini GR, Libby P (2018) Atherosclerosis and inflammation: overview and updates. Clin Sci (Lond) 132(12):1243–1252. https://doi.org/10.1042/CS20180306
5. Bartekova M, Radosinska J, Jelemensky M, Dhalla NS (2018) Role of cytokines and inflammation in heart function during health and disease. Heart Fail Rev 23(5):733–758. https://doi.org/10.1007/s10741-018-9716-x
6. Wu MY, Yiang GT, Liao WT et al (2018) Current mechanistic concepts in ischemia and reperfusion injury. Cell Physiol Biochem 46(4):1650–1667. https://doi.org/10.1159/000489241
7. Zaman S, Wang R, Gandhi V (2014) Targeting the apoptosis pathway in hematologic malignancies. Leuk Lymphoma 55(9):1980–1992. https://doi.org/10.3109/10428194.2013.855307
8. Cotter TG (2009) Apoptosis and cancer: the genesis of a research field. Nat Rev Cancer 9(7):501–507. https://doi.org/10.1038/nrc2663
9. Lev N, Melamed E, Offen D (2003) Apoptosis and Parkinson's disease. Prog Neuropsychopharmacol Biol Psychiat 27(2):245–250
10. Mines MA, Beurel E, Jope RS (2011) Regulation of cell survival mechanisms in Alzheimer's disease by glycogen synthase kinase-3. Int J Alzheimers Dis 2011:861072. https://doi.org/10.4061/2011/861072
11. Roulston A, Marcellus RC, Branton PE (1999) Viruses and apoptosis. Annu Rev Microbiol 53:577–628
12. Wang L, Wang FS, Gershwin ME (2015) Human autoimmune diseases: a comprehensive update. J Intern Med 278(4):369–395. https://doi.org/10.1111/joim.12395
13. Mihaljevic O, Zivancevic-Simonovic S, Milosevic-Djordjevic O et al (2018) Apoptosis and genome instability in children with autoimmune diseases. Mutagenesis 33(5–6):351–357. https://doi.org/10.1093/mutage/gey037
14. Teringova E, Tousek P (2017) Apoptosis in ischemic heart disease. J Transl Med 15(1):87. https://doi.org/10.1186/s12967-017-1191-y
15. Olivetti G, Quaini F, Sala R et al (1996) Acute myocardial infarction in humans is associated with activation of programmed myocyte cell death in the surviving portion of the heart. J Mol Cell Cardiol 28(9):2005–2016
16. Saraste A, Pulkki K, Kallajoki M et al (1997) Apoptosis in human acute myocardial infarction. Circulation 95(2):320–323
17. Baldi A, Abbate A, Bussani R et al (2002) Apoptosis and post-infarction left ventricular remodeling. J Mol Cell Cardiol 34(2):165–174
18. Lee P, Sata M, Lefer DJ et al (2003) Fas pathway is a critical mediator of cardiac myocyte death and MI during ischemia-reperfusion in vivo. Am J Physiol Heart Circ Physiol 284(2):H456–H463
19. Liu XM, Yang ZM, Liu XK (2017) Fas/FasL induces myocardial cell apoptosis in myocardial ischemia-reperfusion rat model. Eur Rev Med Pharmacol Sci 21(12):2913–2918
20. Li Y, Takemura G, Kosai K et al (2004) Critical roles for the Fas/Fas ligand system in postinfarction ventricular remodeling and heart failure. Circ Res 95(6):627–636
21. Fertin M, Bauters A, Pinet F, Bauters C (2012) Circulating levels of soluble Fas ligand and left ventricular remodeling after acute myocardial infarction (from the REVE-2 study). J Cardiol 60(2):93–97. https://doi.org/10.1016/j.jjcc.2012.03.001
22. Nilsson L, Szymanowski A, Swahn E, Jonasson L (2013) Soluble TNF receptors are associated with infarct size and ventricular dysfunction in ST-elevation myocardial infarction. PLoS One 8(2):e55477. https://doi.org/10.1371/journal.pone.0055477
23. Osmancik P, Teringova E, Tousek P et al (2013) Prognostic value of TNF-related apoptosis inducing ligand (TRAIL) in acute coronary syndrome patients. PLoS One 8(2):e53860. https://doi.org/10.1371/journal.pone.0053860

24. Tekin D, Xi L, Kukreja RC (2006) Genetic deletion of Fas receptors or Fas ligands does not reduce infarct size after acute global ischemia-reperfusion in isolated mouse heart. Cell Biochem Biophys 44(1):111–117
25. Sahinarslan A, Boyaci B, Kocaman SA et al (2012) The relationship of serum soluble Fas ligand (sFasL) level with the extent of coronary artery disease. Int J Angiol 21(1):29–34. https://doi.org/10.1055/s-0032-1306418
26. Kawakami H, Shigematsu Y, Ohtsuka T et al (1998) Increased circulating soluble form of Fas in patients with dilated cardiomyopathy. Jpn Circ J 62(12):873–876
27. Tsutamoto T, Wada A, Maeda K et al (2001) Relationship between plasma levels of cardiac natriuretic peptides and soluble Fas: plasma soluble Fas as a prognostic predictor in patients with congestive heart failure. J Card Fail 7(4):322–328
28. Niessner A, Hohensinner PJ, Rychli K et al (2009) Prognostic value of apoptosis markers in advanced heart failure patients. Eur Heart J 30(7):789–796. https://doi.org/10.1093/eurheartj/ehp004
29. Fan Q, Huang ZM, Boucher M et al (2013) Inhibition of Fas-associated death domain-containing protein (FADD) protects against myocardial ischemia/reperfusion injury in a heart failure mouse model. PLoS One 8(9):e73537. https://doi.org/10.1371/journal.pone.0073537
30. Saini HK, Xu YJ, Zhang M, Liu PP, Kirshenbaum LA, Dhalla NS (2005) Role of tumour necrosis factor-alpha and other cytokines in ischemia-reperfusion-induced injury in the heart. Exp Clin Cardiol 10(4):213–222
31. Zhang M, Xu YJ, Saini HK et al (2005) TNF-alpha as a potential mediator of cardiac dysfunction due to intracellular Ca^{2+}-overload. Biochem Biophys Res Commun 327(1):57–63
32. Zhang M, Xu YJ, Saini HK et al (2005) Pentoxifylline attenuates cardiac dysfunction and reduces TNF-alpha level in ischemic-reperfused heart. Am J Physiol Heart Circ Physiol 289(2):H832-839
33. Zhou CN, Yao W, Gong YN et al (2019) 22-oxacalcitriol protects myocardial ischemia-reperfusion injury by suppressing NF-κB/TNF-α pathway. Eur Rev Med Pharmacol Sci 23(12):5495–5502. https://doi.org/10.26355/eurrev_201906_18219
34. Huang XW, Pan MD, Du PH, Wang LX (2018) Arginase-2 protects myocardial ischemia-reperfusion injury via NF-κB/TNF-α pathway. Eur Rev Med Pharmacol Sci 22(19):6529–6537. https://doi.org/10.26355/eurrev_201810_16067
35. Meldrum DR (1998) Tumor necrosis factor in the heart. Am J Physiol Regul Integr Comp Physiol 274:R577–R595
36. Bajaj G, Sharma RK (2006) TNF-alpha-mediated cardiomyocyte apoptosis involves caspase-12 and calpain. Biochem Biophys Res Commun 345(4):1558–1564
37. Jarrah AA, Schwarskopf M, Wang ER et al (2018) SDF-1 induces TNF-mediated apoptosis in cardiac myocytes. Apoptosis 23(1):79–91. https://doi.org/10.1007/s10495-017-1438-3
38. Asgeri M, Pourafkari L, Kundra A et al (2015) Dual effects of tumor necrosis factor alpha on myocardial injury following prolonged hypoperfusion of the heart. Immunol Investig 44(1):23–35
39. Monden Y, Kubota T, Inoue T et al (2007) Tumor necrosis factor-alpha is toxic via receptor 1 and protective via receptor 2 in a murine model of myocardial infarction. Am J Physiol Heart Circ Physiol 293(1):H743–H753
40. Kleinbongard P, Heusch G, Schulz R (2010) TNF alpha in atherosclerosis, myocardial ischemia/reperfusion and heart failure. Pharmacol Ther 127(3):295–314. https://doi.org/10.1016/j.pharmthera.2010.05.002
41. Jeremias I, Kupatt C, Martin-Villalba A et al (2000) Involvement of CD95/Apo1/Fas in cell death after myocardial ischemia. Circulation 102(8):915–920
42. LeBlanc HN, Ashkenazi A (2003) Apo2L/TRAIL and its death and decoy receptors. Cell Death Differ 10:66–75
43. Di Bartolo BA, Cartland SP, Prado-Lourenco L et al (2015) Tumor necrosis factor-related apoptosis-inducing ligand (TRAIL) promotes angiogenesis and ischemia-induced neovascularization via NADPH oxidase 4 (NOX4) and nitric oxide-dependent mechanisms. J Am Heart Assoc 4(11)pii: e002527. https://doi.org/10.1161/JAHA.115.002527

44. Nakajima H, Yanase N, Oshima K et al (2003) Enhanced expression of the apoptosis inducing ligand TRAIL in mononuclear cells after myocardial infarction. Jpn Heart J 44(6):833–844
45. Jiang Y, Chen X, Fan M et al (2017) TRAIL facilitates cytokine expression and macrophage migration during hypoxia/reoxygenation via ER stress-dependent NF-κB pathway. Mol Immunol 82:123–136. https://doi.org/10.1016/j.molimm.2016.12.023
46. Toffoli B, Bernardi S, Candido R et al (2012) TRAIL shows potential cardioprotective activity. Investig New Drugs 30(3):1257–1260
47. Secchiero P, Candido R, Corallini F et al (2006) Systemic tumor necrosis factor-related apoptosisinducing ligand delivery shows antiatherosclerotic activity in apolipoprotein E-null diabetic mice. Circulation 114(14):1522–1530
48. Secchiero P, Corallini F, Ceconi C et al (2009) Potential prognostic significance of decreased serum levels of TRAIL after acute myocardial infarction. PLoS One 4(2):e4442. https://doi.org/10.1371/journal.pone.0004442
49. Volpato S, Ferrucci L, Secchiero P et al (2011) Association of tumor necrosis factor-related apoptosis-inducing ligand with total and cardiovascular mortality in older adults. Atherosclerosis 215(2):452–458. https://doi.org/10.1016/j.atherosclerosis.2010.11.004
50. Deftereos S, Giannopoulos G, Kossyvakis C et al (2012) Association of soluble tumour necrosis factor-related apoptosis-inducing ligand levels with coronary plaque burden and composition. Heart 98(3):214–218. https://doi.org/10.1136/heartjnl-2011-300339
51. Dessein PH, Lopez-Mejias R, Ubilla B et al (2015) TNF-related apoptosis-inducing ligand and cardiovascular disease in rheumatoid arthritis. Clin Exp Rheumatol 33(4):491–497
52. Teringova E, Kozel M, Knot J et al (2018) Relationship between TRAIL and left ventricular ejection fraction in patients with ST-elevation myocardial infarction treated with primary percutaneous coronary intervention. Biomed Res Int 2018:3709084. https://doi.org/10.1155/2018/3709084
53. Luz A, Santos M, Magalhães R et al (2017) Soluble TNF-related apoptosis induced ligand (sTRAIL) is augmented by post-conditioning and correlates to infarct size and left ventricle dysfunction in STEMI patients: a substudy from a randomized clinical trial. Heart Vessels 32(2):117–125. https://doi.org/10.1007/s00380-016-0851-9
54. Kruidering M, Evan GI (2000) Caspase-8 in apoptosis: the beginning of "the end"? IUBMB Life 50(2):85–90
55. Stephanou A, Brar B, Liao Z et al (2001) Distinct initiator caspases are required for the induction of apoptosis in cardiac myocytes during ischaemia versus reperfusion injury. Cell Death Differ 8(4):434–435
56. Scarabelli TM, Stephanou A, Pasini E et al (2002) Different signaling pathways induce apoptosis in endothelial cells and cardiac myocytes during ischemia/reperfusion injury. Circ Res 90(6):745–748
57. Roubille F, Combes S, Leal-Sanchez J et al (2007) Myocardial expression of a dominant-negative form of Daxx decreases infarct size and attenuates apoptosis in an in vivo mouse model of ischemia/reperfusion injury. Circulation 116(23):2709–2717
58. Lu X, Moore PG, Liu H, Schaefer S (2011) Phosphorylation of ARC is a critical element in the antiapoptotic effect of anesthetic preconditioning. Anesth Analg 112(3):525–531. https://doi.org/10.1213/ANE.0b013e318205689b
59. Palee S, Weerateerangkul P, Chinda K et al (2013) Mechanisms responsible for beneficial and adverse effects of rosiglitazone in a rat model of acute cardiac ischaemia-reperfusion. Exp Physiol 98(5):1028–1037. https://doi.org/10.1113/expphysiol.2012.070433
60. Fauconnier J, Meli AC, Thireau J et al (2011) Ryanodine receptor leak mediated by caspase-8 activation leads to left ventricular injury after myocardial ischemia-reperfusion. Proc Natl Acad Sci USA 108(32):13258–13263. https://doi.org/10.1073/pnas.1100286108
61. Liang Y, Lin Q, Zhu J et al (2014) The caspase-8 shRNA-modified mesenchymal stem cells improve the function of infarcted heart. Mol Cell Biochem 397(1–2):7–16. https://doi.org/10.1007/s11010-014-2165-5
62. Levine B, Sinha S, Kroemer G (2008) Bcl-2 family members: dual regulators of apoptosis and autophagy. Autophagy 4(5):600–606

63. Chen Z, Chua CC, Ho YS et al (2001) Overexpression of Bcl-2 attenuates apoptosis and protects against myocardial I/R injury in transgenic mice. Am J Physiol Heart Circ Physiol 280:H2313–H2320
64. Misao J, Hayakawa Y, Ohno M et al (1996) Expression of bcl-2 protein, an inhibitor of apoptosis, and Bax, an accelerator of apoptosis, in ventricular myocytes of human hearts with myocardial infarction. Circulation 94:1506–1512
65. Xing K, Fu X, Jiang L et al (2015) Cardioprotective effect of anisodamine against myocardial ischemia injury and its influence on cardiomyocytes apoptosis. Cell Biochem Biophys 73(3):707–716. https://doi.org/10.1007/s12013-015-0642-4
66. Li M, Xue L, Sun H, Xu S (2016) Myocardial protective effects of L-Carnitine on ischemia-reperfusion injury in patients with rheumatic valvular heart disease undergoing cardiac surgery. J Cardiothorac Vasc Anesth 30(6):1485–1493. https://doi.org/10.1053/j.jvca.2016.06.006
67. Hu Q, Luo W, Huang L et al (2016) Apoptosis-related microRNA changes in the right atrium induced by remote ischemic perconditioning during valve replacement surgery. Sci Rep 7(6):18959. https://doi.org/10.1038/srep18959
68. Zhu LP, Tian T, Wang JY et al (2018) Hypoxia-elicited mesenchymal stem cell-derived exosomes facilitates cardiac repair through miR-125b-mediated prevention of cell death in myocardial infarction. Theranostics 8(22):6163–6177. https://doi.org/10.7150/thno.28021
69. Liu Y, Yang L, Yin J et al (2018) MicroRNA-15b deteriorates hypoxia/reoxygenation-induced cardiomyocyte apoptosis by downregulating Bcl-2 and MAPK3. J Investig Med 66(1):39–45. https://doi.org/10.1136/jim-2017-000485
70. Wen Z, Mai Z, Zhu X et al (2020) Mesenchymal stem cell-derived exosomes ameliorate cardiomyocyte apoptosis in hypoxic conditions through microRNA144 by targeting the PTEN/AKT pathway. Stem Cell Res Ther 11(1):36. https://doi.org/10.1186/s13287-020-1563-8
71. Kura B, Kalocayova B, Devaux Y, Bartekova M (2020) Potential clinical implications of miR-1 and miR-21 in heart disease and cardioprotection. Int J Mol Sci 21(3)pii:700. https://doi.org/10.3390/ijms21030700.
72. Honda HM, Korge P, Weiss JN (2005) Mitochondria and ischemia/reperfusion injury. Ann N Y Acad Sci 1047:248–258
73. Scorrano L, Ashiya M, Buttle K et al (2002) A distinct pathway remodels mitochondrial cristae and mobilizes cytochrome c during apoptosis. Dev Cell 2(1):55–67
74. Korge P, Honda HM, Weiss JN (2003) Effects of fatty acids in isolated mitochondria: implications for ischemic injury and cardioprotection. Am J Physiol Heart Circ Physiol 285(1):H259–H269
75. Borutaite V, Budriunaite A, Morkuniene R, Brown GC (2001) Release of mitochondrial cytochrome c and activation of cytosolic caspases induced by myocardial ischaemia. Biochim Biophys Acta 1537(2):101–109
76. Kuznetsov AV, Schneeberger S, Seiler R et al (2004) Mitochondrial defects and heterogeneous cytochrome c release after cardiac cold ischemia and reperfusion. Am J Physiol Heart Circ Physiol 286(5):H1633–H1641
77. Lundberg KC, Szweda LI (2006) Preconditioning prevents loss in mitochondrial function and release of cytochrome c during prolonged cardiac ischemia/reperfusion. Arch Biochem Biophys 453(1):130–134
78. Skemiene K, Rakauskaite G, Trumbeckaite S et al (2013) Anthocyanins block ischemia-induced apoptosis in the perfused heart and support mitochondrial respiration potentially by reducing cytosolic cytochrome c. Int J Biochem Cell Biol 45(1):23–29. https://doi.org/10.1016/j.biocel.2012.07.022
79. Dzhanashiya PKh, Vladytskaya OV, Salibegashvili NV (2004) Efficiency and mechanisms of the antioxidant effect of standard therapy and refracterin in the treatment of chronic heart failure in elderly patients with postinfarction cardiosclerosis. Bull Exp Biol Med 138(4):412–414

80. Qin F, Liang MC, Liang CS (2005) Progressive left ventricular remodeling, myocyte apoptosis, and protein signaling cascades after myocardial infarction in rabbits. Biochim Biophys Acta 1740(3):499–513
81. Zeng C, Jiang W, Zheng R et al (2018) Cardioprotection of tilianin ameliorates myocardial ischemia-reperfusion injury: role of the apoptotic signaling pathway. PLoS One 13(3):e0193845. https://doi.org/10.1371/journal.pone.0193845
82. Liu D (2018) Effects of procyanidin on cardiomyocyte apoptosis after myocardial ischemia reperfusion in rats. BMC Cardiovasc Disord 18(1):35. https://doi.org/10.1186/s12872-018-0772-x
83. Huang LH, Li J, Gu JP et al (2018) Butorphanol attenuates myocardial ischemia reperfusion injury through inhibiting mitochondria-mediated apoptosis in mice. Eur Rev Med Pharmacol Sci 22(6):1819–1824. https://doi.org/10.26355/eurrev_201803_14601
84. Condorelli G, Roncarati R, Ross J Jr et al (2001) Heart-targeted overexpression of caspase 3 in mice increases infarct size and depresses cardiac function. Proc Natl Acad Sci USA 98:9977–9982
85. Holly TA, Drincic A, Byun Y et al (1999) Caspase inhibition reduces myocyte cell death induced by myocardial ischemia and reperfusion in vivo. J Mol Cell Cardiol 31(9):1709–1715
86. Pavo N, Lukovic D, Zlabinger K et al (2017) Intrinsic remote conditioning of the myocardium as a comprehensive cardiac response to ischemia and reperfusion. Oncotarget 8(40):67227–67240. https://doi.org/10.18632/oncotarget.18438
87. Liu Q (2014) Lentivirus mediated interference of caspase-3 expression ameliorates the heart function on rats with acute myocardial infarction. Eur Rev Med Pharmacol Sci 18(13):1852–1858
88. Hu S, Yan G, Xu H et al (2014) Hypoxic preconditioning increases survival of cardiac progenitor cells via the pim-1 kinase-mediated anti-apoptotic effect. Circ J 78(3):724–731
89. You L, Pan YY, An MY et al (2019) The cardioprotective effects of remote ischemic conditioning in a rat model of acute myocardial infarction. Med Sci Monit 25:1769–1779. https://doi.org/10.12659/MSM.914916
90. Bartekova M, Šimončíková P, Fogarassyová M, Ivanová M et al (2015) Quercetin improves postischemic recovery of heart function in doxorubicin-treated rats and prevents doxorubicin-induced matrix metalloproteinase-2 activation and apoptosis induction. Int J Mol Sci 16(4):8168–8185. https://doi.org/10.3390/ijms16048168
91. Li C, Wang T, Zhang C et al (2016) Quercetin attenuates cardiomyocyte apoptosis via inhibition of JNK and p38 mitogen-activated protein kinase signaling pathways. Gene 577(2):275–280. https://doi.org/10.1016/j.gene.2015.12.012
92. Wang H, Zhao YT, Zhang S et al (2017) Irisin plays a pivotal role to protect the heart against ischemia and reperfusion injury. J Cell Physiol 232(12):3775–3785. https://doi.org/10.1002/jcp.25857
93. Liu BF, Chen Q, Zhang M, Zhu YK (2019) MiR-124 promotes ischemia-reperfusion induced cardiomyocyte apoptosis by targeting sphingosine kinase 1. Eur Rev Med Pharmacol Sci 23(16):7049–7058. https://doi.org/10.26355/eurrev_201908_18747
94. Tan H, Qi J, Fan BY et al (2018) MicroRNA-24-3p attenuates myocardial ischemia/reperfusion injury by suppressing RIPK1 expression in mice. Cell Physiol Biochem 51(1):46–62. https://doi.org/10.1159/000495161
95. Chen Z, Su X, Shen Y et al (2019) MiR322 mediates cardioprotection against ischemia/reperfusion injury via FBXW7/notch pathway. J Mol Cell Cardiol 133:67–74. https://doi.org/10.1016/j.yjmcc.2019.05.020
96. Saini U, Gumina RJ, Wolfe B et al (2013) Preconditioning mesenchymal stem cells with caspase inhibition and hyperoxia prior to hypoxia exposure increases cell proliferation. J Cell Biochem 114(11):2612–2623. https://doi.org/10.1002/jcb.24609
97. Knaapen MW, Davies MJ, De Bie M et al (2001) Apoptotic versus autophagic cell death in heart failure. Cardiovasc Res 51(2):304–312
98. Chapman JG, Magee WP, Stukenbrok HA et al (2002) A novel nonpeptidic caspase-3/7 inhibitor, (S)-(+)-5-[1-(2-methoxymethylpyrrolidinyl)-sulfonyl]isatin reduces myocardial ischemic injury. Eur J Pharmacol 456(1–3):59–68

99. Zhao L, Zhuang J, Wang Y et al (2019) Propofol ameliorates H9c2 cells apoptosis induced by oxygen glucose geprivation and reperfusion injury via inhibiting high levels of mitochondrial fusion and fission. Front Pharmacol 10:61. https://doi.org/10.3389/fphar.2019.00061
100. Fischer UM, Tossios P, Huebner A et al (2004) Myocardial apoptosis prevention by radical scavenging in patients undergoing cardiac surgery. J Thorac Cardiovasc Surg 128(1):103–108
101. Aslanabadi N, Shirzadi HR, Asghari-Soufi H et al (2015) A pilot randomized trial of pentoxifylline for the reduction of periprocedural myocardial injury in patients undergoing elective percutaneous coronary intervention. Eur J Clin Pharmacol 71(2):143–149. https://doi.org/10.1007/s00228-014-1782-y
102. Mansourian S, Bina P, Fehri A et al (2015) Preoperative oral pentoxifylline in case of coronary artery bypass grafting with left ventricular dysfunction (ejection fraction equal to/less than 30%). Anatol J Cardiol 15(12):1014–1019. https://doi.org/10.5152/akd.2014.5883
103. Wolfe F, Michaud K (2004) Heart failure in rheumatoid arthritis: rates, predictors, and the effect of anti-tumor necrosis factor therapy. Am J Med 116(5):305–311
104. Dixon WG, Watson KD, Lunt M et al (2007) Reduction in the incidence of myocardial infarction in patients with rheumatoid arthritis who respond to anti-tumor necrosis factor alpha therapy: results from the British society for rheumatology biologics register. Arthritis Rheum 56(9):2905–2912
105. Barnabe C, Martin BJ, Ghali WA (2011) Systematic review and meta-analysis: anti-tumor necrosis factor α therapy and cardiovascular events in rheumatoid arthritis. Arthritis Care Res 63(4):522–529
106. Wu JJ, Poon KY, Bebchuk JD (2013) Association between the type and length of tumor necrosis factor inhibitor therapy and myocardial infarction risk in patients with psoriasis. J Drugs Dermatol 12(8):899–903
107. Armstrong AW (2013) Do TNF inhibitors reduce the risk of myocardial infarction in psoriasis patients? JAMA 309(19):2043–2044
108. Mann DL, McMurray JJ, Packer M et al (2004) Targeted anticytokine therapy in patients with chronic heart failure: results of the randomized Etanercept Worldwide Evaluation (RENEWAL). Circulation 109(13):1594–1602
109. Padfield GJ, Din JN, Koushiappi E et al (2013) Cardiovascular effects of tumour necrosis factor α antagonism in patients with acute myocardial infarction: a first in human study. Heart 99(18):1330–1335
110. Chung ES, Packer M, Lo KH et al (2003) Randomized, double-blind, placebo-controlled, pilot trial of infliximab, a chimeric monoclonal antibody to tumor necrosis factor-α, in patients with moderate-to-severe heart failure: results of the anti-TNF therapy against congestive heart failure (ATTACH) trial. Circulation 107(25):3133–3140

Chapter 4
Autophagy in Cardiac Physiology and Pathology

Tania Zaglia and Loren J. Field

Abstract Adult cardiomyocytes have only a limited capacity for regeneration, with intrinsic renewal rates of approximately 1% per year reported for humans and rodents. Individual cardiomyocytes are thus quite long-lived and must be able to continuously adapt to physiologic and pathophysiologic stresses. Stress adaptation often entails the renewal of intracellular constituents, as exemplified by the changes in enzymatic activity which accompanies shifts in oxygenation and metabolism, occurring during perinatal development; the modification in myofiber content in response to altered cardiac workload; and the renewal of proteins and organelles which takes place in response to reactive oxygen species-induced damage. Multiple proteolytic pathways have evolved to promote the efficient degradation of intracellular constituents, which is an essential step for their ultimate renewal. This chapter is focused on the role of the Autophagy in the heart. We begin with a description of types and molecular regulation of Autophagy, and then briefly summarize the importance of Autophagy in pathophysiologic regulation in other cell types. This is followed by a more detailed description of the role of Autophagy in cardiomyocytes during physiologic (i.e. perinatal development and physiologic hypertrophy), as well as pathophysiologic (i.e. cardiac atrophy, hypertrophy, proteinopathies and aging) conditions.

Keywords Autophagy · Cardiomyocytes · Cell death · Postnatal development · Cardiac diseases · Ageing

---

T. Zaglia (✉)
Department of Cardiac, Thoracic, Vascular Sciences and Public Health, University of Padova, via Giustiniani 2, 35128, Padova, Italy
e-mail: tania.zaglia@unipd.it

Department of Biomedical Sciences, University of Padova, Via Ugo Bassi 58/B, 35122, Padova, Italy

Veneto Institute of Molecular Medicine, Via Orus 2, 35129, Padova, Italy

L. J. Field
The Krannert Institute of Cardiology and the Wells Center for Pediatric Research, Indiana University School of Medicine, Indianapolis, USA

© Springer Nature Switzerland AG 2022

L. A. Kirshenbaum (ed.), *Biochemistry of Apoptosis and Autophagy*,
Advances in Biochemistry in Health and Disease 18,
https://doi.org/10.1007/978-3-030-78799-8_4

Introduction

Autophagy is a housekeeping cellular process critically involved in the turnover of proteins and organelles. Autophagy is a multi-step mechanism whereby a series of reactions culminate in the delivery of the material to be degraded into lysosomes. The identification of this complex and highly dynamic cellular phenomenon stemmed from the serendipitous discovery of lysosomes in the early 1950 by de Duve and colleagues, who were utilizing differential centrifugation to establish the sub-cellular localization of glucose-6-phosphatase [1–3]. Ultrastructural analyses revealed that lysosomes were sites of focal cytoplasmic degradation which facilitate protein and organelle turnover in response to injury [4], as well as during normal physiologic processes [5]. de Duve subsequently coined the term 'Autophagy' to describe this process. A number of groups have exploited genetically-tractable organisms to identify genes which are required for Autophagy (reviewed in [6]). Once the genes were identified, there was rapid progress in dissecting the roles which Autophagy plays in physiology and pathophysiology. The importance of this pathway has been recognized by two Nobel Prizes in *Physiology or Medicine*: the 1974 Prize was awarded to de Duve, Claude and Palade for the discovery of the lysosome (as well as of other organelles), and the 2016 Prize was awarded to Ohsumi for the identification of Autophagy regulatory genes in yeast. At the time being, advancement in the understanding of the molecular basis underlying organization and function of Autophagy, has placed it at the center of a wide range of cellular processes, including metabolism, growth, division, and remarkably, cell death. As a result, cellular biologists, regardless of the study topic, cannot dribble considering the state of Autophagy in their research. However, despite such burst of popularity, a veil of mystery lingers on Autophagy, leaving the verdict on whether '*…it is friend or foe…*' open for future judgement.

This chapter focuses on the role of Autophagy in the heart, and in particular the role which it plays in response to physiologic and pathophysiologic stresses in cardiomyocytes.

The chapter is organized in the following paragraphs:

1. The necessity to maintain cardiomyocyte health
2. Biochemistry and regulation of Autophagy
    - 2.1 *The different types of Autophagy*
    - 2.2 *Regulation of Autophagy in response to stress conditions*
3. The role of Autophagy in physiology and pathology
4. Autophagy and cell death
    - 4.1 *Revisiting an old debate: does autophagic cell death exist in vivo?*
    - 4.2 *The Autophagy-apoptosis interplay*
5. Cardiac Autophagy in physiology and pathology
    - 5.1 *The necessary role of Autophagy throughout cardiomyocyte life*

The Necessity to Maintain Cardiomyocyte Health

The heart continuously pumps oxygenated, nutrient enriched blood to all cells of the body. Cardiomyocytes account for 85% of the cardiac mass, and constitute '*the fundamental work unit of the heart*' [7]. To fulfil this role, cardiomyocytes exhibit a unique cytoarchitecture characterized by the precise arrangement of myofibrils and organelles, and maintain a well-organized three-dimensional microenvironment with neighbouring cardiomyocytes and interstitial cells. Most studies have shown that adult ventricular cardiomyocytes have a very limited capacity for renewal, which occurs predominately via the proliferation of pre-existing cells at a rate of ~ 1% per year (reviewed in [8]). Indeed, the vast majority of cardiomyocytes stop dividing shortly after birth and the number of contractile cells observed in the neonatal heart remains approximately the same throughout the entire lifespan of the organism, supporting the notion that the mammalian heart is largely '*a postmitotic organ*' [9, 10]. The increase in cardiac mass, occurring during postnatal life, results from hypertrophic cardiomyocyte growth (which is in part driven by neuroendocrine activity) [11] and is accompanied by the proliferation of interstitial cells (i.e. cardiac fibroblasts) and blood vessel constituents [12]. In mice, postnatal hypertrophic growth is characterized by the formation of bi-nucleated cardiomyocytes, which is largely completed by postnatal day 14 (Fig. 4.1). The inability of a large number of ventricular cardiomyocytes to re-enter the cell cycle limits the regenerative capacity of the heart following tissue damage. This implies that those cardiomyocytes which

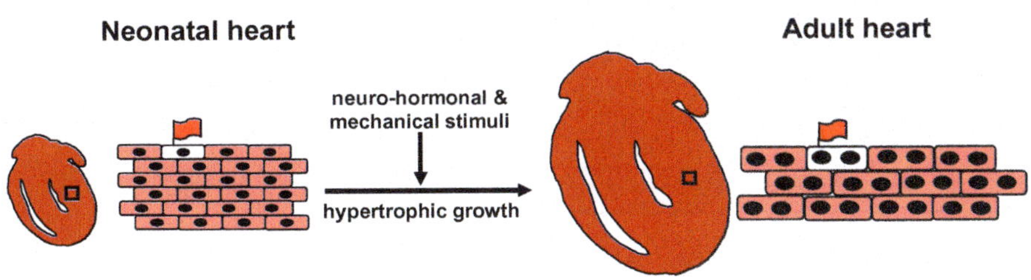

Fig. 4.1 Heart postnatal development. During development, cardiac mass increases through cardiomyocyte proliferation (hyperplastic growth). Few days after birth, cardiomyocytes stop divide and postnatal heart enlargement continues with physiologic hypertrophy of preexisting cells until the adult phenotype is reached (see red flag)

die, as a consequence of a pathologic event, are predominantly substituted by non-contractile scar tissue. Although hypertrophic growth of the surviving myocardium can partially compensate for loss of myocardial mass, ablation of a large volume of myocardial tissue ultimately leads to an irreversible decrease of cardiac performance [13] (Fig. 4.2). While acute management strategies and pharmacological treatment have significantly reduced early post-infarction mortality, pathologic remodelling of the ventricular myocardium, with subsequent progression to chronic heart failure (HF), remains a relevant clinical problem, and is the main cause of death in industrialized countries. To date, HF affects nearly 6 million people in the United States, with 670,000 new cases diagnosed each year; in most instances heart transplantation remains the only therapeutic option [14]. Thus, reconstitution of the myocardial mass

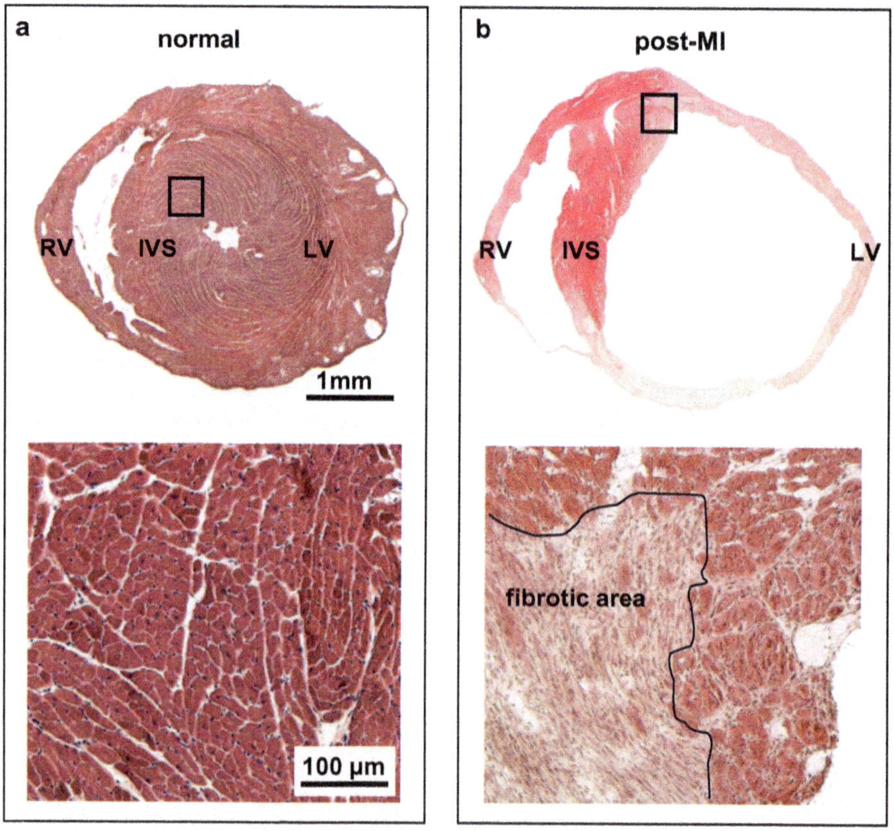

Fig. 4.2 Cardiac remodeling after myocardial infarction. **a–b** Haematoxilin–eosin staining on ventricular sections from normal **a** and 30-day post-MI, **b** mouse hearts. Top panels in **a** and **b** show heart sections from the mid portion of the ventricles, while bottom panels illustrate high magnification details of the region boxed above. The ischemic insult leads to the irreversible loss of cardiomyocytes and subsequent scar formation (see the fibrotic area evidenced in **b**). RV, right ventricle; IVS, interventricular septum; LV, left ventricle

lost upon damage remains a major goal of cardiovascular research, as this holds the potential for true regeneration of diseased hearts [15]. Several approaches have been attempted to attain this goal, including transplantation of *bona fide* cardiomyocyte donor cells, transplantation or mobilization of donor cells with presumed cardiomyogenic activity, directed differentiation of non-cardiomyocyte cells to cardiomyocytes, and gene- or growth factor-based induction of cardiomyocyte proliferation (reviewed in [16, 17]). An alternative and perhaps more achievable approach is that of developing strategies to block and/or reverse the molecular changes which transpire during adverse post-injury remodelling, and in so doing, prevent and/or delay HF progression. Towards this end, it is important to keep in mind that *'terminally differentiated'* cardiomyocytes are endowed with a high capacity for plasticity, which allows for their continuous adaptation to intrinsic and environmental stresses encountered during normal daily activities, ageing and injury [18]. As such, cardiomyocytes require to maintain the equilibrium between the synthesis, folding, assembly and ultimately removal of structural and signalling proteins to preserve their proper function during the entire lifespan. Removal of cellular constituents, unfolded/misfolded proteins and damaged organelles is mediated by the Ubiquitin Proteasome System (UPS) and the Autophagy/Lysosome System (ALS/Autophagy), which together constitute the primary intracellular proteolytic machineries [19, 20]. This chapter focuses on Autophagy and on its role in cardiomyocyte physiology and pathology.

Biochemistry and Regulation of Autophagy

The Different Types of Autophagy

Autophagy is an evolutionary conserved, non-selective catabolic process responsible for the degradation of cellular cytoplasmic components and organelles during basal states and stress conditions. In mammalian cells, three types of Autophagy have been described, which employ different mechanisms to deliver components to the lysosomes for degradation; they are: **(i) macroautophagy**, **(ii) chaperone-mediated autophagy (CMA)** and **(iii) microautophagy** (Fig. 4.3). The majority of data accrued thus far on the role of Autophagy in muscle have focused on macroautophagy. This process mediates the degradation of a wide spectrum of substrates, including misfolded proteins, oligomers, protein aggregates and damaged organelles (i.e. mitochondria, ribosomes). The first step of macroautophagy is the formation of a double membrane (the *phagophore*) which encloses and insulates cytoplasmic components, giving rise to the *autophagosome.* Once established, autophagosomes move along microtubules until they reach and fuse with lysosomes, enabling the degradation of the engulfed material by acid hydrolases [21, 22]. Although the majority of the data on the mechanisms underlying Autophagy have used yeast as a model system, it has been demonstrated that yeast and mammals share most Autophagy signalling pathways [23], and Autophagy-related genes (ATGs) have

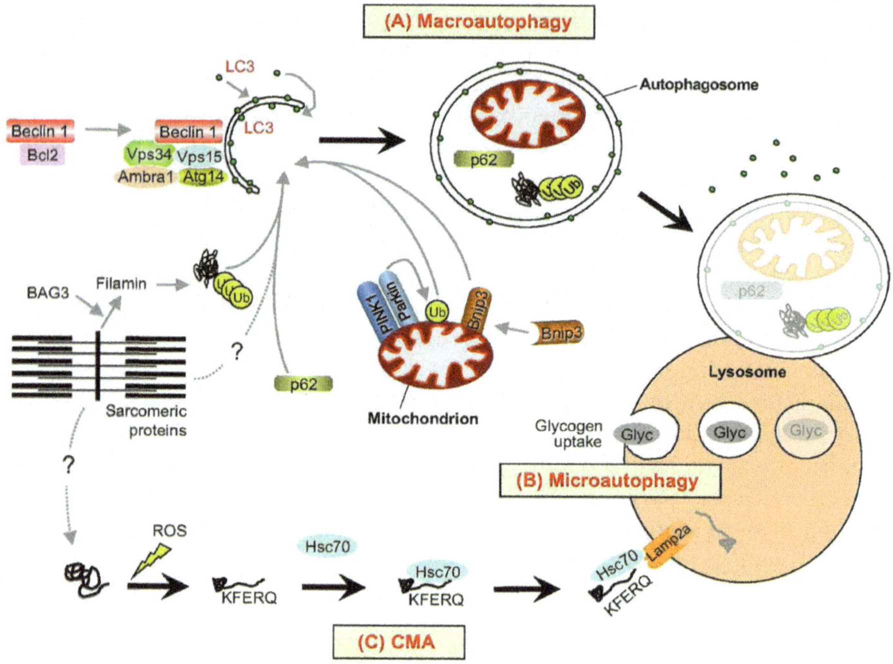

Fig. 4.3 Schematic of macroAutophagy, microAutophagy and chaperone-mediated Autophagy (CMA) pathways in protein and organelle degradation. Reproduced with permission from Bonaldo and Sandri [21]

been identified as common effectors of key steps of the process. Among these, the ATG1/ATG13 kinase complex induces Autophagy, while ATG6 (whose mammalian homolog is Beclin 1) triggers the formation of the phagophore by interacting with VPS15 and VPS34. ATG12 and ATG8 (LC3 in mammals) subsequently promote the engulfment of cytoplasmic material and formation of the autophagosome [24]. In CMA, complexes of chaperone/co-chaperone proteins recognize selective cytosolic proteins which are subsequently targeted to the lysosome surface and, upon interaction with Lysosome Associated Protein (LAMP) type-2a, are translocated into the organelle lumen where degradation takes place [25]. Finally, in microautophagy, lysosomes directly engulf cytosolic fractions via membrane invaginations, but the mechanisms regulating this process are still poorly understood, particularly in the myocardium [21].

Regulation of Autophagy in Response to Stress Conditions

Although autophagy-mediated degradation of proteins, protein aggregates and damaged organelles contributes to the maintenance of cell function and survival under basal conditions, it also plays a key survival role in response to cellular stress. Nutrient deprivation (i.e. starvation) is, for example, a well-established trigger of Autophagy [26]. During starvation, reduction in the availability of growth factors activates Autophagy to ensure adequate energy supply and preserve cell homeostasis, through protein kinase Gcn2 (which is activated by uncharged tRNA molecules) and other amino acid signaling pathways, involving class III phosphatidylinositol 3 (PI3)-kinase and Beclin 1 [27, 28]. Interestingly, the amino acids produced by Autophagy-mediated protein breakdown activate a negative feedback mechanism by both interfering with Erk1/2 signaling [29] and activating mTOR [30]. TOR belongs to a family of conserved serine/threonine protein kinases and constitutes the main intracellular sensor of nutrient-derived signals. In yeast, the accumulation of nitrogen sources activates TOR, which in turn inhibits the expression of genes required for adaptation to cell starvation, including those related to Autophagy [31]. In addition, TOR also mediates physiologic repression of Autophagy by cooperating with the Ras/cAMP/PKA signalling pathway (which has a central role in the regulation of cell metabolism) and inhibits autophagosome biogenesis [32]. Other important intracellular molecules which impinge on Autophagy activity include ATP (whose decrease induces Autophagy directly or via AMPK-dependent sensing of TOR [33, 34]) and cytosolic calcium (whose increase has been shown to activate Autophagy via TOR inhibition [35]). In addition, the transcription factor FoxO3 (a key mediator of muscle atrophy downstream of Akt/PKB) has been shown to increase the expression of several Autophagy-related genes via TOR-independent signalling pathway [36, 37].

Autophagy is also induced in other situations unfavourable for cell function, including oxidative stress or hypoxia, both of which are associated with the Endoplasmic Reticulum (ER) stress response [38]. Excessive accumulation of unfolded proteins during the ER stress response induces the so-called Unfolded Protein Response (UPR), which attempts to restore normal ER function, via transcriptional and post-transcriptional mechanisms, including activation of ATG genes mediated by the eIF2α/ATF4 signalling pathway. In fact, Autophagy is the major degradation system activated during the UPR [39–41].

Activation of the ER stress response can also occur as a consequence of a primary dysfunction of the UPS (the other main intracellular proteolytic mechanism). In this case, Autophagy induction functions as a compensatory mechanism to remove aggregates of UPS protein substrates [42]. Conversely, reduced Autophagy is associated with increased proteasome activity [43, 44]. This suggests that the UPS and the ALS, which have previously been considered two distinct processes, may indeed be cross-regulated. However, the molecular mechanisms underlying such interplay have only currently started to be uncovered. The recent demonstration that the ESCR-TIII protein CHMP2B (a key modulator of Autophagy) is targeted, in cardiomyocytes, by the muscle specific ubiquitin ligase Atrogin1, provides an example of the molecular mechanism underlying UPS/ALS cross-regulation. Indeed, genetic ablation of Atrogin1, by altering the turnover and function of CHMP2B, blocked

Autophagy. Given that ER stress activation in these animals could not be compensated by increased autophagic flux, cell proteotoxicity was amplified, culminating in cardiomyocyte death, replacement fibrosis and decreased cardiac performance [42] (Fig. 4.4).

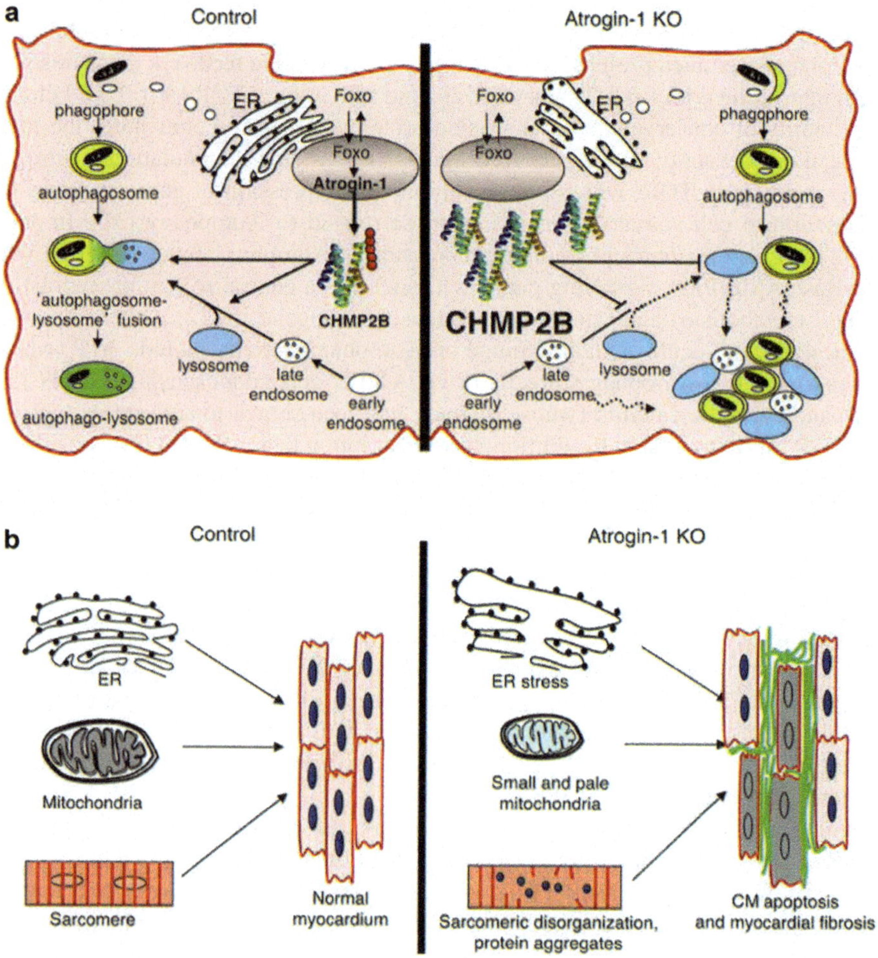

Fig. 4.4 Cross-talk between UPS and Autophagy. **a** In the heart, the muscle specific ubiquitin ligase Atrogin1 regulates the turnover of the ESCRIII protein CHMP2B, which is essential for Autophagy flux, as it mediates endosome maturation and autophagosome/lysosome fusion. **b** The impaired turnover of CHMP2B compromises endosome maturation and autophagosome/lysosome fusion, resulting in the block of the autophagic flux activation of markers of ER stress, expansion of ER cisternae, and modifications of mitochondria morphology, all of which lead to cardiomyocyte apoptosis, resulting in a diffuse myocardial interstitial fibrosis. Reproduced with permission from Zaglia et al. [42]

The Role of Autophagy in Physiology and Pathology

Autophagy regulates multiple key cellular activities. A large body of evidence indicates that the induction of Autophagy following short-term nutrient deprivation serves as a pro-survival mechanism. For example, yeast with defective Autophagy cannot sustain nitrogen starvation [45]. Moreover, mice lacking Atg5 die after birth due to their inability to adapt the energy metabolism in the early postnatal phases [46]. As indicated above, Autophagy also impacts on intracellular protein quality control and ER homeostasis. The cytoprotective effects of Autophagy extend to cell-specific functions. For example, antigen presentation and removal of intracellular infectious pathogens (i.e. bacteria and viruses) both rely on Autophagy, suggesting a role in innate and adaptative immunity, respectively [47, 48]. Finally, several lines of evidence suggest that induction of Autophagy may alleviate the ageing process. In support of this, caloric restriction and enhancement of basal Autophagy activity increased longevity in several animal species and *Drosophila* [49, 50] (this concept will be expanded in paragraph 5.5 below). Moreover, knockdown of Atg7 and Atg12 significantly reduced the life-expectancy in *C. elegans* [51].

Given the multiple roles played in several aspects of cellular biology, it is not surprising that alterations in Autophagy are implicated in diverse pathologies affecting post-mitotic cells of the nervous system and the myocardium. In the former, it has been demonstrated that several neurodegenerative disorders (including Huntington's and Parkinson's disease) are characterized by intracellular accumulation of protein aggregates which compromise neuronal function and viability [20, 52, 53] (Fig. 4.5). Moreover, mutations in CHMP2B and mSnf7-2, two members of the ESCRTIII complex required for endosome maturation and autophagosome-lysosome fusion, lead to frontotemporal dementia and amyotrophic lateral sclerosis, respectively, both of which are characterized by neuronal accumulation of ubiquitinated proteins [54, 55]. Conversely, pharmacologic induction of Autophagy in murine models of Huntington's disease increased protein aggregate removal, thereby reducing disease progression [56, 57]. In muscle, defective Autophagy is associated with myopathies, such as Danon's disease, an inherited disorder resulting from mutations in the lysosomal protein LAMP2 (which is required for autophagosome-lysosome fusion). LAMP2 mutations result in the intracellular accumulation of autophagic vacuoles, which are the histopathological hallmark of the disease [58, 59]. Due to its housekeeping function in cardiac cells, alterations in Autophagy are continuously being identified in several inherited or acquired cardiovascular disorders. However, the exact role of Autophagy in disease onset and progression is not completely elucidated, and the current view is that Autophagy may exert distinct roles, depending on the type, severity and stage of a given disease. The role of Autophagy is also largely debated in cancer. Here, dysfunctional Autophagy has been shown to promote: cancer initiation, likely by interfering with DNA repair mechanisms and promoting chromosomal instability, and tumor progression, by contrasting the relative nutrient deprivation of the rapidly proliferating neoplastic cells, in the

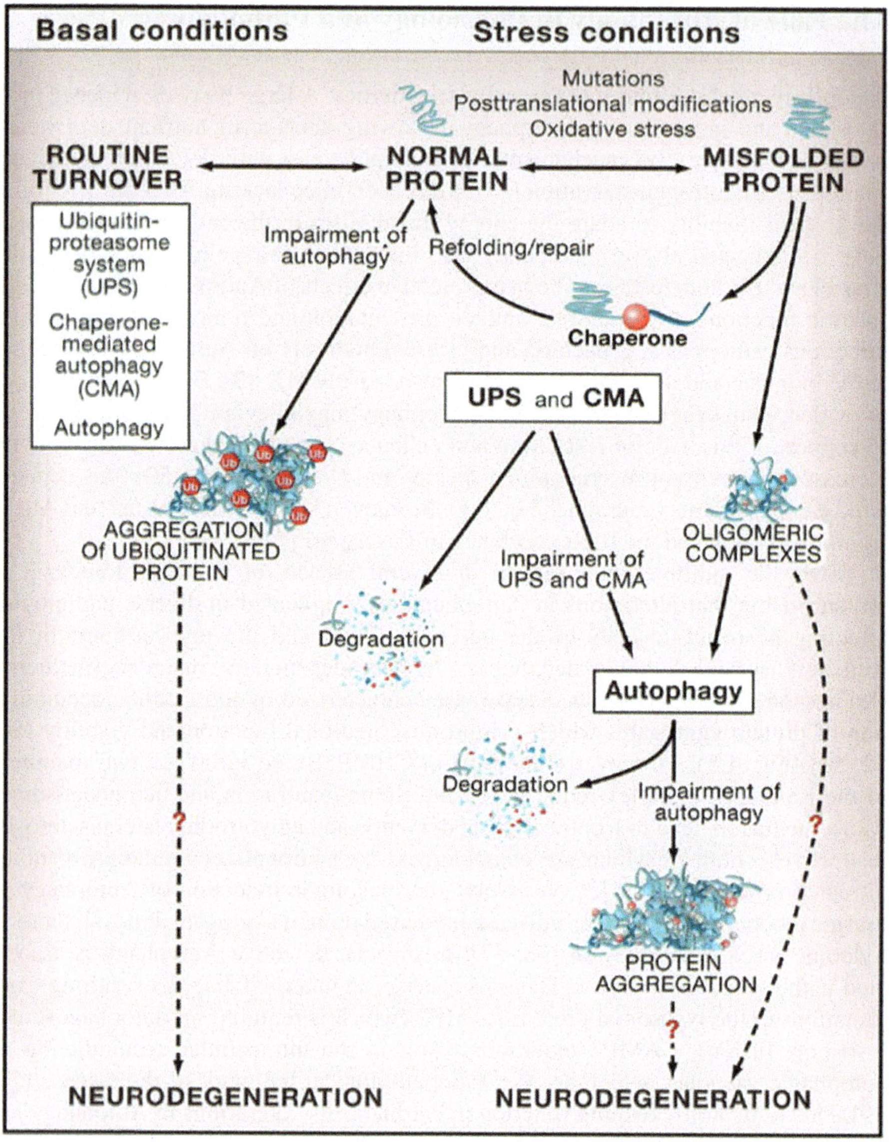

Fig. 4.5 Cellular mechanisms cooperating in protein quality control. Correct cellular protein turnover is guaranteed by the cooperation of the UPS, CMA, and Autophagy. In post-mitotic cells (neurons, cardiomyocytes) Autophagy impairment compromises removal of protein aggregates, leading to cellular degeneration. If primary defects occur in UPS or CMA, Autophagy represents the sole route for the removal of abnormal and potentially toxic proteins. Dysfunctional Autophagy causes the formation of protein aggregates and cell proteotoxicity. Reproduced with permission from Levine and Kroemer [20]

hypoxic tumor microenvironment [60–63]. At the time being, the molecular mechanisms underlying Autophagy-dependency in cancer are incompletely understood and require further research for the identification of novel and more efficient therapeutics.

Autophagy and Cell Death

Although the data presented above support the notion that correct Autophagy is required to preserve cell survival, several lines of evidence indicate that, in some circumstances, Autophagy may contribute to cell loss thorough a mechanism known as '*autophagic cell death*'.

Revisiting an Old Debate: Does Autophagic Cell Death Exist In Vivo?

It is generally thought that cell death may occur through three different modalities: apoptosis (also known as type 1 cell death); autophagic cell death (ACD; type 2 cell death) and necrosis (type 3 cell death). The distinction among these modalities is based mainly on the histopathological cellular phenotype and the molecular effectors of cellular degeneration. ACD is characterized by the presence of cytoplasmic vacuolization in the absence of chromatin condensation (the latter being a hallmark of apoptosis) [64]. Morphologic evidence for ACD was first described more than 40 years ago by Richard Lockshin [65] and was revisited in the 1990s with the discovery of Autophagy-related genes and the observation that cell death can occur in a caspase-independent manner [62]. Circumstantial evidence supporting the presence of ACD in vivo came from the identification of autophagic vacuoles in myocytes of patients affected by Danon's disease. However, more recent studies indicate that autophagosome accumulation in the disease is secondary to defects in autophagosome maturation and fusion with lysosomes [58, 59], thus changing the paradigm of 'cell death caused by enhanced Autophagy' to 'cell death caused by impaired Autophagy'. This interpretation agrees well with several in vitro and in vivo studies showing that knocking-down of ATG genes causes increased cell death during starvation, exposure to cytotoxic agents or cell infection. Moreover, evidence indicating that Autophagy helps maintain cell energy homeostasis and degrade intracellular toxic aggregates further supports a predominantly protective role in mammalian cells, and that ACD may indicate cell death 'accompanied by Autophagy'.

To date, most data supporting the existence of ACD have been obtained from studies using *Drosophila melanogaster* and nematodes [66, 67]. Evidence indicating the presence of ACD in mammals is less convincing, due in part to off-target activities of the chemical agents used to suppress Autophagy (i.e. chloroquine and 3-methyladenine) [68], incomplete inhibition when using gene silencing approaches

[69], and potential cross-talk with apoptotic pathways when employing constitutive ablation of ATG genes [70]. Thus, definitive evidence demonstrating that Autophagy is a direct effector of cell death in mammals under baseline conditions is lacking. Autophagy may nonetheless contribute to cell loss in response to stress. In support of this, it has been demonstrated that genetic ablation of Beclin1 (which is required for the sequestration of autophagic proteins to a pre-autophagosomal structure) reduces maladaptive Autophagy and cardiac dysfunction in pressure-overload dependent HF [71]. However, given that several key ALS mediators also participate to Autophagy-independent signaling pathways (i.e. caspase activation), it is prudent to use a certain degree of caution when interpreting these results.

The Autophagy-Apoptosis Interplay

Although the existence of ACD is still somewhat controversial, it is well accepted that there is considerable overlap between signalling pathways regulating Autophagy and apoptosis. For example, several studies demonstrated that many molecules are regulators of both apoptosis and Autophagy (i.e. p53, Akt, Jun, BH3-only proteins, oncogenes) [72], albeit with opposite consequences. The initial signalling via these proteins can activate Autophagy, with the ensuing removal of pro-apoptotic cytosolic proteins (including the kinase SRC and caspase 8 [73, 74]), and degradation of damaged mitochondria, which would otherwise initiate apoptosis [75]. These processes collectively attenuate cell death. Conversely, when the intensity and duration of stressor stimuli overcome the threshold of cell tolerance, apoptosis-associated signalling cascades are activated and suppress Autophagy, e.g. via caspase-dependent degradation of key ALS components (such as beclin1 and ATG3 [76, 77]). In addition, cleavage of Autophagy mediators generates protein fragments with pro-apoptotic function, thus initiating a positive feedback mechanism to accelerate cell death [78]. Of note, the interplay between Autophagy and apoptosis regulation differs in the context of single cell *versus* multicellular tissues. For example, soluble factors released by apoptotic cells can promote Autophagy in the neighbouring ones via a paracrine mechanism, thus increasing their tolerance to injury and restricting the extension of the tissue damage. In the context of myocardial ischemia, this phenomenon has been shown to reduce the area of cardiomyocyte death [72]. All these data underscore the complex nature of the interplay between Autophagy and apoptosis in preventing cell death and maintaining cell and tissue integrity.

Cardiac Autophagy in Physiology and Pathology

The preservation of cell structure and function is of paramount importance in post-mitotic cells, such as cardiomyocytes. Although cardiomyocytes are largely terminally differentiated, they exhibit a high degree of plasticity and are able to continuously adapt to changes in perfusional demand associated with physiologic and pathologic stresses. Given the absence of robust cardiomyocyte renewal, the adult heart is dependent upon cytoprotective mechanisms to ensure protein quality control as well as the integrity of cellular structures and organelles (particularly mitochondria, which play a central role in energy production and apoptosis regulation). Autophagy constitutes one such mechanism. Here, we will focus on the role of cardiac Autophagy in postnatal life and in the adult myocardium under baseline, physiologic and pathologic conditions.

The Necessary Role of Autophagy Throughout Cardiomyocyte Life

The sudden interruption of the trans-placental nutrient supply at birth results in transient, but severe, starvation in all cells of the body. Activation of Autophagy is essential for cell survival during the perinatal window, as evidenced by the observation that mice lacking key Autophagy mediators (such as Atg5 and Atg7) die by postnatal day 5 [46, 79]. In the heart, Autophagy activity reaches its maximum level at 3–6 h after birth, and remains transiently elevated even when nutrient supply is restored. Increased Autophagy and protein breakdown in perinatal life is essential for the recycling of amino acids as energy sources, and for the mobilization of glycogen stores which are particularly abundant in fetal cardiomyocytes. Indeed, electron microscopy analyses of neonatal mouse and rat hearts revealed the presence of autophagosomes which often contain only glycogen granules. Autophagy-dependent gluconeogenesis is essential for supplying energy to cardiomyocytes, thereby ensuring their survival and supporting their contractile activity during perinatal life [80]. Autophagy also modulates the postnatal developmental gene program by facilitating the removal of fetal-specific intracellular components, and thus plays a role in the 'fetal-to-neonatal' metabolic transition (i.e. carbohydrates and low fat in fetal life *versus* low carbohydrates and high fat in neonatal life). Perinatal life is also associated with the transition from a state of relative hypoxia to an oxygen-enriched environment, which occurs concomitantly with the establishment of autonomous respiration [81]. Consequently, this phase of development is characterized by increased reactive oxygen species (ROS) production which, in turn, has also been linked to ER stress activation [82, 83]. As both oxidative and ER stress are associated with ALS induction, it is likely that Autophagy represents a key mechanism protecting neonatal hearts from ROS-dependent cell damage. The housekeeping function of Autophagy in cardiomyocyte maturation during normal physiologic conditions is strongly supported by the observation that deficiency in LAMP2 results in accumulation of autophagic vacuoles in cardiomyocytes resulting in cell proteotoxicity, decreased cardiac performance, and death during adolescence [59, 84]. Once the heart has completed post-natal development, and mature cardiomyocytes have lost the regenerative potential, removal of damaged proteins and organelles by Autophagy becomes a fundamental requirement for the maintenance of normal function. This concept is readily supported by numerous studies wherein cardiomyocyte-restricted genetic manipulation of Autophagy circuits in otherwise normal animals results in cardiomyocyte death, leading to adverse myocardial remodeling and ultimately to HF [85, 86]. Altogether, these examples support the essential role of regulated Autophagy for the maintenance of heart homeostasis from birth to the adulthood.

Autophagy During Cardiac Atrophy

It is well accepted that the fully developed heart is able to increase its mass, e.g. when facing a chronic elevation in workload. What is less commonly recognized is that the opposite adaptation can also take place via a process known as cardiac atrophy, which is characterized by decreased cardiomyocyte size, heart weight and activity [87–89]. Metabolic unloading (i.e. starvation, nutrient deprivation) is the major cause of cardiac atrophy. Studies in rabbits [87] and rats [90] subjected to prolonged fasting revealed that starvation-induced cardiac atrophy results mainly from the increased protein degradation mediated by the UPS and ALS. Interestingly, metabolic unloading-dependent atrophy is typically reversible upon re-establishment of nutrient availability. Clinically relevant conditions associated with nutrient deprivation-mediated cardiac atrophy include malabsorption and metabolic dysfunction, dietary deficiency, loss of nutrients via urinary or digestive tract dysfunction, and cachexia (the latter of which is responsible for one third of cancer deaths). Cardiac atrophic remodelling during cachexia was initially described by Burch and colleagues in 1968, and more recent studies revealed a correlation between the severity of cardiac atrophy and decreased life-span in tumor-bearing rats [91, 92]. Other studies, using a tumor-induction model in mice, revealed that cancer cachexia leads to both skeletal and cardiac muscle wasting, albeit through different molecular mechanisms. While skeletal muscle atrophy is mediated predominately by UPS activation during cachexia, cardiac atrophy is mediated mainly by Autophagy activation [93]. β2-adrenoceptor agonists (i.e. formoterol; clenbuterol) which enhance protein synthesis and inhibit proteolysis can be used to counteract cachexia-induced skeletal muscle atrophy [94]. Unfortunately, these drugs have a number of adverse side effects when administered chronically, including cardiotoxicity, which may add relevant co-morbidities, such as HF or arrhythmias [95, 96].

In addition to metabolic factors, mechanical/circulatory unloading and the consequent decreased cardiac workload (which is often incurred by prolonged bedrest or space flights) is also associated to reduction in heart size. Indeed, three weeks of bedrest is sufficient to induce a significant decrease in cardiac mass in human subjects

[97]. Similarly, short-term space flight results in rapid cardiac atrophic remodelling in humans [89]. To date, the molecular mechanisms underlying cardiac atrophy are poorly understood, due in part to the paucity of spontaneous animal models and the complexity of achieving heart unloading experimentally. Heterotopic heart transplantation is a commonly used surgical experimental model of unloading-dependent cardiac atrophy. This model (initially developed in dogs and subsequently adapted to rats and mice) entails transplantation of a donor heart such that the great vessels are anastomosed to the abdominal aorta and inferior vena cava of a recipient animal. In such configuration, the aortic blood of the recipient animal perfuses retrogradely the coronary tree of the heterotopic heart, leaving the left ventricle chamber almost excluded from the circulation and thus physiologically unloaded [98]. Atrophic remodelling is mediated by the activation of multiple proteolytic pathways including the ALS, as evidenced by the induction of key Autophagy markers such as LC3, Atg5 and Atg12 [99]. In human hearts, Autophagy markers, which are elevated in HF, decrease following the relative mechanical unloading upon left ventricular assist device implantation [100]. These observations support that Autophagy may play, in the same pathologic condition, either adaptive or maladaptive roles which depend on the disease stage. Unfortunately, the implication of such finding is limited due to the plethora of parallel signaling events which underlie HF progression. Finally, it has recently been shown that in addition to metabolic and mechanical stimuli, the ablation of neuro-hormonal signalling (as for example, catecholamines released by cardiac sympathetic neurons) also results in atrophic remodelling of the heart. The initial atrophic response to denervation (which is mediated by increased UPS activity) is subsequently followed by Autophagy activation accompanied by decreased protein synthesis [101]. This process is important as it closely mimics the condition of the transplanted heart (in which atrophic remodelling has not been addressed in detail). In line with these results, chronic pharmacologic therapy with β-blockers, in rabbits [98] or mice (*Zaglia, unpublished*), provokes heart atrophy.

Autophagy During Cardiac Hypertrophy

Cardiac hypertrophy is characterized by increased heart mass due to cardiomyocyte enlargement, which is accompanied with a profound reorganization of their intracellular architecture. The structural, functional and molecular changes occurring during cardiomyocyte remodelling differ based on the type, magnitude and duration of the extrinsic stimuli. For example, increasing cardiac workload with chronic exercise training leads to physiologic hypertrophy, which is characterized by the presence of increased cardiomyocyte volume, deposition of new sarcomeres, increased myocardial vascularization and improved heart performance. Alternatively, augmenting the pressure gradient between the ventricle and aorta (as in hypertension and aortic coartation) or the cardiac volume load (as in valvular disease) results in pathologic hypertrophy, which also features cardiomyocyte growth, but, in contrast with the physiologic hypertrophy, occurs with decreased capillary density, reactivation of the

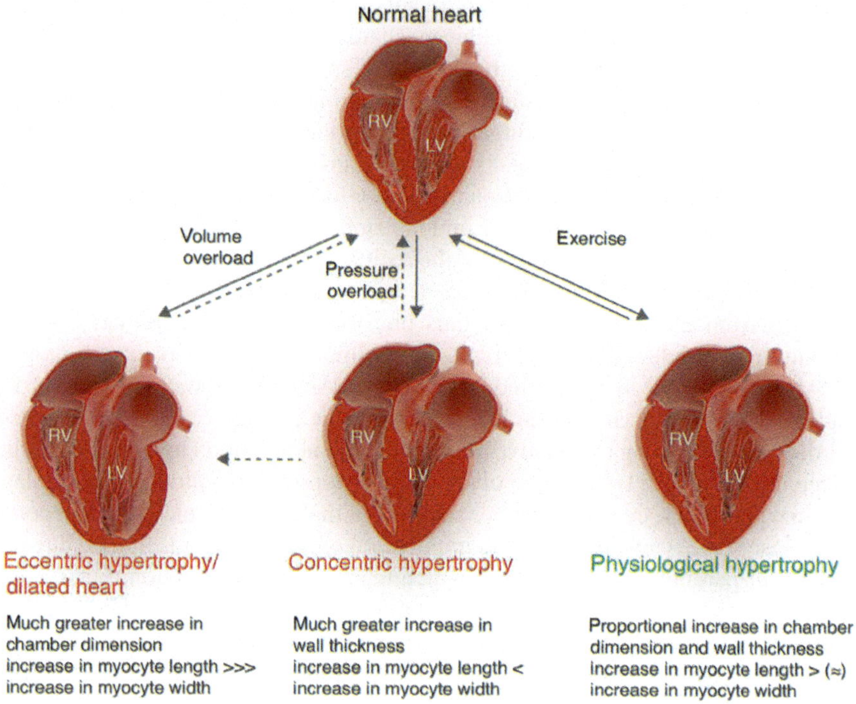

Fig. 4.6 Hypertrophic remodeling of the myocardium. Reproduced from Chung and Leinwand [137]

fetal gene program, cardiomyocyte apoptosis and ultimately leads to replacement myocardial fibrosis. While physiologic cardiac hypertrophy is not a risk factor for HF, pathologic hypertrophy progresses toward maladaptive myocardial remodelling, ultimately resulting in cardiac dysfunction and failure [18] (Fig. 4.6). *Which is the role of autophagy in physiologic versus pathologic hypertrophy?*

Exercise, a primary stimulator of physiologic hypertrophy, is associated to energy deprivation, moderately increased oxidative stress and unfolded protein response, all of which are known to induce Autophagy [102–104]. Most studies have addressed the effects of endurance exercise on the regulation of Autophagy-related genes in skeletal muscles, both in animal models and, more interestingly, humans [105]. Recently, exercise-induced Autophagy has also been described in the heart. For example, repetitive cycles of running have been demonstrated to significantly increase the number of autophagosomes in cardiomyocytes, as reported using a transgenic mouse model expressing the autophagosome protein LC3 fused with GFP. Such induction results from the disruption of the Beclin1/BCL-2 complex; BCL-2 is an anti-apoptotic protein which inhibits Autophagy activation by binding to the BH3 domain of Beclin1 [106].

Pathologic hypertrophy is also associated with modulation of ALS pathway activity. For example, pressure overload is characterized by a dramatic induction of ALS in the early phases, followed by a moderate reduction in more advanced disease stages. This supports the notion that '*...autophagic activity carries out different functions depending on disease stage and severity...*' [107]. The observation that decreased Autophagy in Beclin1 haploinsufficient mice significantly reduced cardiomyocyte death and systolic dysfunction [71] suggested that Autophagy may, in the long term, contribute to the transition from adaptive to maladaptive hypertrophy and HF. However, other studies came to the seemingly opposite conclusion that decreased Autophagy is indeed detrimental in maladaptive hypertrophy, as evidenced by the observation that cardiac-specific deficiency of Atg5 accelerates progression to HF upon pressure overload and β-adrenergic stress [86]. The apparent contradiction between these studies has been commented by *Rothermel and Hill*, who suggested that the degree of Autophagy inhibition (totally ablated in $Atg5^{flox/flox}$;MLC2v-Cre^{+} mice, while only partially reduced in the Beclin1 haploinsufficient mice) contributed to the differential response to stress [71]. Collectively, these observations have led to a consensus view on the '*dual role of Autophagy*' in heart diseases and on the fact that '*...depending on the amplitude and duration of the autophagic response and the initial level of basal Autophagy flux, Autophagy may elicit either harmful or beneficial effects...*' [108]. As a consequence, it has been hypothesized that a precise '*...window of optimal Autophagy action...*' exists, beyond which cell viability is compromised [108] (Fig. 4.7). In light of this, it has been proposed that in the early phases of cardiac hypertrophy, Autophagy exerts a beneficial role by removing intracellular aggregates of unfolded/misfolded proteins which cannot be removed by the overloaded UPS. Indeed, it has been shown both in animal models and human hearts that Autophagy activation parallels the intracellular accumulation of ubiquitinated proteins [109], thus leading to the inclusion of pressure-overload hypertrophy into the category of 'cardiac proteinopathies' (see paragraph 5.4 below). Over time, increased ROS production and mitochondrial calcium overload can induce Autophagy beyond the 'beneficial' range, reaching therefore the maladaptive levels which characterize the failing phase of hypertrophy. Thus, modulation of ALS pathways appears a promising therapeutic mechanism to be exploited in the setting of pathologic hypertrophy, however, additional preclinical work is clearly needed to determine the strategy, the optimal time window and the degree for safe and efficient Autophagy manipulation.

Autophagy and Cardiac Proteinopathy

Protein quality control (PQC) is a housekeeping mechanism to avoid intracellular accumulation of damaged proteins and consequent cell proteotoxicity. PQC is dependent upon the cooperation between chaperones (which promote correct protein folding), the UPS (which mediates ubiquitination of proteins to be degraded by the proteasome) and Autophagy (which removes protein aggregates that cannot be

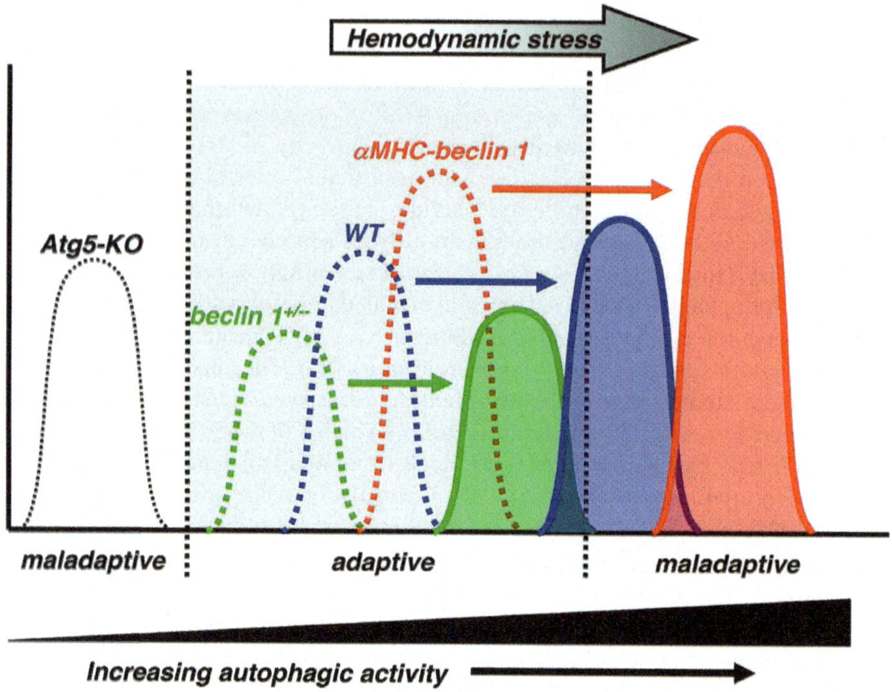

Fig. 4.7 Illustration of the dual nature of Autophagy in pathologic cardiac hypertrophy. Basal activation of Autophagy is required for normal cellular function. Too little or too much Autophagy can each be maladaptive. Hemodynamic stress triggers Autophagy, which, depending on the amplitude and duration of autophagic activation and the initial level of basal autophagic flux, may elicit either harmful or beneficial effects. Reproduced from Rothermel and Hill [108]

destroyed by the UPS). As indicated above, UPS and Autophagy are functionally interconnected and influence each other. Thus perturbation of UPS activity causes aggregation of misfolded proteins in soluble oligomers and inclusion bodies, which, unless they are removed by Autophagy, may lead to cardiomyocyte death by a mechanism known as proteotoxicity (Fig. 4.5) [110–112]. UPS dysfunction has been described in several cardiac disorders, including hypertrophic and dilated cardiomyopathies, HF, and familial desmin-related cardiomyopathy (DRC). The latter is a *bona fide* cardiac proteinopathy, caused by mutations in genes encoding for desmin or desmin-related proteins, like αB-crystallin (CryAB), which binds both desmin and actin and functions as a chaperone protein [113–115]. Transgenic mouse models expressing the R120G-missense mutation of CryAB [116] or the D7-des protein [117]) recapitulate well the DRC phenotype. Indeed, these animals develop intracellular desmin aggregates, sarcomeric disarrangement, cardiomyocyte hypertrophy, and increased cell death which ultimately leads to contractile dysfunction and increased mortality. Cardiomyocytes from these mice display increased autophagosome content, indicating that Autophagy is spontaneously activated. Moreover,

Autophagy inhibition worsened intracellular accumulation of ubiquitinated protein aggregates, accelerating cardiomyocyte death and disease progression [118, 119]. These data support a compensatory role of Autophagy as an adaptive and protective response to counteract cell proteotoxicity following primary or secondary UPS dysfunction. Although the signalling mechanisms responsible for proteotoxicity-dependent activation of Autophagy have not been determined, these data suggest that enhancement of Autophagy may delay disease progression. This was exploited by *Robbins and colleagues*, who demonstrated the beneficial role of exercise-dependent Autophagy induction in preserving cardiomyocyte integrity and cardiac function in a mouse model of DRC [120]. In further support of the key role of Autophagy in this disease mechanism, amelioration of the DRC phenotype was achieved by exploiting the heat-shock protein family of chaperones, which, by improving correct protein folding, prevent from protein aggregation and consequent Autophagy overload [121].

Autophagy and Cardiac Ageing

Ageing is accompanied by a significant increase in the incidence of heart diseases, featuring increased cardiomyocyte death which in turn results in adverse myocardial remodelling with diastolic dysfunction, increased myocardial stiffness and decreased adaptation to extrinsic neuro-hormonal stimuli [122–124]. Several theories have been proposed to explain the molecular mechanisms which contribute to cardiac ageing, including the argument that age-related cardiomyocyte damage/death is a consequence of increased intracellular oxidative stress. In support of this, it is well accepted that ROS production increases during ageing due to progressive mitochondrial dysfunction [125]. ROS activate AMPK, which in turn induces Atg4 transcription and consequently ALS activity [82, 126]. ALS activation should negatively regulate ROS production by removing damaged mitochondria via mitophagy. In support of the existence of this negative feedback loop, genetic ablation of Atg5 results in increased age-related accumulation of damaged mitochondria and cardiac dysfunction [127]. Conversely, ROS promote the intra-lysosomal formation of undegradable aggregates (i.e. lipofuscin), which in turn alter lysosome pH resulting in the inhibition of autophagosome/lysosome fusion [128, 129]. Thus, aging-dependent declines in Autophagy function can generate a positive feedback loop with progressively increased ROS content. Although the notion that Autophagy activity declines during ageing is generally accepted, the underlying mechanism is debated. Strain- and species-dependent variations, as well as the use of different Autophagy molecular read-outs as surrogate markers for its activity, have undoubtedly contributed to the lack of a consensus mechanistic view. Utilization of 'Autophagy flux' as the gold standard for Autophagy activity quantification would go a long way in clarifying the situation [130]. Given that experimental Autophagy activation can extend lifespan in multiple species, and given that physiologic interventions associated with Autophagy activation (i.e. caloric restriction, endurance exercise) can also extend lifespan [50,

131–136], understanding the mechanisms which give rise to age-related reduction of this proteolytic mechanism is clearly warranted.

Concluding Remarks

The data reviewed here clearly demonstrate that Autophagy plays a pivotal role in regulating cardiomyocyte homeostasis in both physiologic and pathophysiologic conditions. Indeed, well-functioning Autophagy is required to maintain cellular health across the wide spectrum of adaptation to physiologic stresses encountered by the heart from birth to old age. While there is a good consensus that this concept is correct, the role of Autophagy in cardiac diseases is still debated, with some studies indicating that too much Autophagy may be detrimental in the response to pathologic stimuli, and contribute to, rather than antagonize, the pathogenesis of cardiac maladaptation. Such divergent observations, which we have attempted to give a balanced view on, argue that Autophagy must operate within a precise window of activity to exert beneficial effects on the post-mitotic cardiomyocytes. Despite intense research efforts of the last decades, the molecular mechanisms regulating Autophagy in health and disease have not been completely elucidated, and additional studies of the role of the ALS are thus warranted. Further, clarification of the degree to which Autophagy exerts a beneficial or maladaptive role in cardiovascular disorders has the potential to develop novel strategies aimed at enhancing or inhibiting such proteolytic machinery to achieve a therapeutic effect.

Acknowledgements We are grateful to Marco Mongillo (University of Padova) for critical reading and discussion, and Nicola Pianca, Ph.D. (University of Padova) for critical reading and technical assistance.

Conflict of Interest Nothing to declare.

References

1. Appelmans F, Wattiaux R, De Duve C (1955) Tissue fractionation studies. 5. The association of acid phosphatase with a special class of cytoplasmic granules in rat liver. Biochem J 3:438–445
2. De Duve C, Pressman BC, Gianetto R et al (1955) Tissue fractionation studies. 6. Intracellular distribution patterns of enzymes in rat-liver tissue. Biochem J 4:604–617
3. Novikoff AB, Beaufay H, De Duve C (1956) Electron microscopy of lysosomerich fractions from rat liver. J Biophys Biochem Cytol 4(Suppl):179–184
4. Buechner, Letterer, Roulet (eds) (1955) Handbook of general physiology. Springer, Berlin-Goettingen-Heidelberg
5. Hruban Z, Spargo B, Swift H et al (1963) Focal cytoplasmic degradation. Am J Pathol 657–683
6. Ohsumi Y (1999) Molecular mechanism of autophagy in yeast, Saccharomyces cerevisiae. Philos Trans R Soc Lond B Biol Sci 1389:1577–1580; discussion 1580–1571

7. Soonpaa MH, Kim KK, Pajak L et al (1996) Cardiomyocyte DNA synthesis and binucleation during murine development. Am J Physiol 5(Pt 2):H2183-2189
8. Soonpaa MH, Rubart M, Field LJ (2013) Challenges measuring cardiomyocyte renewal. Biochim Biophys Acta 4:799–803
9. Woodcock EA, Matkovich SJ (2005) Cardiomyocytes structure, function and associated pathologies. Int J Biochem Cell Biol 9:1746–1751
10. Foglia MJ, Poss KD (2016) Building and re-building the heart by cardiomyocyte proliferation. Development 5:729–740
11. O'Connell TD, Ishizaka S, Nakamura A et al (2003) The alpha(1A/C)- and alpha(1B)-adrenergic receptors are required for physiological cardiac hypertrophy in the double-knockout mouse. J Clin Invest 11:1783–1791
12. Maillet M, van Berlo JH, Molkentin JD (2013) Molecular basis of physiological heart growth: fundamental concepts and new players. Nat Rev Mol Cell Biol 1:38–48
13. Sutton MG, Sharpe N (2000) Left ventricular remodeling after myocardial infarction: pathophysiology and therapy. Circulation 25:2981–2988
14. Bui AL, Horwich TB, Fonarow GC (2011) Epidemiology and risk profile of heart failure. Nat Rev Cardiol 1:30–41
15. Laflamme MA, Murry CE (2005) Regenerating the heart. Nat Biotechnol 7:845–856
16. Rubart M, Field LJ (2008) Stem cell differentiation: cardiac repair. Cells Tissues Organs 1–2:202–211
17. Rubart M, Field LJ (2006) Cardiac regeneration: repopulating the heart. Annu Rev Physiol 29–49
18. Shimizu I, Minamino T (2016) Physiological and pathological cardiac hypertrophy. J Mol Cell Cardiol 245–262
19. Willis MS, Bevilacqua A, Pulinilkunnil T et al (2014) The role of ubiquitin ligases in cardiac disease. J Mol Cell Cardiol 43–53
20. Levine B, Kroemer G (2008) Autophagy in the pathogenesis of disease. Cell 1:27–42
21. Bonaldo P, Sandri M (2013) Cellular and molecular mechanisms of muscle atrophy. Dis Model Mech 1:25–39
22. Mizushima N (2007) Autophagy: process and function. Genes Dev 22:2861–2873
23. Delorme-Axford E, Guimaraes RS, Reggiori F et al (2015) The yeast Saccharomyces cerevisiae: an overview of methods to study autophagy progression. Methods 3–12
24. Wesselborg S, Stork B (2015) Autophagy signal transduction by ATG proteins: from hierarchies to networks. Cell Mol Life Sci 24:4721–4757
25. Arias E, Cuervo AM (2011) Chaperone-mediated autophagy in protein quality control. Curr Opin Cell Biol 2:184–189
26. Kadowaki M, Karim MR, Carpi A et al (2006) Nutrient control of macroAutophagy in mammalian cells. Mol Aspects Med 5–6:426–443
27. Talloczy Z, Jiang W, Virgin HWt et al (2002) Regulation of starvation- and virus-induced autophagy by the eIF2alpha kinase signaling pathway. Proc Natl Acad Sci U S A 1:190–195
28. Kanazawa T, Taneike I, Akaishi R et al (2004) Amino acids and insulin control autophagic proteolysis through different signaling pathways in relation to mTOR in isolated rat hepatocytes. J Biol Chem 9:8452–8459
29. Pattingre S, Bauvy C, Codogno P (2003) Amino acids interfere with the ERK1/2-dependent control of macroAutophagy by controlling the activation of Raf-1 in human colon cancer HT-29 cells. J Biol Chem 19:16667–16674
30. Jewell JL, Russell RC, Guan KL (2013) Amino acid signalling upstream of mTOR. Nat Rev Mol Cell Biol 3:133–139
31. Kim YC, Guan KL (2015) mTOR: a pharmacologic target for autophagy regulation. J Clin Invest 1:25–32
32. Schmelzle T, Beck T, Martin DE et al (2004) Activation of the RAS/cyclic AMP pathway suppresses a TOR deficiency in yeast. Mol Cell Biol 1:338–351
33. Meijer AJ, Dubbelhuis PF (2004) Amino acid signalling and the integration of metabolism. Biochem Biophys Res Commun 2:397–403

34. Meijer AJ, Codogno P (2007) AMP-activated protein kinase and autophagy. Autophagy 3:238–240
35. Hoyer-Hansen M, Bastholm L, Szyniarowski P et al (2007) Control of macroAutophagy by calcium, calmodulin-dependent kinase kinase-beta, and Bcl-2. Mol Cell 2:193–205
36. Mammucari C, Milan G, Romanello V et al (2007) FoxO3 controls autophagy in skeletal muscle in vivo. Cell Metab 6:458–471
37. Zhao J, Brault JJ, Schild A et al (2007) FoxO3 coordinately activates protein degradation by the autophagic/lysosomal and proteasomal pathways in atrophying muscle cells. Cell Metab 6:472–483
38. Lee WS, Yoo WH, Chae HJ (2015) ER Stress and autophagy. Curr Mol Med 8:735–745
39. Glembotski CC (2008) The role of the unfolded protein response in the heart. J Mol Cell Cardiol 3:453–459
40. B'Chir W, Maurin AC, Carraro V et al (2013) The eIF2alpha/ATF4 pathway is essential for stress-induced autophagy gene expression. Nucleic Acids Res 16:7683–7699
41. Jiang Q, Li F, Shi K et al (2014) Involvement of p38 in signal switching from autophagy to apoptosis via the PERK/eIF2alpha/ATF4 axis in selenite-treated NB4 cells. Cell Death Dis e1270
42. Zaglia T, Milan G, Ruhs A et al (2014) Atrogin-1 deficiency promotes cardiomyopathy and premature death via impaired autophagy. J Clin Invest 6:2410–2424
43. Korolchuk VI, Menzies FM, Rubinsztein DC (2010) Mechanisms of cross-talk between the ubiquitin-proteasome and autophagy-lysosome systems. FEBS Lett 7:1393–1398
44. Su H, Wang X (2011) p62 Stages an interplay between the ubiquitin-proteasome system and autophagy in the heart of defense against proteotoxic stress. Trends Cardiovasc Med 8:224–228
45. Levine B, Klionsky DJ (2004) Development by self-digestion: molecular mechanisms and biological functions of autophagy. Dev Cell 6:463–477
46. Kuma A, Hatano M, Matsui M et al (2004) The role of autophagy during the early neonatal starvation period. Nature 7020:1032–1036
47. Deretic V (2005) Autophagy in innate and adaptive immunity. Trends Immunol 10:523–528
48. Levine B, Deretic V (2007) Unveiling the roles of autophagy in innate and adaptive immunity. Nat Rev Immunol 10:767–777
49. Madeo F, Zimmermann A, Maiuri MC et al (2015) Essential role for autophagy in life span extension. J Clin Invest 1:85–93
50. Simonsen A, Cumming RC, Brech A et al (2008) Promoting basal levels of autophagy in the nervous system enhances longevity and oxidant resistance in adult Drosophila. Autophagy 2:176–184
51. Hars ES, Qi H, Ryazanov AG et al (2007) Autophagy regulates ageing in C. elegans. Autophagy 2:93–95
52. Komatsu M, Waguri S, Chiba T et al (2006) Loss of autophagy in the central nervous system causes neurodegeneration in mice. Nature 7095:880–884
53. Ogata M, Hino S, Saito A et al (2006) Autophagy is activated for cell survival after endoplasmic reticulum stress. Mol Cell Biol 24:9220–9231
54. Lee JA, Gao FB (2008) Roles of ESCRT in autophagy-associated neurodegeneration. Autophagy 2:230–232
55. Lee JA, Beigneux A, Ahmad ST et al (2007) ESCRT-III dysfunction causes autophagosome accumulation and neurodegeneration. Curr Biol 18:1561–1567
56. Ravikumar B, Vacher C, Berger Z et al (2004) Inhibition of mTOR induces autophagy and reduces toxicity of polyglutamine expansions in fly and mouse models of Huntington disease. Nat Genet 6:585–595
57. Sarkar S, Perlstein EO, Imarisio S et al (2007) Small molecules enhance autophagy and reduce toxicity in Huntington's disease models. Nat Chem Biol 6:331–338
58. Lacoste-Collin L, Garcia V, Uro-Coste E et al (2002) Danon's disease (X-linked vacuolar cardiomyopathy and myopathy): a case with a novel Lamp-2 gene mutation. Neuromuscul Disord 9:882–885

59. Tanaka Y, Guhde G, Suter A et al (2000) Accumulation of autophagic vacuoles and cardiomyopathy in LAMP-2-deficient mice. Nature 6798:902–906
60. Mathew R, Kongara S, Beaudoin B et al (2007) Autophagy suppresses tumor progression by limiting chromosomal instability. Genes Dev 11:1367–1381
61. Dorsey FC, Steeves MA, Prater SM et al (2009) Monitoring the autophagy pathway in cancer. Methods Enzymol 251–271
62. Maiuri MC, Zalckvar E, Kimchi A et al (2007) Self-eating and self-killing: crosstalk between autophagy and apoptosis. Nat Rev Mol Cell Biol 9:741–752
63. Kenific CM, Debnath J (2015) Cellular and metabolic functions for autophagy in cancer cells. Trends Cell Biol 1:37–45
64. Kroemer G, Galluzzi L, Vandenabeele P et al (2009) Classification of cell death: recommendations of the nomenclature Committee on Cell Death 2009. Cell Death Differ 1:3–11
65. Lockshin RA, Zakeri Z (2001) Programmed cell death and apoptosis: origins of the theory. Nat Rev Mol Cell Biol 7:545–550
66. Nelson C, Baehrecke EH (2014) Autophagy and cell death in the fly. Methods Enzymol 181–199.
67. Denton D, Xu T, Kumar S (2015) Autophagy as a pro-death pathway. Immunol Cell Biol 1:35–42
68. Chatterjee T, Muhkopadhyay A, Khan KA et al (1998) Comparative mutagenic and genotoxic effects of three antimalarial drugs, chloroquine, primaquine and amodiaquine. Mutagenesis 6:619–624
69. McManus MT, Sharp PA (2002) Gene silencing in mammals by small interfering RNAs. Nat Rev Genet 10:737–747
70. Yousefi S, Perozzo R, Schmid I et al (2006) Calpain-mediated cleavage of Atg5 switches autophagy to apoptosis. Nat Cell Biol 10:1124–1132
71. Zhu H, Tannous P, Johnstone JL et al (2007) Cardiac autophagy is a maladaptive response to hemodynamic stress. J Clin Invest 7:1782–1793
72. Marino G, Niso-Santano M, Baehrecke EH et al (2014) Self-consumption: the interplay of autophagy and apoptosis. Nat Rev Mol Cell Biol 2:81–94
73. Amir M, Zhao E, Fontana L et al (2013) Inhibition of hepatocyte autophagy increases tumor necrosis factor-dependent liver injury by promoting caspase-8 activation. Cell Death Differ 7:878–887
74. Sandilands E, Serrels B, Wilkinson S et al (2012) Src-dependent autophagic degradation of Ret in FAK-signalling-defective cancer cells. EMBO Rep 8:733–740
75. Kim I, Lemasters JJ (2011) Mitochondrial degradation by autophagy (mitophagy) in GFP-LC3 transgenic hepatocytes during nutrient deprivation. Am J Physiol Cell Physiol 2:C308-317
76. Oral O, Oz-Arslan D, Itah Z et al (2012) Cleavage of Atg3 protein by caspase-8 regulates autophagy during receptor-activated cell death. Apoptosis 8:810–820
77. Luo S, Rubinsztein DC (2010) Apoptosis blocks Beclin 1-dependent autophagosome synthesis: an effect rescued by Bcl-xL. Cell Death Differ 2:268–277
78. Wirawan E, Vande Walle L, Kersse K et al (2010) Caspase-mediated cleavage of Beclin-1 inactivates Beclin-1-induced autophagy and enhances apoptosis by promoting the release of proapoptotic factors from mitochondria. Cell Death Dis e18
79. Komatsu M, Waguri S, Ueno T et al (2005) Impairment of starvation-induced and constitutive autophagy in Atg7-deficient mice. J Cell Biol 3:425–434
80. Viragh S, Szabo E, Challice CE (1982) Glycogen-containing lysosomes and glycogen loss in the cardiocytes of embryonic and neonatal mice. Adv Myocardiol 553–561
81. Saugstad OD (2005) Oxidative stress in the newborn—a 30-year perspective. Biol Neonate 3:228–236
82. Scherz-Shouval R, Shvets E, Fass E et al (2007) Reactive oxygen species are essential for autophagy and specifically regulate the activity of Atg4. Embo j 7:1749–1760
83. Scherz-Shouval R, Elazar Z (2007) ROS, mitochondria and the regulation of autophagy. Trends Cell Biol 9:422–427

84. Nishino I, Fu J, Tanji K et al (2000) Primary LAMP-2 deficiency causes X-linked vacuolar cardiomyopathy and myopathy (Danon disease). Nature 6798:906–910
85. Pattison JS, Robbins J (2011) Autophagy and proteotoxicity in cardiomyocytes. Autophagy 10:1259–1260
86. Nakai A, Yamaguchi O, Takeda T et al (2007) The role of autophagy in cardiomyocytes in the basal state and in response to hemodynamic stress. Nat Med 5:619–624
87. Samarel AM, Parmacek MS, Magid NM et al (1987) Protein synthesis and degradation during starvation-induced cardiac atrophy in rabbits. Circ Res 6:933–941
88. Kolar F, MacNaughton C, Papousek F et al (1993) Systolic mechanical performance of heterotopically transplanted hearts in rats treated with cyclosporin. Cardiovasc Res 7:1244–1247
89. Perhonen MA, Franco F, Lane LD et al (2001) Cardiac atrophy after bed rest and spaceflight. J Appl Physiol 2:645–653
90. Crie JS, Sanford CF, Wildenthal K (1980) Influence of starvation and refeeding on cardiac protein degradation in rats. J Nutr 1:22–27
91. Burch GE, Phillips JH, Ansari A (1968) The cachetic heart. A clinico-pathologic, electrocardiographic and roentgenographic entity. Dis Chest 5:403–409
92. Zhou X, Wang JL, Lu J et al (2010) Reversal of cancer cachexia and muscle wasting by ActRIIB antagonism leads to prolonged survival. Cell 4:531–543
93. Cosper PF, Leinwand LA (2011) Cancer causes cardiac atrophy and Autophagy in a sexually dimorphic manner. Cancer Res 5:1710–1720
94. Toledo M, Busquets S, Penna F et al (2016) Complete reversal of muscle wasting in experimental cancer cachexia: additive effects of activin type II receptor inhibition and beta-2 agonist. Int J Cancer 8:2021–2029
95. Mettauer B, Rouleau JL, Burgess JH (1985) Detrimental arrhythmogenic and sustained beneficial hemodynamic effects of oral salbutamol in patients with chronic congestive heart failure. Am Heart J 4:840–847
96. Molenaar P, Chen L, Parsonage WA (2006) Cardiac implications for the use of beta2-adrenoceptor agonists for the management of muscle wasting. Br J Pharmacol 6:583–586
97. Goldstein MA, Edwards RJ, Schroeter JP (1992) Cardiac morphology after conditions of microgravity during COSMOS 2044. J Appl Physiol (2 Suppl):94s–100s
98. Kadner A, Chen RH, Adams DH (2000) Heterotopic heart transplantation: experimental development and clinical experience. Eur J Cardiothorac Surg 4:474–481
99. Razeghi P, Volpini KC, Wang ME et al (2007) Mechanical unloading of the heart activates the calpain system. J Mol Cell Cardiol 2:449–452
100. Kassiotis C, Ballal K, Wellnitz K et al (2009) Markers of Autophagy are downregulated in failing human heart after mechanical unloading. Circulation 11(Suppl):S191-197
101. Zaglia T, Milan G, Franzoso M et al (2013) Cardiac sympathetic neurons provide trophic signal to the heart via beta2-adrenoceptor-dependent regulation of proteolysis. Cardiovasc Res 2:240–250
102. de Lange P, Moreno M, Silvestri E et al (2007) Fuel economy in food-deprived skeletal muscle: signaling pathways and regulatory mechanisms. Faseb j 13:3431–3441
103. Wu J, Ruas JL, Estall JL et al (2011) The unfolded protein response mediates adaptation to exercise in skeletal muscle through a PGC-1alpha/ATF6alpha complex. Cell Metab 2:160–169
104. Sahlin K, Shabalina IG, Mattsson CM et al (2010) Ultraendurance exercise increases the production of reactive oxygen species in isolated mitochondria from human skeletal muscle. J Appl Physiol 4:780–787
105. Jamart C, Benoit N, Raymackers JM et al (2012) Autophagy-related and autophagy-regulatory genes are induced in human muscle after ultraendurance exercise. Eur J Appl Physiol 8:3173–3177
106. He C, Bassik MC, Moresi V et al (2012) Exercise-induced BCL2-regulated autophagy is required for muscle glucose homeostasis. Nature 7382:511–515
107. Rothermel BA, Hill JA (2007) Myocyte autophagy in heart disease: friend or foe? Autophagy 6:632–634

108. Rothermel BA, Hill JA (2008) Autophagy in load-induced heart disease. Circ Res 12:1363–1369
109. Tannous P, Zhu H, Nemchenko A et al (2008) Intracellular protein aggregation is a proximal trigger of cardiomyocyte autophagy. Circulation 24:3070–3078
110. Wang C, Wang X (2015) The interplay between autophagy and the ubiquitin-proteasome system in cardiac proteotoxicity. Biochim Biophys Acta 2:188–194
111. Su H, Li J, Zhang H et al (2015) COP9 signalosome controls the degradation of cytosolic misfolded proteins and protects against cardiac proteotoxicity. Circ Res 11:956–966
112. Willis MS, Patterson C (2013) Proteotoxicity and cardiac dysfunction–Alzheimer's disease of the heart? N Engl J Med 5:455–464
113. Bennardini F, Wrzosek A, Chiesi M (1992) Alpha B-crystallin in cardiac tissue. Association with actin and desmin filaments. Circ Res 2:288–294
114. Nicholl ID, Quinlan RA (1994) Chaperone activity of alpha-crystallins modulates intermediate filament assembly. Embo J 4:945–953
115. Wang K, Spector A (1996) alpha-crystallin stabilizes actin filaments and prevents cytochalasin-induced depolymerization in a phosphorylation-dependent manner. Eur J Biochem 1:56–66
116. Wang X, Osinska H, Klevitsky R et al (2001) Expression of R120G-alphaB-crystallin causes aberrant desmin and alphaB-crystallin aggregation and cardiomyopathy in mice. Circ Res 1:84–91
117. Wang X, Osinska H, Dorn GW 2nd et al (2001) Mouse model of desmin-related cardiomyopathy. Circulation 19:2402–2407
118. Bence NF, Sampat RM, Kopito RR (2001) Impairment of the ubiquitin-proteasome system by protein aggregation. Science 5521:1552–1555
119. Tannous P, Zhu H, Johnstone JL et al (2008) Autophagy is an adaptive response in desmin-related cardiomyopathy. Proc Natl Acad Sci U S A 28:9745–9750
120. Bhuiyan MS, Pattison JS, Osinska H et al (2013) Enhanced autophagy ameliorates cardiac proteinopathy. J Clin Invest 12:5284–5297
121. Sanbe A, Daicho T, Mizutani R et al (2009) Protective effect of geranylgeranylacetone via enhancement of HSPB8 induction in desmin-related cardiomyopathy. PLoS One 4:e5351
122. Shirakabe A, Ikeda Y, Sciarretta S et al (2016) Aging and autophagy in the heart. Circ Res 10:1563–1576
123. Nakayama H, Nishida K, Otsu K (2016) Macromolecular degradation systems and cardiovascular aging. Circ Res 10:1577–1592
124. Linton PJ, Gurney M, Sengstock D et al (2015) This old heart: cardiac aging and autophagy. J Mol Cell Cardiol 44–54
125. Terman A, Kurz T, Navratil M et al (2010) Mitochondrial turnover and aging of long-lived postmitotic cells: the mitochondrial-lysosomal axis theory of aging. Antioxid Redox Signal 4:503–535
126. Li L, Chen Y, Gibson SB (2013) Starvation-induced autophagy is regulated by mitochondrial reactive oxygen species leading to AMPK activation. Cell Signal 1:50–65
127. Taneike M, Yamaguchi O, Nakai A et al (2010) Inhibition of autophagy in the heart induces age-related cardiomyopathy. Autophagy 5:600–606
128. Zheng K, Li Y, Wang S et al (2016) Inhibition of autophagosome-lysosome fusion by ginsenoside Ro via the ESR2-NCF1-ROS pathway sensitizes esophageal cancer cells to 5-fluorouracil-induced cell death via the CHEK1-mediated DNA damage checkpoint. Autophagy 9:1593–1613
129. Jaishy B, Zhang Q, Chung HS et al (2015) Lipid-induced NOX2 activation inhibits autophagic flux by impairing lysosomal enzyme activity. J Lipid Res 3:546–561
130. Mizushima N, Yoshimori T (2007) How to interpret LC3 immunoblotting. Autophagy 6:542–545
131. Harrison DE, Strong R, Sharp ZD et al (2009) Rapamycin fed late in life extends lifespan in genetically heterogeneous mice. Nature 7253:392–395

132. Lee IH, Cao L, Mostoslavsky R et al (2008) A role for the NAD-dependent deacetylase Sirt1 in the regulation of autophagy. Proc Natl Acad Sci U S A 9:3374–3379
133. Matsushima S, Sadoshima J (2015) The role of sirtuins in cardiac disease. Am J Physiol Heart Circ Physiol 9:H1375-1389
134. Alvers AL, Wood MS, Hu D et al (2009) Autophagy is required for extension of yeast chronological life span by rapamycin. Autophagy 6:847–849
135. Speakman JR, Mitchell SE (2011) Caloric restriction. Mol Aspects Med 3:159–221
136. Sujkowski A, Bazzell B, Carpenter K et al (2015) Endurance exercise and selective breeding for longevity extend Drosophila healthspan by overlapping mechanisms. Aging (Albany NY) 8:535–552
137. Chung E, Leinwand LA (2014) Pregnancy as a cardiac stress model. Cardiovasc Res 4:561–570

Chapter 5
Caspase Signaling Pathways as Convenors of Stress Adaptation

Charis Putinski and Lynn A. Megeney

Abstract Caspase proteins are a group of proteases that manage intermediate and late stages of programmed cell death/apoptosis. Despite the core cell death function inherent to caspases, these proteins also maintain distinct nonapoptotic functions across most metazoan organisms. Here, we review the role of caspases in cell differentiation and stress adaptation, and the evolution of nonapoptotic activity of these proteases and related factors. We also discuss how caspases integrate other cell death signaling pathways to manage and guide stress response in the mammalian heart, independent of cell death.

Keywords Caspase · Nonapoptotic · Cell differentiation · Cardiac hypertrophy

Introduction

Eukaryotic cells retain a remarkable capacity to adjust and remodel in response to environmental demands. This innate stress sensing capacity is generally divided into two distinct outcomes, the cell adapts and survives, or the cell dies, often through a regulated form of death. Typically, stress induced survival and death are treated as divergent cell fates, yet the morphologic and biochemical alterations for each stress response are remarkably similar. This general overlap in the biology of cell stress has led to the hypothesis that successful adaptation and cell death may utilize similar proteins and biochemical pathways [1, 2]. This supposition derived from earlier observations where investigators first demonstrated that cell death pathways were

C. Putinski · L. A. Megeney (✉)
Ottawa Hospital Research Institute, Sprott Centre for Stem Cell Research, Regenerative Medicine Program, Ottawa Hospital, Ottawa, ON K18L6, Canada
e-mail: lmegeney@ohri.ca

Department of Cellular and Molecular Medicine, University of Ottawa, Ottawa, ON K18M5, Canada

L. A. Megeney
Department of Medicine, University of Ottawa, Ottawa, ON K18M5, Canada

© Springer Nature Switzerland AG 2022
L. A. Kirshenbaum (ed.), *Biochemistry of Apoptosis and Autophagy*,
Advances in Biochemistry in Health and Disease 18,
https://doi.org/10.1007/978-3-030-78799-8_5

conserved and required components of cell differentiation, across a broad species range [3–7]. More recently, the evolutionary role of cell death proteins as essential stress sensors and survival factors were confirmed in yeast models, where death centric proteases were shown to limit accumulation of stress-induced toxic protein aggregates [8–11].

In this chapter, we discuss the diversity of cell death regulatory mechanisms with a specific emphasis on caspase protease-mediated signaling pathways. Here, we examine the emerging intersection between caspase signaling and autophagy as it relates to stress remodeling and how caspase proteases may alter a wide variety of stress adaptations including cell hypertrophy and cell atrophy.

Cell Death and Apoptosis

There are two main types of cell death -necrosis and apoptosis- which differ in mechanism and morphology. Necrosis occurs when there is severe damage to cell membranes resulting in enzymes leak from lysosomes, which initiates digestion of the cell. Here, necrotic cells appear enlarged due to swelling which is rapidly followed by cell rupture and release of cellular contents. Autophagy is another form of lysosomal mediated cell death which is activated under conditions such as nutrient deprivation, stress, infection and cancer. However, unlike necrosis, autophagy may also occur as a managed catabolic process, in which cells survive by lysosomal degradation of their own cellular components.

Apoptosis is a highly regulated form of cell death characterized by nuclear dissolution, without complete loss of membrane integrity. Here, DNA fragmentation is evident and apoptotic cells shrink and form cytoplasmic buds which dissociate into apoptotic bodies. These fragments are phagocytised, eliminating the induction of an inflammatory response. Cell injury due to infections, DNA damage and the accumulation of misfolded proteins are examples of pathological states that induce an apoptotic cell death response. Canonical apoptotic cell death pathways derive from an intrinsic source originating from the mitochondria or an extrinsic source mediated by pro-death ligand/receptor interactions. One class of proteins known to mediate apoptosis is the caspase protease family. Caspases are cysteinyl aspartate-specific proteases and they have been well characterized for their role in cell death progression and the apoptotic caspase cascade. During cell death progression, activated caspases cleave numerous substrates converging on the activation of nucleases that degrade DNA and other enzymes, which promote destruction of the cytoskeleton and nucleoproteins. These enzymes cleave their substrates by hydrolyzing peptide bonds on the carboxyl end of an aspartic acid residue within their recognition sequence [12]. Caspases are engaged through the sequential activation and inactivation of numerous pro- and anti-apoptotic proteins.

Although there is a strong preference to consider apoptosis in a strictly negative connotation (as a disease inducing mechanism), apoptosis is also a vital process that maintains organism homeostasis. Specifically, one subtype of apoptosis, referred to as programmed cell death (PCD), is as prominent as cell division in shaping the developing organism. PCD controls cell numbers, provides proper structural sculpting of developing tissues, which together drives the basic morphogenic process. In developing amphibians/frogs, the transition from the tadpole stage to the adult is coincident with a PCD directed removal of the tail and a portion of the intestine. The formation of digits in higher vertebrates is another well defined example of PCD controlled tissue sculpting. Apoptosis also serves as a protective mechanism by eliminating defective cells in developing and adult organisms. A classic example in this regard is the directed apoptotic process that eliminates self-reactive lymphocytes which otherwise could instigate the development of autoimmunity [13]. More recently, the apoptotic machinery and its constituent proteins have been demonstrated to be involved in various non-death functions, including cellular differentiation and stress remodeling.

Caspase Signaling in Death and Nondeath Cell Function

Caspase proteases serve a vital role in apoptosis progression. Caspases are synthesized as inactive zymogens composed of three domains: an N terminal prodomain and the p20 and p10 domains. The active enzyme is a heterotetramer composed of two p20/p10 heterodimers and two active sites [12]. Mechanisms of caspase activation include processing by upstream caspases, induced proximity or the association with a regulatory subunit. Caspases are activated by proteolytic processing of the inactive zymogen between p20 and p10, and between the prodomain and p20. Each cleavage event occurs at the carboxyl end of an aspartic acid residue. This caspase cascade mechanism is evident in the activation of downstream effector caspases 3, 6 and 7 by action of initiator caspases. Induced proximity refers to caspase activation induced by high concentrations of the pro-enzyme allowing various pro-caspases to mutually cleave and activate one another. This occurs during the activation of caspase 8 where ligand binding allows recruitment of several pro-caspase 8 molecules. Activation of caspase 9 requires association with regulatory cofactor apoptosis activating compound 1 (Apaf-1). This leads to formation of the caspase 9 holoenzyme, commonly referred to as the apoptosome.

As noted previously, the caspase cascade is initiated by either intrinsic or extrinsic signals. Death signals can arise from an intrinsic source which originates from the mitochondria. Mitochondria contain several pro-apoptotic proteins, including cytochrome *c*, and anti-apoptotic proteins. Mitochondrial permeability, controlled by these pro- and anti-apoptotic proteins, determines the fate of the cell. During stress or nutrient deprivation, sensors activate pro-apoptotic proteins Bak and Bax which dimerize and insert into the mitochondrial membrane. This forms pores which allow cytochrome *c* and other mitochondrial proteins to be released into the cytosol. This same fate can ensue by action of other sensors which inhibit anti-apoptotic proteins Bcl-2 and Bcl-x_L. The binding of cytochrome *c* and Apaf-1 results in apoptosome formation and subsequent initiator caspase 9 activation. The extrinsic pathway involves signals being received through a death receptor, such as Fas of the tumor necrosis factor (TNF) family of receptors followed by initiator caspase 8 activation. Regardless of the pathway utilized, the downstream activation of effector caspase 3 or 7 results in subsequent proteolytic processing of target proteins and cellular dissolution/death [14, 15].

A growing body of literature has provided evidence that apoptotic caspase signaling mechanisms possess non-death functions in addition to their role in cell death progression. Primarily, it has been demonstrated that caspase-dependent apoptosis pathways act as essential drivers of cell differentiation. One of the earliest observations in support of the non-death function of caspase activity was in skeletal muscle biology. Here, the upregulation of caspase 3 activity was synonymous with skeletal muscle differentiation and the inhibition of caspase 3 led to a significant reduction in the formation of mature myofibers [4]. Terminal lens fiber cell differentiation is accompanied by a series of morphological and caspase inspired biochemical events that resemble those that occur during apoptosis [16]. Caspase 3, 9, 7 and 2 are transiently activated during erythrocyte differentiation, and the inhibition of these caspases results in a dramatic reduction in erythroid differentiation [3]. More recently, it has also been demonstrated that cardiomyocyte differentiation is a caspase-dependent process [17].

Interestingly, caspases have also been linked to earlier steps in cell differentiation programs, specifically at the stage of stem cell self-renewal versus commitment to differentiation. A core set of transcription factors, including Oct4, Sox2 and Nanog, are involved in maintaining stem cells in a self-renewing state [18–20]. Fujita et al. reported that caspase 3-mediated cleavage of Nanog in embryonic stem cells (ESCs) is required for ESCs to stop the self-renewal process and initiate differentiation. Defects in differentiation were noted in stem cells lacking the caspase 3 gene [21]. Similarly, a role for caspase 3 activity in muscle stem/progenitor cell self-renewal has been observed [22]. Here, caspase 3 cleavage of the paired-box transcription factor Pax7 was reported to be crucial in terminating the self-renewal process allowing for initiation of the myogenic differentiation program. These observations demonstrate that caspase signaling acts at multiple levels to limit stem cell self-renewal and engage differentiation programs, a biologic control that is likely conserved across all cell lineages.

These non-apoptotic roles of death centric proteases are consistent with an ancient and conserved physiologic role, which are evident in organisms such as Drosophila and C. Elegans. Genetic based reporter systems in Drosophila have shown widespread caspase activation in otherwise healthy progenitor cells [23]. Similarly, studies in the worm have shown that caspases manage stem cell differentiation independent of cell death [24]. Observations in the yeast model *S. cerevisiae* have established that the functional caspase equivalent, the metacaspase Yca1, has defined non-death functions. Here, Lee et al. [25] identified that the deletion or inactivation of the single yeast metacaspase Yca1 resulted in reduced cell growth and viability. It was further determined that Yca1 is essential in limiting protein aggregation within the cell, rather than solely promoting cell death [8, 9, 11]. Interestingly, cell death signal pathways have also been shown to be active remodeling agents in cardiomyocytes and essential for the induction of cardiac hypertrophy [26, 27]. The involvement of caspases as active cellular stress sensing agents is an emergent area of investigation, spanning a diverse biology and will be the focus of the subsequent chapter components.

Stress Adaptations and the Involvement of Caspase Activity

Stress adaptations are reversible changes in the size, number, morphology, metabolic activity or function of cells in response to changes in their environment. These adaptations can be physiological which include cellular reactions to normal stimulation by endogenous chemical mediators or hormones. For instance, during pregnancy physiological enlargement of the breasts and uterus occurs, which reverts back to a basal state size postpartum. Pathological adaptations are those alterations that allow cells to evade injury by modulating their structure and function, yet if severe and/or prolonged, will lead to detrimental outcomes. Cardiac hypertrophy induced by hypertension is an example of such pathological adaptation and caspase signaling may be central to this clinically relevant alteration (Fig. 5.1).

Hypertrophy

Hypertrophy is typically defined as an increase in cell size with a concomitant increase in organ size. Hypertrophy can be characterized as physiological or pathological in reference to many organ systems. In the cardiac system, physiological hypertrophy of the heart is reversible while prolonged irreversible pathological hypertrophy can lead to injury and cell death. Physiological cardiac hypertrophy occurs in response to developmental maturation, pregnancy or exercise, and is adaptive. This is exemplified by the increased heart mass and individual cardiac muscle cell size seen in endurance sports athletes, induced by increased workload. Cardiac enlargement observed in hypertension and aortic valve diseases are maladaptive and detrimental examples of pathological hypertrophy.

As previously mentioned, caspase-dependent signaling has been demonstrated to be an initiating factor in pathological cardiac hypertrophy [26, 27]. Here, the disruption of the intrinsic cell death pathway lead to significant reduction of cardiomyocyte hypertrophy while the treatment of cardiomyocytes with a caspase activating compound lead to induction of a hypertrophy response. This demonstrated that caspase-dependent signaling is both required and sufficient to promote pathological cardiac hypertrophy [26]. Overall, these studies demonstrated that cell death pathways behave as active remodeling agents in cardiomyocytes, independent of inducing apoptosis (Fig. 5.1b). A role for caspase-activated DNase (CAD) in the regulation of pathological cardiac hypertrophy has also been observed [28]. CAD is a caspase 3 activated endonuclease that plays an important role in the induction of DNA fragmentation during apoptosis. This study found that disruption of CAD

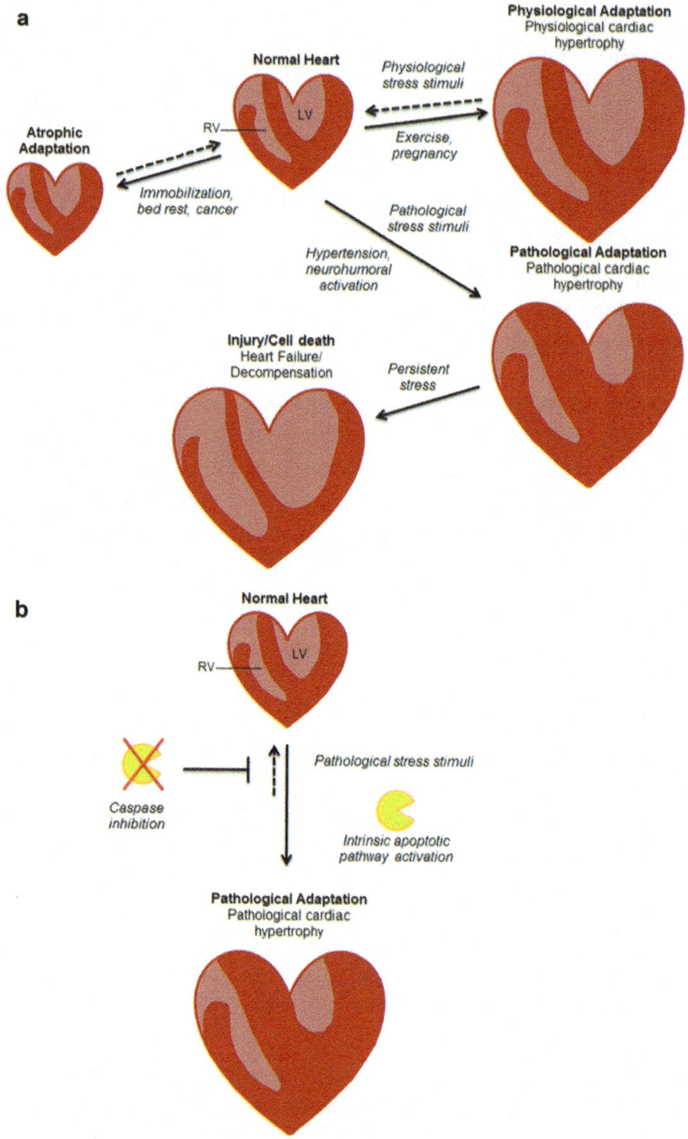

Fig. 5.1 Cardiac stress adaptation and a role for caspase-dependent pathways in stress remodeling. **a** Stress adaptations can be reversible and categorized as physiological, while others can be pathological, often progressing to injury and cell death if prolonged. A reduction (atrophy) or enlargement (hypertrophy) in cell size and organ mass can result depending on the stimuli. This can be exemplified by the remodeling of the myocardium during different types of environmental demands. **b** The intrinsic apoptotic pathway and the associated caspases are activated during pathologic and physiologic cardiac hypertrophy. The phenotypic divergence in each form of hypertrophy is due in part to variable caspase 3 activity. Here, the kinase CK2 phosphorylates caspase 3 substrates inhibiting their cleavage, while also directly supressing cleavage activation of caspase 3 [26, 33]

led to reduced pressure overload-induced cardiac hypertrophy, fibrosis, and cardiac dysfunction [28].

No matter the mechanism of hypertrophy, a limit is reached where enlargement of the myocardium can no longer compensate for the increased load. This results in degenerative changes in myocardial fibers, a loss of contractile capabilities and eventual ventricular dilation and cardiac failure. Therefore, although cardiac hypertrophy is initially adaptive it often transits to a myopathic response concurrent with an increased incidence of caspase-mediated cell death. Interestingly, caspase activity can be localized to promote dismantling of specific sub-cellular structures during remodeling and differentiation without ushering in cell death [29, 30]. During hypertrophic remodeling, caspase activity may be involved in the reorganization and remodeling of the cytoskeleton and contractile proteins [31]. For instance, caspase 3 has been shown to target α-actin, α-actinin, and cardiac troponin T prior to the onset of apoptosis during β-adrenergic stimulation [31]. More recently, it has been demonstrated that caspase 3 targets the actin remodeling protein gelsolin, to release a highly active fragment, which alone engages the pathologic hypertrophy phenotype [32]. Indeed, this caspase gelsolin interaction spurs release of gelsolin itself to propagate hypertrophy in neighboring cardiomyocytes [32], a mechanism that is consistent with the decompensation that eventually occurs in the context of pathologic hypertrophy.

While these observations are consistent with caspase activation giving rise to a pathologic form of cardiac growth, this does not preclude the role for caspase activity in physiologic or beneficial forms of cardiac adaptation. Indeed, one may envision a scenario where muted or greatly reduced levels of caspase activity may institute the more subtle cardiac growth that occurs during endurance exercise or pregnancy. Consistent with this nuanced model of caspase activation, a recent study has shown that induction of physiologic hypertrophy is in fact dependent on activation of caspase 3 activity. Here, investigators demonstrated that the cytokine cardiotrophin 1 (CT1) can induce beneficial remodeling of individual cardiomyocytes, that translates into an organ level beneficial adaptation, similar to endurance exercise training [33]. Unlike pathologic hypertrophy agonists, CT1 promotes a very attenuated activation of the mitochondrial/intrinsic cell death pathway (Fig. 5.1b). Nevertheless, this temporally restricted activation of caspase signaling is essential for CT1 to remodel the heart [33]. How this temporal control of caspase activity translates into such distinct forms of remodeling remains unknown. A reasonable supposition is that extended activation of caspase 3 during pathologic settings will target substrate(s) such as gelsolin, which are not targeted or available during attenuated caspase activation patterns.

Atrophy

In contrast to hypertrophy, atrophy is the reduction in cell size through loss of cell substance. Cell atrophy typically underpins the reductions in tissue and organ mass that occur during disease pathology and aging. Cardiac and skeletal muscle are dramatically impacted by atrophy which can arise from reduced workload due to extended bed rest, immobilization or cancer, loss of innervation, malnutrition, aging, and reduced endocrine stimulation or blood supply. Resulting cellular morphology is the same in both physiological and pathological atrophy, where smaller cells are observed and their function is diminished due to decreased protein synthesis and increased protein degradation. For example, diaphragm atrophy can occur during prolonged mechanical ventilation [34]. Evidence for cardiac atrophic remodeling in cancer-induced cachexia in mice has also been reported [35] and the ubiquitin–proteasome pathway is known to have a dominant role in this degradation process [36, 37]. Interestingly, both caspase 3 and calpain proteases have been implicated in muscle atrophy and the breakdown of structural proteins without ensuing cell death [31, 38, 39]. Evidence has suggested that both caspase 3 and calpain play a role in respiratory muscle atrophy caused by inactivity [40]. Similarly, both proteases have been shown to be required for disuse limb muscle atrophy [41]. Here, the inhibition of either caspase 3 or calpain was sufficient to protect against immobilization-induced limb muscle atrophy. Although the mechanisms of action have not been completely elucidated, structural or sarcomeric proteins have been reported to be targets of caspase 3 during such negative remodeling events, independent of cell death [31]. In addition to targeting structural proteins, caspase proteases may augment cell atrophy by engaging transcriptional responses that govern this form of cell stress. Nuclear factor kappa B (NF-κB) is a major transcription factor involved in various forms of muscle atrophy [42]. Interestingly, caspases have been demonstrated to activate NF-κB in response to specific stimuli [43]. Additionally, the activation of initiator caspase 8 has been observed in skeletal muscle during catabolic stimuli, such endotoxin, and during sarcopenia providing further support for a role of caspases in muscle proteolysis and atrophy [44–46]. Collectively, these studies support a temporally precise use of caspase inhibition in the treatment of muscle atrophy, bearing in mind that satellite cell mediated growth and repair may be adversely affected.

Caspase Signaling: At the Intersection Between Proteostasis and Autophagy

Caspase activity has been implicated as a pathogenic signal across a broad spectrum of neurodegenerative disease. Within this array of pathologies, the role of caspase activity as a causative disease agent has been most intensely studied in patients with amyotrophic lateral sclerosis (ALS) and frontotemporal lobar degeneration with ubiquitinated inclusions (FTLD-U) [47–51]. Evidence supports that activated caspases are involved in the cleavage of transactive response DNA-binding protein-43 (TDP-43), a protein known to be involved in abnormal aggregation [52–54]. Despite the propensity to treat caspase activation as a disease intensifying vector, this protease may in fact restrain or limit ALS pathology. Suzuki et al. demonstrated that caspase-mediated TDP-43 cleavage induced by endoplasmic reticulum (ER)-stress and staurosporine treatment generates C-terminal fragments (CTFs) of TDP-43, including

CTF35 and CTF27 through cleavage at Asp89 and Asp169, respectively. Interestingly, in contrast to full-length TDP-43, CTF27 did not induce cell death while it aggregated in neuronal cells. CTF35 was determined to be less effective than TDP-43 at inducing a cell death response. Heightened cell death-inducing capabilities were alternatively observed in a non-cleavable TDP-43 mutant [55]. This study supports a role for caspases in reducing toxicity through promoting TDP-43 cleavage. Other studies have also reported a similar mechanism for caspase-mediated fragmentation to that observed with TDP-43 [56, 57].

Most recently, two intriguing studies have shown that the proteostatic function of caspase 3 in neurons is not limited to resolving TDP43 aggregates. Here, investigators have shown that expression of a caspase 3 cleavage resistant mutation of the Tau protein (the insoluble plaque forming constituent that gives rise to Alzheimers pathology) leads to the development of overt brain abnormalities reminiscent of early stage Alzhemiers [58]. Secondly, caspase activity has been shown to participate with the ubiquitin ligase CHIP to disperse Tau protein aggregates [59]. Indeed, the ability of caspase 3 to disperse toxic protein aggregates is reminiscent of metacaspase function in yeast models [8, 9, 11]. While metacaspase and caspase proteases are distinct in many respects [60], the functional conservation across this larger and diverse protease family suggests there is a common role in protein quality control.

Caspase activation has also been linked to control of autophagy. Autophagy is a process where cells self digest either as a pathologic response to a particular disease condition or as a survival mechanism. In the latter case, autophagy involves lysosomal digestion during nutrient deprivation. Here, portions of the cytosol and intracellular organelles are sequestered into autophagic vacuoles formed from the ribosome-free regions of the rough endoplasmic reticulum (RER). Autophagolysosomes are formed when these vacuoles fuse with lysosomes resulting in digestion of the cellular components by lysosomal enzymes. If prolonged, autophagy can lead to cell death induction by apoptosis. Interestingly, autophagy and apoptosis are intricately linked to ensure strict regulation of cellular homeostasis. Caspases have been reported to cleave various autophagy-related proteins (Atg) and the resulting cleavage fragments serve different roles [61]. For instance, Atg Beclin-1 is cleaved by caspase 3 in Alzheimer's disease resulting in impaired Beclin-1 functionality. This suggests that caspase-mediated Beclin-1 cleavage may have a role in the dysfunction of autophagy in neurodegenerative diseases [62]. Proteasome inhibition has also been shown to induce autophagy [63]. Furthermore, proliferating T cells have been observed to express activated caspase 8 on autophagosomal membranes [64]. One study showed a link between autophagic signaling and caspase 8 activation in response to proteasomal pathway blockade. Here, caspase 8 activation was dependent on the induction of autophagy and was independent of death ligands [65]. More recently, effector caspases have been linked to non-death control of the autophagy response, through targeting and cleaving key mitochondrial control proteins that moderate autophagic flux [66, 67]. Again, these observations suggest that caspase enzymes can be deployed as proteostatic control mechanisms during cell stress, and that this coordinated protease activity averts cell death rather than inducing it.

Mechanisms that Restrain Caspase Activity: A Key Step in the Non-lethal Deployment of Caspase Proteases

The level and duration of caspase activation is a primary determinant in the death versus non-death role for this protease family. In broad terms unrestrained caspase activation is a prototypical cell death signal, whereas spatially and/or temporally restrained caspase activation is associated with non-death outcomes. For example, low levels of transient caspase activity were reported during pathological cardiac adaptation, whereas high levels of long standing caspase activation led to cell death [26]. A similar mechanism was noted during neurohormonal agonist induced hypertrophy where adenoviral expression of wild-type Gαq induced hypertrophy after 24 h and constitutively active mutant Gαq rapidly progressed to cell death [68]. Furthermore, the Ras GTPase activating protein (AP) (RasGAP) can respond to varied levels of caspase activity. Here, low levels of apoptotic-stimuli lead to partial RasGAP cleavage and do not reduce cell viability while high levels of capase activity result in RasGAP promoted cell death [69, 70]. Another example is the transient caspase 3 activation observed during neural stem cell differentiation, which peaks at 24 h and is reduced thereafter [6]. Temporal caspase 3 activated mammalian sterile twenty-like kinase (MST1) has been demonstrated to induce myoblast differentiation while prolonged MST1 activation leads to accelerated apoptosis progression [4]. One study also demonstrated that the degree of caspase activity during lens cell differentiation was lower than that observed during apoptosis [71].

The non-death deployment of caspase activity and its modification of cell stress likely reside with two distinct regulatory mechanisms. First is the sub-cellular localization of cleaved caspase molecules. Caspase activity can be confined to specific sub-cellular compartments and depending on its location apoptosis may or may not be induced. Punctuate cytoplasmic active caspase 3 is observed in maturing megakaryocytes (MKs) during platelet formation and more diffuse caspase 3 staining is observed in apoptotic MKs [72]. Caspase 8 has also been shown to have different spatial expression during T-cell activation. Here, caspase 8 was demonstrated to be activated in aggregates within membrane lipid rafts. Conversely, during apoptosis high levels of diffuse cytosolic caspase 8 expression was observed [73]. These studies suggest that active caspases can be sequestered to specific sub-cellular compartments to regulate a non-death process, providing access to a limited and specific pool of substrates [7, 74]. This sequestration may derive from protein/protein interactions that restrain and position caspases in the required sub-cellular localization. An interaction between caspase 3 and the small heat shock protein αB-crystallin has also been observed. This physical association was shown to be involved in directing caspase to specific sub-cellular locations and restraining its proteolytic activity in both skeletal muscle and cardiomyocytes [75, 76]. Caspases can also be regulated by a family of proteins called inhibitors of apoptosis (IAPs). The X-linked inhibitor of apoptosis protein (XIAP) is the most potent IAP, although eight different human IAPs have been observed. XIAP can inhibit activities of caspase 3, 7 and 9 [77]. Interestingly, XIAP contains E3 ubiquitin ligase activity allowing it to further decrease caspase activity by promoting ubiquitination of caspase [77]. For example, XIAP has been observed to physically interact with and inhibit caspase 3 activity in cardiomyocytes [78, 79]. XIAP can also modify caspase activity during myoblast differentiation [80, 81]. Again, the capacity to localize and direct caspase function to a specific subcellular compartment appears to have been an early evolutionary adaptation in eukaryotes, as evidenced by the yeast metacaspase, Yca1, which colocalizes with chaperone remodeling proteins to control protein aggregation during cell stress [8].

A second locus of control that appears to be essential to the non-death deployment of caspase function is the post-translational modification of the protease itself, as well as its constituent substrates. As caspase proteases cleave their substrates at specific recognition sequences, modifications to these peptides or to the flanking regions can profoundly alter the targeting capacity of the protease. Of interest, a number of kinases have been demonstrated to alter caspase activity, which impacts the resulting cellular response and phenotype [82–84]. A well studied example is the phosphorylation of peptides in close proximity to the caspase 3 cleavage site. Here, casein kinase 2 (CK2) produces a steric inhibition on caspase 3 cleavage by phosphorylating neighboring serine residues [85]. These phosphorylation sites strongly overlap with known caspase 3 cleavage sites within various known caspase targets, including the autocatalysis site on caspase 3 itself [84, 85]. CK2-mediated phosphorylation of Pax7 has been demonstrated to inhibit caspase cleavage of Pax7 and promote satellite cell self-renewal of myofibers [22]. One study also demonstrated that point mutations preventing the phosphorylation of Pax7 result in reduced Pax7 in proliferating muscle progenitors due to coincident Pax7 ubiquitin and caspase-mediated degradation [86]. Additionally, altering the essential aspartic acid residues within the caspase cleavage site of Pax7 abolished the caspase-mediated cleavage event, forcing satellite cell self renewal and limiting the induction of the myogenic differentiation program [22].

This CK2 mediated suppression of caspase 3 is also the mechanism by which CT1 manages the induction of beneficial physiologic hypertrophy [33]. For example, CT1 induces a rapid activation of the intrinsic cell death pathway, followed by an equally rapid decline in this same signaling conduit. Loss or inhibition of CK2 during CT1 exposure leads to extended caspase 3 activation and development of a cell phenotype that mirrors pathologic hypertrophy [33]. As such, caspase activation can govern both pathologic and remedial outcomes in the heart dependent, a divergent response that is managed by fine tuning caspase activity (Fig. 1b).

Conclusions

When cells are presented with stressful stimuli they can adapt and remodel, or if the demand is too severe, undergo injury and subsequent cell death. The resulting cell fate is determined by a plethora of factors including the type, duration and extent of the stress, and the adaptive capabilities of the target cell. Cell adaptations can be reversible, referred to as physiological adaptations, while other responses can be pathological and have irreversible effects on overall cellular function and morphology. Divergent morphologies can result depending on the stimuli. For instance, cardiac hypertrophy occurs in athletes training for endurance sports due to increased workload. This is exemplified by a physiological enlargement of heart mass and cardiac cell size. Conversely, a reduction in cell size occurs in atrophic remodeling. This form of stress adaptation is common during extended bed rest or diminished workload.

In this chapter, we covered the role of the apoptotic signaling, and specifically the involvement of caspase proteases, in the adaptation to stress. Caspases participate in cellular functions that extend beyond cell death, including cellular differentiation and stress remodeling. The timing, level and sub-cellular localization of caspase activation during stress can have a significant impact on cell fate. Here, we have explored the emerging role of caspase signaling pathways as convenors of stress adaptation. For instance, a role for the intrinsic apoptotic pathway and associated caspases has been observed during cardiac hypertrophy [26]. Interestingly, caspases may be involved in the morphological changes associated with this pathological adaptation through the cleavage of specific substrates and contractile protein components. Further, caspases have been shown to participate in dispersing toxic aggregates, such as TDP-43, involved in atrophic diseases [55]. Based on these findings, targeting the components of apoptotic pathways, including caspases, may serve as a means to inhibit and/or reverse the detrimental effects of injurious stimuli. Modifying caspases, by for instance silencing their functionality or sequestering active caspases and/or their substrates to specific sub-cellular locations, may be a potential method of altering cellular adaptive capabilities from a pathological to a more beneficial physiological state. However, robust caspase inhibition as a therapeutic strategy must be approached with caution, as there is overwhelming experimental data confirming a beneficial role(s) for this protease family, i.e. CT1 mediated beneficial cardiac remodeling. As such, targeting of caspase function must be conducted in a judicious manner as blockade of proteolytic activity may have unintended outcomes that limit normal function or induce additional pathology.

References

1. Dick SA, Megeney LA (2013) Cell death proteins: an evolutionary role in cellular adaptation before the advent of apoptosis. BioEssays 35:974–983
2. Shrestha A, Megeney LA (2012) The non-death role of metacaspase proteases. Front Oncol 2:78
3. Zermati Y et al (2001) Caspase activation is required for terminal erythroid differentiation. J Exp Med 193:247–254
4. Fernando P, Kelly JF, Balazsi K, Slack RS, Megeney LA (2002) Caspase 3 activity is required for skeletal muscle differentiation. Proc Natl Acad Sci U S A 99:11025–11030
5. Arama E, Agapite J, Steller H (2003) Caspase activity and a specific cytochrome C are required for sperm differentiation in Drosophila. Dev Cell 4:687–697
6. Fernando P, Brunette S, Megeney LA (2005) Neural stem cell differentiation is dependent upon endogenous caspase 3 activity. FASEB J Off Publ Fed Am Soc Exp Biol 19:1671–1673

7. Fernando P, Megeney LA (2007) Is caspase-dependent apoptosis only cell differentiation taken to the extreme? FASEB J Off Publ Fed Am Soc Exp Biol 21:8–17
8. Lee REC, Brunette S, Puente LG, Megeney LA (2010) Metacaspase Yca1 is required for clearance of insoluble protein aggregates. Proc Natl Acad Sci U S A 107:13348–13353
9. Shrestha A, Puente LG, Brunette S, Megeney LA (2013) The role of Yca1 in proteostasis. Yca1 regulates the composition of the insoluble proteome. J Proteomics 81:24–30
10. Hill SM, Hao X, Liu B, Nystrom T (2014) Life-span extension by a metacaspase in the yeast Saccharomyces cerevisiae. Science 344:1389–1392
11. Shrestha A, Brunette S, Stanford WL, Megeney LA (2019) The metacaspase Yca1 maintains proteostasis through multiple interactions with the ubiquitin system. Cell Discovery 5:6
12. Earnshaw WC, Martins LM, Kaufmann SH (1999) Mammalian caspases: structure, activation, substrates, and functions during apoptosis. Annu Rev Biochem 68:383–424
13. Opferman JT, Korsmeyer SJ (2003) Apoptosis in the development and maintenance of the immune system. Nat Immunol 4:410–415
14. Cory S (1998) Cell death throes. Proc Natl Acad Sci U S A 95:12077–12079
15. McIlroy D, Sakahira H, Talanian RV, Nagata S (1999) Involvement of caspase 3-activated DNase in internucleosomal DNA cleavage induced by diverse apoptotic stimuli. Oncogene 18:4401–4408
16. Bassnett S (2002) Lens organelle degradation. Exp Eye Res 74:1–6
17. Abdul-Ghani M et al (2011) Wnt11 promotes cardiomyocyte development by caspase-mediated suppression of canonical Wnt signals. Mol Cell Biol 31:163–178
18. Chambers I et al (2003) Functional expression cloning of Nanog, a pluripotency sustaining factor in embryonic stem cells. Cell 113:643–655
19. Yu J et al (2007) Induced pluripotent stem cell lines derived from human somatic cells. Science 318:1917–1920
20. Yamanaka S (2008) Induction of pluripotent stem cells from mouse fibroblasts by four transcription factors. Cell Prolif 41(Suppl 1):51–56
21. Fujita J et al (2008) Caspase activity mediates the differentiation of embryonic stem cells. Cell Stem Cell 2:595–601
22. Dick SA et al (2015) Caspase 3 cleavage of Pax7 inhibits self-renewal of satellite cells. Proc Natl Acad Sci U S A 112:E5246–E5252
23. Baena-Lopez LA et al (2018) Novel initiator caspase reporters uncover previously unknown features of caspase activating cells. Development 145:1–11
24. Weaver BP, Weaver YM, Mitani S, Han M (2017) Coupled caspase and N-End rule ligase activities allow recognition and degradation of pluripotency factor LIN-28 during non-apoptotic development. Dev Cell 41:665–673
25. Lee REC, Puente LG, Kaern M, Megeney LA (2008) A non-death role of the yeast metacaspase: Yca1p alters cell cycle dynamics. PLoS One 3:e2956
26. Putinski C et al (2013) Intrinsic-mediated caspase activation is essential for cardiomyocyte hypertrophy. Proc Natl Acad Sci U S A 110:E4079–E4087
27. Balakumar P, Singh M (2006) The possible role of caspase-3 in pathological and physiological cardiac hypertrophy in rats. Basic Clin Pharmacol Toxicol 99:418–424
28. Gao L et al (2015) Novel role for caspase-activated DNase in the regulation of pathological cardiac hypertrophy. Hypertens 65:871–881
29. Abdul-Ghani M, Megeney LA (2008) Rehabilitation of a contract killer: caspase-3 directs stem cell differentiation. Cell Stem Cell 2:515–516
30. Fuchs Y, Steller H (2011) Programmed cell death in animal development and disease. Cell 147:742–758
31. Communal C et al (2002) Functional consequences of caspase activation in cardiac myocytes. Proc Natl Acad Sci U S A 99:6252–6256
32. Putinski C, Abdul-Ghani M, Brunette S, Burgon PG, Megeney LA (2018) Caspase cleavage of gelsolin is an inductive cue for pathologic cardiac hypertrophy. J Am Heart Assoc 7:e010404
33. Abdul-Ghani M et al (2017) Cardiotrophin 1 stimulates beneficial myogenic and vascular remodeling of the heart. Cell Res 27:1195–2015

34. Petrof BJ, Hussain SN (2016) Ventilator-induced diaphragmatic dysfunction: what have we learned? Curr Opin Crit Care 22:67–72
35. Tian M et al (2010) Cardiac alterations in cancer-induced cachexia in mice. Int J Oncol 37:347–353
36. Hasselgren PO, Fischer JE (1997) The ubiquitin-proteasome pathway: review of a novel intracellular mechanism of muscle protein breakdown during sepsis and other catabolic conditions. Ann Surg 225:307–316
37. Mitch WE, Goldberg AL (1996) Mechanisms of muscle wasting. The role of the ubiquitin-proteasome pathway. N Engl J Med 335:1897–1905
38. Du J et al (2004) Activation of caspase-3 is an initial step triggering accelerated muscle proteolysis in catabolic conditions. J Clin Invest 113:115–123
39. Huang J, Forsberg NE (1998) Role of calpain in skeletal-muscle protein degradation. Proc Natl Acad Sci U S A 95:12100–12105
40. Nelson WB, Smuder AJ, Hudson MB, Talbert EE, Powers SK (2012) Cross-talk between the calpain and caspase-3 proteolytic systems in the diaphragm during prolonged mechanical ventilation. Crit Care Med 40:1857–1863
41. Talbert EE, Smuder AJ, Min K, Kwon OS, Powers SK (2013) Calpain and caspase-3 play required roles in immobilization-induced limb muscle atrophy. J Appl Physiol 114:1482–1489
42. Li H, Malhotra S, Kumar A (2008) Nuclear factor-kappa B signaling in skeletal muscle atrophy. J Mol Med (Berl) 86:1113–1126
43. Lamkanfi M, Festjens N, Declercq W, Vanden Berghe T, Vandenabeele P (2007) Caspases in cell survival, proliferation and differentiation. Cell Death Differ 14:44–55
44. Baker DJ, Hepple RT (2006) Elevated caspase and AIF gene expression correlate with progression of sarcopenia during aging in male F344BN rats. Exp Gerontol 41:1149–1156
45. Supinski GS, Ji X, Callahan LA (2010) p38 Mitogen-activated protein kinase modulates endotoxin-induced diaphragm caspase activation. Am J Respir Cell Mol Biol 43:121–127
46. Supinski GS, Ji X, Callahan LA (2009) The JNK MAP kinase pathway contributes to the development of endotoxin-induced diaphragm caspase activation. Am J Physiol Regul Integr Comp Physiol 297:R825–R834
47. Martin LJ (1999) Neuronal death in amyotrophic lateral sclerosis is apoptosis: possible contribution of a programmed cell death mechanism. J Neuropathol Exp Neurol 58:459–471
48. Li M et al (2000) Functional role of caspase-1 and caspase-3 in an ALS transgenic mouse model. Science 288:335–339
49. Inoue H et al (2003) The crucial role of caspase-9 in the disease progression of a transgenic ALS mouse model. EMBO J 22:6665–6674
50. Wootz H, Hansson I, Korhonen L, Napankangas U, Lindholm D (2004) Caspase-12 cleavage and increased oxidative stress during motoneuron degeneration in transgenic mouse model of ALS. Biochem Biophys Res Commun 322:281–286
51. Su JH et al (2000) DNA damage and activated caspase-3 expression in neurons and astrocytes: evidence for apoptosis in frontotemporal dementia. Exp Neurol 163:9–19
52. Igaz LM et al (2009) Expression of TDP-43 C-terminal Fragments in Vitro Recapitulates Pathological Features of TDP-43 Proteinopathies. J Biol Chem 284:8516–8524
53. Nonaka T, Kametani F, Arai T, Akiyama H, Hasegawa M (2009) Truncation and pathogenic mutations facilitate the formation of intracellular aggregates of TDP-43. Hum Mol Genet 18:3353–3364
54. Zhang Y-J et al (2009) Aberrant cleavage of TDP-43 enhances aggregation and cellular toxicity. Proc Natl Acad Sci U S A 106:7607–7612
55. Suzuki H, Lee K, Matsuoka M (2011) TDP-43-induced death is associated with altered regulation of BIM and Bcl-xL and attenuated by caspase-mediated TDP-43 cleavage. J Biol Chem 286:13171–13183
56. Gamas P et al (2009) Inhibition of imatinib-mediated apoptosis by the caspase-cleaved form of the tyrosine kinase Lyn in chronic myelogenous leukemia cells. Leukemia 23:1500–1506
57. Hashimoto Y, Niikura T, Ito Y, Nishimoto I (2000) Multiple mechanisms underlie neurotoxicity by different types of Alzheimer's disease mutations of amyloid precursor protein. J Biol Chem 275:34541–34551

58. Biundo F et al (2017) Abolishing Tau cleavage by caspase at Aspartate(421) causes memory/synaptic plasticity deficits and pre-pathological Tau alterations. Transl Psychiatry 7:e1198
59. Ravalin M et al (2019) Specificity for latent C termini links the E3 ubiqutin ligase CHIP to caspases. Nat Chem Biol 15:786–794
60. Minina EA et al (2020) Classification and nomenclature of metacaspases and paracaspases: no more confusion with caspases. Mol Cell 77:927–929
61. Rubinstein AD, Kimchi A (2012) Life in the balance—a mechanistic view of the crosstalk between autophagy and apoptosis. J Cell Sci 125:5259–5268
62. Rohn TT et al (2011) Depletion of Beclin-1 due to proteolytic cleavage by caspases in the Alzheimer's disease brain. Neurobiol Dis 43:68–78
63. Pandey UB et al (2007) HDAC6 rescues neurodegeneration and provides an essential link between autophagy and the UPS. Nature 447:859–863
64. Bell BD et al (2008) FADD and caspase-8 control the outcome of autophagic signaling in proliferating T cells. Proc Natl Acad Sci U S A 105:16677–16682
65. Laussmann MA et al (2011) Proteasome inhibition can induce an autophagy-dependent apical activation of caspase-8. Cell Death Differ 18:1584–1597
66. DeVorkin L et al (2014) The Drosophila effector caspase Dcp-1 regulates mitochondrial dynamics and autophagic flux via SesB. J Cell Biol 205:477–492
67. DeVorkin L, Gorski SM (2014) A mitochondrial-associated link between an effector caspase and autophagic flux. Autophagy 10:1866–1867
68. Adams JW et al (1998) Enhanced Galphaq signaling: a common pathway mediates cardiac hypertrophy and apoptotic heart failure. Proc Natl Acad Sci U S A 95:10140–10145
69. Yang JY, Widmann C (2001) Antiapoptotic signaling generated by caspase-induced cleavage of RasGAP. Mol Cell Biol 21:5346–5358
70. Yang J-Y et al (2004) Partial cleavage of RasGAP by caspases is required for cell survival in mild stress conditions. Mol Cell Biol 24:10425–10436
71. Weber GF, Menko AS (2005) The canonical intrinsic mitochondrial death pathway has a non-apoptotic role in signaling lens cell differentiation. J Biol Chem 280:22135–22145
72. De Botton S et al (2002) Platelet formation is the consequence of caspase activation within megakaryocytes. Blood 100:1310–1317
73. Koenig A, Russell JQ, Rodgers WA, Budd RC (2008) Spatial differences in active caspase-8 defines its role in T-cell activation versus cell death. Cell Death Differ 15:1701–1711
74. Feinstein-Rotkopf Y, Arama E (2009) Can't live without them, can live with them: roles of caspases during vital cellular processes. Apoptosis 14:980–995
75. Kamradt MC, Chen F, Sam S, Cryns VL (2002) The small heat shock protein alpha B-crystallin negatively regulates apoptosis during myogenic differentiation by inhibiting caspase-3 activation. J Biol Chem 277:38731–38736
76. Chis R et al (2012) alpha-Crystallin B prevents apoptosis after H_2O_2 exposure in mouse neonatal cardiomyocytes. Am J Physiol Heart Circ Physiol 303:H967–H978
77. O'Riordan MXD, Bauler LD, Scott FL, Duckett CS (2008) Inhibitor of apoptosis proteins in eukaryotic evolution and development: a model of thematic conservation. Dev Cell 15:497–508
78. Potts MB, Vaughn AE, McDonough H, Patterson C, Deshmukh M (2005) Reduced Apaf-1 levels in cardiomyocytes engage strict regulation of apoptosis by endogenous XIAP. J Cell Biol 171:925–930
79. Piacentino V et al (2012) X-linked inhibitor of apoptosis protein-mediated attenuation of apoptosis, using a novel cardiac-enhanced adeno-associated viral vector. Hum Gene Ther 23:635–646
80. Murray TVA et al (2008) A non-apoptotic role for caspase-9 in muscle differentiation. J Cell Sci 121:3786–3793
81. Smith MI, Huang YY, Deshmukh M (2009) Skeletal muscle differentiation evokes endogenous XIAP to restrict the apoptotic pathway. PLoS One 4:e5097
82. Kurokawa M, Kornbluth S (2009) Caspases and kinases in a death grip. Cell 138:838–854

83. Desagher S et al (2001) Phosphorylation of bid by casein kinases I and II regulates its cleavage by caspase 8. Mol Cell 8:601–611
84. Turowec JP et al (2014) An unbiased proteomic screen reveals caspase cleavage is positively and negatively regulated by substrate phosphorylation. Mol Cell Proteomics 13:1184–1197
85. Duncan JS et al (2011) A peptide-based target screen implicates the protein kinase CK2 in the global regulation of caspase signaling. Sci Signal 4:ra30
86. Gonzalez N et al (2016) Ck2-dependent phosphorylation is required to maintain Pax7 protein levels in proliferating muscle progenitors. PLoS One 11:e0154919

Chapter 6
Cross Talk Between Apoptosis and Autophagy in Regulating the Progression of Heart Disease

Niketa Sareen, Lorrie A. Kirshenbaum, and Sanjiv Dhingra

Abstract Apoptosis, the programmed cell death is prominent in multiple cardiac pathologies eventually resulting in heart failure. Autophagy, on the other hand plays an important role in recycling nutrients by causing the degradation of intracellular components not required by the cardiomyocytes, during cellular stress. Both apoptosis and autophagy play significant roles in cardiac development and disease. Absence of apoptosis in the developing embryo results in various congenital heart defects; however, excessive apoptosis in heart has been linked to negative ventricular remodeling. Similarly, desired levels of cellular proteins are maintained by basal levels of autophagy, and abnormal changes in the autophagic pathway have detrimental effects on myocytes. Though the mechanisms of each of these pathways differ, there is some dialogue that is always active amongst these two processes through common signaling pathways. Therefore, the interaction of proteins specific to autophagy with the apoptotic proteins might result in activation or inactivation of the process. This chapter summarizes mechanisms of both the pathways in cardiomyocytes and the molecules that link these pathways to regulate cell death.

Keywords Apoptosis · Autophagy · Heart failure · Crosstalk · Cardiac development

Heart failure results in body's insufficiency to meet the demands of required blood flow or pressure leading to symptoms like breathlessness, fatigue and death [1]. Moreover, it is also considered as an end result of various complications including hypertension, valvular heart disease and myocardial infarction [2]. Among several reported mechanisms, the death of cardiac myocytes has been suggested to be a leading cause of various cardiovascular diseases [3]. How exactly cardiac cells die is still an unresolved and ongoing problem. However, studies from various laboratories

N. Sareen · L. A. Kirshenbaum · S. Dhingra (✉)
Department of Physiology and Pathophysiology, Rady Faculty of Health Sciences, St. Boniface Hospital Albrechtsen Research Centre, Institute of Cardiovascular Sciences, Max Rady College of Medicine, University of Manitoba, 351 Tache Avenue, R226, Winnipeg R-3028-2, Canada
e-mail: sdhingra@sbrc.ca

© Springer Nature Switzerland AG 2022

L. A. Kirshenbaum (ed.), *Biochemistry of Apoptosis and Autophagy*, Advances in Biochemistry in Health and Disease 18,
https://doi.org/10.1007/978-3-030-78799-8_6

have shown apoptosis, necrosis and autophagy to be the processes through which cells die under different cardiac insults [4]. These processes can occur independently, parallel or sequentially to one another depending on the time, extent and type of injury. Although the cells undergoing necrosis, apoptosis or autophagy meet similar fate which is "cell death", but the signaling pathways involved in these processes are different. Cellular death by necrosis involves various characteristic changes in the cell including swelling of cytoplasm and organelles such as mitochondria due to compromised functioning of the plasma membrane. Necrotic cells release cellular contents in the surrounding which sets off an inflammatory response [1]. Unlike necrosis which is considered unprogrammed process, apoptosis is a form of cell death which is programmed and involves organized events such as membrane blebbing, shrinking of cellular cytoplasm, chromatin condensation and fragmentation of cytoplasm and nuclei followed by their packaging into apoptotic bodies to be later phagocytosed by nearby macrophages. In contrast to necrosis, the phagocytosis of the apoptotic bodies prevents the release of inflammatory contents into extracellular space thus preventing the inflammatory response during apoptosis [1, 5].

Different from necrosis and apoptosis, autophagy is a mechanism to recycle the cellular organelles, proteins and lipids by lysosomal degradation, releasing the biochemicals in circulation and thus helping the cells when there are inadequate nutrients [4]. Autophagosomes are specific structures that are formed during autophagy where proteins and organelles are sequestered for degradation [6]. Both apoptosis and autophagy are well organized processes and are reported to be involved in cardiac complications leading to heart failure.

Apoptosis in Heart Failure

Term apoptosis as a form of cell death was first defined by Ker et al. in1971 [7]. However, the first indication of apoptosis in heart can be accounted back to mid-90s where it was reported to play a role in the ischemic death of cardiac myocytes following acute myocardial infarction (MI) [8]. There is enormous amount of literature indicating the role of apoptosis in heart failure. Various cardiac complications like congenital heart disease, arrythmogenic right ventricular cardiomyopathy, valvular heart disease and other cardiomyopathies which result in heart failure have been linked to apoptosis [9]. Additionally, increased levels of apoptotic markers like Bax, Caspases along with decreased levels of anti-apoptotic markers like Bcl_2, $Bcl\text{-}X_L$ were observed following an MI [10]. Overall, apoptosis has been linked to pathological left ventricular remodeling leading to heart failure [11].

Cardiomyocyte death during stress conditions can be a result of apoptosis that require caspases as a death signal or alternatively it can be caspase independent mechanisms that do not involve caspases for the execution of the apoptotic death of cardiomyocytes [12]. Apoptosis can occur through extrinsic or the death receptor pathway which involves death receptor molecules on the surface of the cardiomyocytes. Alternatively, it can occur via activation of the intrinsic pathway involving mitochondria and endoplasmic reticulum during intracellular stress including hypoxia, nutrient inadequacy, DNA damage or cellular toxicity which further cause caspase activation.

Extrinsic Pathway of Apoptosis

This pathway is also known as the death receptor pathway as it involves the binding of a ligand to the death receptors located on the cell membrane. The domains of these receptors include extracellular domain which interacts with the ligand; transmembrane domain; and intracellular death domain which recruits caspases after ligand binding [13]. Binding of a ligand to the extracellular domain of one of the following receptors including TNFR (Tumor Necrosis Factor Receptor), Fas or TRAIL-R (TNF-related apoptosis inducing ligand), leads to the activation of the receptor after oligomerization [5]. This ligand-receptor complex then recruits FADD (Fas Associated Death Domain), which further binds to the initiator pro caspase 8; later activating it to caspase 8 by cleavage. Caspase 8 activation leads to the activation of executioner caspases 3, 6 and 7 [14]. Instead, the release of cytochrome c from the mitochondria is another mechanism by which caspase 8 transmits apoptotic signals. Caspase 8 is known for its ability to cleave Bid (BH3 interacting-domain death agonist) thus forming tBid (truncated Bid), which is then transported to mitochondria triggering cytochrome c (cyt c) release. The release of cyt c from mitochondria later activates caspase 9 and 3 which participates in apoptosis by cleaving the cellular proteins [15].

Cardiomyocytes express both Fas and TNFR1 receptors. However, increased level of TNFα has been observed in failing human heart [13]. In addition, transgenic mice overexpressing cardio-specific TNFα develop dilated cardiomyopathy [12, 16]. Another study conducted by Fan et al. [17], highlights the role of Fas/FADD pathway towards cardiomyocyte apoptosis after ischemia reperfusion injury. According to this study, the animal models lacking FADD had demonstrated reduction in apoptosis of cardiomyocytes along with decreased scar size in mouse models of ischemia reperfusion (I/R) injury [17]. Similarly, in another study use of vanadyl sulfate to reduce FasL levels led to a reduction in apoptotic cell death in the ischemia reperfusion (I/R) in vivo models [18], highlighting the role of Fas mediated apoptosis in heart failure.

Intrinsic Pathway of Apoptosis

Mitochondria are responsible for an uninterrupted and effective supply of ATP to the contracting cardiac myocytes [13]. However, under stress conditions, mitochondria can also contribute towards cardiomyocyte apoptosis. The intrinsic apoptotic pathway involves mitochondria and is activated when the cardiomyocytes encounter oxidative stress, DNA damage, nutrient or survival factor deficiency or other physical/chemical toxin exposure [19]. Bcl_2 family proteins are the major controlling factors of intrinsic apoptotic pathway. Mitochondrial integrity is sustained under normal conditions by anti-apoptotic members of Bcl_2 family (Bcl_2 and Bcl-X_L) which prevents the release of pro-apoptotic factors from mitochondria. However, stress activates proapoptotic members of Bcl_2 family such as Bax, Bak, Bnip3, Nix/Bnip3L, Bad, Bid, Noxa and Puma [20, 21]. Bax and Bak proteins undergo oligomerization and consequently form a channel for the release of cytochrome c (cyt c) from the inner mitochondrial membrane to the cytosol. In the cytosol cyt c is associated with Apaf-1(Apoptosis protease activating factor-1) and ATP to form a complex, which further recruits and activates procaspase 9 and apoptosis executioner caspases-3, 6, and 7 [22, 23]. Moreover, in addition to cyt c, factors like Smac/Diablo are released by mitochondria which interfere with and prevent the action of apoptosis inhibitor proteins including XIAP (X linked Inhibitor of Apoptosis Protein), cIAP1 and cIAP2 [23].

The balance between anti-apoptotic and pro-apoptotic proteins regulates mitochondrial health and cell survival. Enhanced expression of Bcl-X_L in cardiac cells has been shown to prevent hypoxia mediated apoptosis by maintaining mitochondrial integrity [24]. In another study, miRNA induced downregulation of Bcl_2 expression was responsible for myocardial apoptosis in hypoxia reoxygenation models [25]. In a pressure overload model, transition from cardiac hypertrophy to heart failure was accompanied by a substantial shift in Bax to Bcl-X_L ratio [26]. Upregulation of oxidative stress in the cells increased the levels of pro-apoptotic protein Bax and mitochondrial damage thus enhancing the rate of I/R injury which was significantly reduced in Bax knock out mice. This suggests a role of Bax in causing mitochondrial damage and I/R injury [13]. Activation of Bax and Bak has been linked to another pro-apoptotic gene Bnip3, which is activated during oxidative stress and further leads to activation of Bax/Bak [13, 27]. Furthermore, myocardial systolic and diastolic dysfunction and left ventricular interstitial fibrosis has been attributed to increased expression of Bnip3 leading to heart failure [28]. In a recent study, it is reported that Bnip3 knockdown prevented doxorubicin induced cardiac cell death [29]. Furthermore, in another study involving rat MI models, induction of MI was accompanied by increased cytochrome c release from mitochondria as well as increased expression of apoptotic markers including Bax, Apaf1 along with caspases 9 and 3 [30].

Endoplasmic Reticulum (ER) Mediated Intrinsic Apoptosis

ER plays an important role in protein production, folding and transportation to golgi apparatus. ER undergoes stress when there is an imbalance in the import and secretion of folded proteins, causing the activation of stress related proteins-inositol requiring protein-1(IRE1), pancreatic eukaryotic translation initiator factor 2α kinase (PERK) and activating transcription factor-6 (ATF6) [31]. This ER trauma is followed by UPR (unfolded protein response) to overcome the stress, however prolonged UPR can lead to apoptosis either by activation of caspase 12 followed by caspase3 activation [13]; through IRP1 or by upregulation of transcription factor CHOP (C/EBP homologous protein) [31] which further leads to the activation of pro-apoptotic proteins like BH3-only protein Puma [32]. Increased levels of Puma have been observed in I/R injury models of cardiomyocytes; Importance of Puma in ER stress is supported by reports which demonstrated reduced ER stress in Puma deficient mice models [13].

Apoptotic Mechanisms Independent of Caspases

Under certain conditions when caspases do not get activated, cells switch to alternate form of apoptosis that involves death factors which are independent of caspases [33]. These factors include apoptosis inducing factor (AIF), endonuclease G (Endo G), serine protease high temperature requirement protein A2 (HtrA2/Omi) and Bnip3 [12]. Recently, a dual role of AIF has been reported, when in mitochondria AIF is involved in cell survival. However, when it translocates to the nucleus, AIF contributes to cell death [34]. Normally, AIF is present in the intermembrane space of mitochondria and is required for oxidative phosphorylation in the cells. However, in the presence of a death stimulus; AIF is translocated to the nucleus thus leading to the DNA fragmentation independent of caspase activation [3]. There have been reports regarding the involvement of AIF in various forms of cardiomyopathies like ischemia reperfusion, oxidative stress, cardiac hypertrophy and heart failure [12]. Similar to AIF, endo G and HtrA2/Omi participate in caspase independent apoptosis. Both of these factors are translocated from mitochondria to the nucleus and cytosol respectively to execute apoptosis in cardiomyocytes [12].

Protective Role of Apoptosis During Cardiac Development

Cell death by apoptosis does not always have negative effects. Several reports indicate the role played by apoptosis during normal cardiac morphogenesis [35]. Various reports indicate that the failure of normal apoptotic signaling in hearts of developing embryo leads to the occurrence of various congenital heart defects. The major phases of heart development involving apoptosis include septa formation between the chambers and valve formation. Moreover, during the transitional phase of heart from fetal to adult, apoptosis has been observed in the interventricular septum and right ventricular wall after birth. In addition to septa formation, in the course of cardiac development, other regions of marked apoptosis include the zones of fusion of atrioventricular cushions. Furthermore, the cells other than cardiac myoblasts, including the cells of aortic and pulmonary valves also undergo apoptosis during the cardiac development. Hence the disruption of apoptosis in any of the above mentioned developmental phases leads to various defects in the developing child such as long QT syndrome, presence of the accessory pathways or other congenital heart defects [36].

Autophagy

Autophagy refers to the process of degrading various intracellular components including organelles as well as other macromolecules which are no longer required by the cells [37]. It is considered to be an evolutionarily conserved mode of cellular organelle degradation [38] and its function is known to be preserved from yeast to humans [37]. Autophagy being considered as a pivotal mechanism for the maintenance of the cellular energy levels during nutrient deprivation and stress conditions; does this by the breakdown of the several important cellular biomolecules including amino acids, fatty acids [1] in addition to damaged organelles such as mitochondria, ER, golgi body. For the process of autophagy to complete, the formation of autophagosomes is one of the key steps. These double membraned vesicles that are reported to be responsible for delivering biomolecules, proteins or organelles to lysosome are the hall mark of autophagy process [39]. Following fusion with lysosomes, autophagosomes are ultimately degraded by lysosomal enzymes. The process of autophagy has been shown to be both essential as well as detrimental for numerous intracellular events. The basal levels of autophagy help in the maintenance of requisite levels of the cellular proteins, on the contrary, increased and abnormal levels of autophagic protein degradation might result in the breakdown of necessary biomolecules ultimately proving fatal for the cell [40].

Mechanisms Underlying Autophagy

Cellular autophagy is divided into three different types including –microautophagy, chaperone mediated autophagy and macroautophagy. Out of these, microautophagy involves non-specific, direct consumption of the cytoplasmic contents followed by lysosomal degradation [38]. Chaperone mediated autophagy, however, is different in its binding of the target protein through the chaperones which are responsible for the delivery of these specific proteins to lysosomes ultimately leading to the degradation of target proteins [40]. Finally, macroautophagy encompasses the formation of an autophagosome which fuses with the lysosome in order to complete the degradation of the cellular components in the cytoplasm [41]. The process of macroautophagy comprises different stages including- induction and nucleation of phagophore which involves the formation of an isolation membrane, followed by expansion of phagophores and engulfment of the biomolecules or organelles to be degraded, forming autophagosomes which ultimately fuses with lysosomes [40]. The formation of the phagophore is stimulated by ULK1 complex which includes Atg13, Ulk 1(Unc-51-like kinase 1), Ulk2 and FIP200 (Focal Adhesion Kinase family interacting protein 200) [40]. Initiation of autophagy is regulated by mTOR, which negatively regulates autophagy by preventing the formation of ULK1 complex under normal growth conditions. However under stress conditions like starvation or hypoxia the reduction in PI3K-Akt and activation of AMPK respectively result in the inhibition of mTOR activity, thus leading to the activation of autophagy [1, 40]. Phagophore nucleation is completed after the formation of VPS34 (Lipid kinase vacuolar protein sorting 34) and Beclin 1 complex [42]. mTOR regulatory step; this is another mode of autophagic regulation through Beclin 1 which acts as a positive regulator of autophagy. BH3 domain of Beclin 1 has the affinity to bind to anti-apoptotic proteins including Bcl_2 and Bcl-X_L; which prevents interaction of Beclin 1 with VPS34 thus preventing autophagy [1]. The nucleation of phagophore is followed by elongation phase. This step involves transformation of cytoplasmic LC3-I to membrane associated LC3-II after binding to phosphatidylethanolamine. For this binding reaction to occur it requires a pair of ubiquitination like reactions. The first reaction is the binding of Atg5 and Atg12 complex to the phagophore membrane in the presence of Atg7 and Atg10. Second reaction requires the presence of Atg3, Atg5, Atg7 and Atg12 and it drives the conversion of LC3-I to LC3-II [40]. The last stage in autophagy is the maturation of autophagosome and its fusion to the lysosomes. Fusion between lysosomes and autophagosomes is facilitated by various proteins including Rab-7 whose action is mediated by lysosomal receptor proteins LAMP1 and LAMP2 [43]. Moreover, another set of proteins, SNARE (Soluble N-ethylmaleimide sensitive factor attachment protein receptor) proteins have also been reported to play an important role in the fusion of autophagosomes with lysosomes [40].

Role of Autophagy in Heart Failure

In heart, the role of autophagy has been controversial and has been reported to have positive as well as negative effects during pathological conditions including hypertension, myocardial ischemia, hypertrophy, dilated cardiomyopathy [44, 45]. Increased levels of autophagy have been reported to play a protective role in various pathological conditions including neurodegenerative diseases like dementia, Parkinson's disease and Huntington's disease where autophagy degrades harmful proteins in the cytoplasm [37]. Additionally, in the heart, autophagy has been shown to be beneficial during I/R injury [43, 46]. In a study by Oka et al. [49], the essential role of mitochondrial autophagy has been indicated in the heart. This study reported that the failure of autophagy to eliminate defective mitochondria may result in the dilated cardiomyopathy and myocarditis after inflammation through the activation of TLR pathway [47]. Furthermore, the absence of Atg5, an autophagic gene which in conjunction with Atg12 plays a role in autophagosome formation, results in dilation of left ventricles in mice [48]. In a recent study by Zhang et al. [51], autophagy has also been suggested to play a role in the differentiation of cardiac progenitor cells [49]. However, several other studies reported negative effects of excessive autophagic process on the cells. For instance, Zhu et al. [50], revealed that autophagy plays a detrimental role in the development of heart failure in a pressure overload mouse model. In this study, increased levels of LC3-II and LAMP-1 were reported in pressure overload mice models compared to controls indicating increased autophagic process [50]. Another study indicated the prevalence of autophagy in the failing hearts where increased levels of cathepsin D and enhanced autophagy was observed [51].

Crosstalk Between Autophagy and Apoptosis

Although the mechanisms of apoptosis and autophagy are different, yet there is some dialogue that always keeps on going among these two processes through common signaling pathways. Several studies have reported a crosstalk between autophagy and apoptosis.

Beclin 1 has a BH3 (Bcl2 Homology 3) domain that interacts with BH3 receptor domain of anti-apoptotic proteins including Bcl_2 and $Bcl\text{-}X_L$. Interaction of Beclin 1 with the anti-apoptotic proteins results in its inactivation thus leading to inhibition of Beclin 1 mediated autophagy [52]. Moreover, binding of Beclin 1 to Bcl_2 and $Bcl\text{-}X_L$ receptors results in the inhibition of anti-apoptotic effect of Bcl_2 and $Bcl\text{-}X_L$ proteins that might further result in the activation of pro apoptotic members such as Bax and Bak [53].

Dapk-(Death associated protein kinase 1) is also another common protein reported to be involved in both autophagy and apoptosis. This kinase protein is responsible for the phosphorylation of BH3 domain of Beclin 1; which further prevents inhibition of Beclin 1 by Bcl_2 and $Bcl\text{-}X_L$, leading to autophagy in the cells [54].

FLIP-(FADD Like IL-Beta-converting enzyme like protein) is an anti-apoptotic protein which also shows anti autophagic response by competing with LC3 for binding site on Atg3. Under autophagic conditions LC3 interacts with Atg3 that is important for elongation of autophagosome and thus for the process of autophagy. Flip binds to Atg3 which results in competitive inhibition of LC3-Atg3 complex and results in the inhibition of autophagy process [55].

Atgs-(Autophagy related proteins) along with their established essential role in autophagy, these proteins can also play a major role in regulating apoptosis under certain conditions. The formation of complex between autophagy proteins Atg3-Atg12 is shown to regulate mitochondria mediated apoptosis. However once the complex is disrupted the mitochondrial autophagy is reduced [41]. Caspase mediated cleavage of Atg4 shows similar effect in inducing apoptosis by mediating mitochondrial potential imbalance [56, 57]. Moreover, there are some indications that ATG5 may bind to FADD which would further stimulate the induction of caspase dependent apoptosis [58]. In addition to this, ATG5 is also involved in the formation of complex with ATG12 during the binding of autophagosome to the lysosomal membrane. However, under some circumstances, calpain mediates cleavage of ATG5 results in the generation of N terminal fragment from ATG5, causing its mitochondrial translocation and cytochrome c release following disruption of mitochondrial membrane potential. This cleavage of ATG5, thus leads to activation of intrinsic apoptotic pathway [59]. Similarly, another study indicated the role of ATG12 in mediating apoptosis by binding to anti apoptotic proteins Bcl_2 and Mcl_1 and causing their inactivation. This inactivation was responsible for mitochondrial cytochrome c release causing apoptosis [60].

BNIP3-(Bcl2 nineteen-kilodalton protein) is a well-known protein responsible for the induction of apoptosis. It has also been known to be responsible for activation of autophagy in cardiac myocytes [61]. The interaction of Bnip3 with Bcl_2 and Bcl-X_L results in the inhibition of Bnip3 function. One of the mechanisms of autophagy induction by Bnip3 is by blocking the interaction of Beclin 1 and anti-apoptotic protein Bcl-X_L [42]. Moreover, upregulation of Bnip3 during ischemia reperfusion injury has been reported to promote autophagy and protect against apoptosis induced myocardial damage [62].

Caspases are known for their role in mediating apoptotic cell death. During cellular apoptosis caspases have been known to breakdown essential cellular proteins, which then lead to appearance of characteristic apoptotic morphological features. However caspases have also been reported to degrade and inactivate some of the autophagy related proteins including Beclin 1, p62, Atg4D, Atg5, Atg7, Atg3 and AMBRA [41, 42]. The cleavage of Beclin 1 by caspase 3, 6 or 9 has been reported to result in the formation of C terminal fragment of Beclin 1, which leads to mitochondrial membrane permeabilization and causes release of cyt c (cytochrome c). The release of cyt c from the mitochondrial membrane is involved in the regulation of mitochondrial apoptotic pathway [42]. Similar to this mechanism, caspases have also been reported to induce mitochondrial apoptotic pathway by cleaving Atg5 protein [63]. Therefore, the above-mentioned research indicates a strong correlation between the two major death pathways and switch between their functions.

Conclusion

One of the major reasons for the development of heart failure following a cardiac injury is the death of cardiomyocytes. Cellular death may occur through either one or multiple pathways which consist of necrosis, apoptosis and autophagy. Of these, necrosis results in the release of the cellular components into cytoplasm leading to inflammation which is characteristic of necrotic cell death. Inflammation is absent in the other two forms of cell death. However, the process of apoptosis results in cell death which is executed by the various caspases in the cell. These caspases may be activated either by intrinsic or extrinsic pathway depending on the involvement of the death receptor; in the mitochondria or the endoplasmic reticulum. Autophagy related cell death mainly is a mode of degradation of unwanted cellular organelles or components which are not needed in the cell. The myocyte death, however, may not be the result of one single pathway working by itself; rather through multiple pathways working in co-ordination. Therefore, various components of apoptosis might play an important role in autophagy or necrosis by activating or deactivating the process. However, in spite of the numerous literature indicating the link between the major death pathways; more studies are required to understand the depth of these pathways in regulating cardiomyocyte death which ultimately results in heart failure. Understanding these pathways will help us to find the molecular targets for development of future therapies.

References

1. Whelan RS, Kaplinskiy V, Kitsis RN (2010) Cell death in the pathogenesis of heart disease: mechanisms and significance. Annu Rev Physiol 72(1):19–44. https://doi.org/10.1146/annurev.physiol.010908.163111
2. Favaloro B, Allocati N, Graziano V, Di Ilio C, De Laurenzi V (2012) Role of apoptosis in disease. Aging 4(5):330–349
3. Chiong M et al (2011) Cardiomyocyte death: mechanisms and translational implications. Cell Death Dis **2**(12):e244. doi: https://doi.org/10.1038/cddis.2011.130
4. He C, Klionsky DJ (2009) Regulation mechanisms and signaling pathways of autophagy. Annu Rev Genet 43:67–93. https://doi.org/10.1146/annurev-genet-102808-114910
5. Duprez L, Wirawan E, Berghe TV, Vandenabeele P (2009) Major cell death pathways at a glance. Microbes Infect 11(13):1050–1062. https://doi.org/10.1016/j.micinf.2009.08.013
6. Nakatogawa H (2020) Mechanisms governing autophagosome biogenesis. Nat Rev Mol Cell Biol 1–20. https://doi.org/10.1038/s41580-020-0241-0
7. Kerr JFR, Wyllie AH, Currie AR (1972) Apoptosis: a basic biological phenomenon with wide-ranging implications in tissue Kinetics. Br J Cancer 26(4):239–257
8. Itoh G et al (1995) DNA fragmentation of human infarcted myocardial cells demonstrated by the nick end labeling method and DNA agarose gel electrophoresis. Am J Pathol 146(6):1325–1331
9. Takemura G, Kanoh M, Minatoguchi S, Fujiwara H (2013) Cardiomyocyte apoptosis in the failing heart—A critical review from definition and classification of cell death. Int J Cardiol 167(6):2373–2386. https://doi.org/10.1016/j.ijcard.2013.01.163
10. Qin F, Liang MC, Liang C (2005) Progressive left ventricular remodeling, myocyte apoptosis, and protein signaling cascades after myocardial infarction in rabbits. Biochem Biophys Acta BBA Mol Basis Dis**1740**(3):499–513. https://doi.org/10.1016/j.bbadis.2004.11.007
11. Park M et al (2009) Apoptosis predominates in nonmyocytes in heart failure. Am J Physiol Heart Circ Physiol 297(2):H785–H791. https://doi.org/10.1152/ajpheart.00310.2009
12. Kim N-H, Kang PM (2010) Apoptosis in cardiovascular diseases: mechanism and clinical implications. Korean Circ J 40(7):299–305. https://doi.org/10.4070/kcj.2010.40.7.299
13. Lee Y, Gustafsson ÅB (2009) Role of apoptosis in cardiovascular disease. Apoptosis 14(4):536–548. https://doi.org/10.1007/s10495-008-0302-x
14. Galluzzi L et al (2012) Molecular definitions of cell death subroutines: recommendations of the nomenclature committee on cell death. Cell Death Differ 19(1):107–120. https://doi.org/10.1038/cdd.2011.96
15. Tsipis A (2012) Heart muscle and apoptosis. In: Veselka J (ed) Cardiomyopathies—From basic research to clinical management. InTech
16. Bryant D et al (1998) Cardiac failure in transgenic mice with myocardial expression of tumor necrosis factor-α. Circulation 97(14):1375–1381. https://doi.org/10.1161/01.CIR.97.14.1375
17. Fan Q et al (2013) Inhibition of fas-associated death domain-containing protein (FADD) protects against myocardial ischemia/reperfusion injury in a heart failure mouse model. PLoS ONE 8(9). https://doi.org/10.1371/journal.pone.0073537
18. Bhuiyan MdS, Takada Y, Shioda N, Moriguchi S, Kasahara J, Fukunaga K (2008) Cardioprotective effect of vanadyl sulfate on ischemia/reperfusion-induced injury in rat heart in vivo is mediated by activation of protein Kinase B and induction of FLICE-inhibitory protein. Cardiovasc Drug Rev 26(1):10–23. https://doi.org/10.1111/j.1527-3466.2008.00039.x

19. Marín-García J (2016) Cell death in the pathogenesis and progression of heart failure. Heart Fail Rev 21(2):117–121. https://doi.org/10.1007/s10741-016-9538-7
20. O'Neill KL, Huang K, Zhang J, Chen Y, Luo X (2016) Inactivation of prosurvival Bcl-2 proteins activates Bax/Bak through the outer mitochondrial membrane. Genes Dev 30(8):973–988. https://doi.org/10.1101/gad.276725.115
21. Carrington EM et al (2017) Anti-apoptotic proteins BCL-2, MCL-1 and A1 summate collectively to maintain survival of immune cell populations both in vitro and in vivo. Cell Death Differ 24(5):878–888. https://doi.org/10.1038/cdd.2017.30
22. Yuan S, Akey CW (2013) Apoptosome structure, assembly, and procaspase activation. Structure 21(4):501–515. https://doi.org/10.1016/j.str.2013.02.024
23. LaCasse EC, Mahoney DJ, Cheung HH, Plenchette S, Baird S, Korneluk RG (2008) IAP-targeted therapies for cancer. Oncogene 27(48):6252–6275. https://doi.org/10.1038/onc.2008.302
24. Reeve JL et al (2007) Distinct mechanisms of cardiomyocyte apoptosis induced by doxorubicin and hypoxia converge on mitochondria and are inhibited by Bcl-xL. J Cell Mol Med 11(3):509–520. https://doi.org/10.1111/j.1582-4934.2007.00042.x
25. Liu L et al (2014) MicroRNA-15b enhances hypoxia/reoxygenation-induced apoptosis of cardiomyocytes via a mitochondrial apoptotic pathway. Apoptosis Int J Program Cell Death 19(1):19–29. https://doi.org/10.1007/s10495-013-0899-2
26. Sharma AK, Dhingra S, Khaper N, Singal PK (2007) Activation of apoptotic processes during transition from hypertrophy to heart failure in guinea pigs. Am J Physiol Heart Circ Physiol 293(3):H1384–H1390. https://doi.org/10.1152/ajpheart.00553.2007
27. Kubli DA, Quinsay MN, Huang C, Lee Y, Gustafsson ÅB (2008) Bnip3 functions as a mitochondrial sensor of oxidative stress during myocardial ischemia and reperfusion. Am J Physiol Heart Circ Physiol 295(5):H2025–H2031. https://doi.org/10.1152/ajpheart.00552.2008
28. Chaanine AH, Gordon RE, Kohlbrenner E, Benard L, Jeong D, Hajjar RJ (2013) Potential role of BNIP3 in cardiac remodeling, myocardial stiffness, and endoplasmic reticulum mitochondrial calcium homeostasis in diastolic and systolic heart failure. Circ Heart Fail 6(3):572–583. https://doi.org/10.1161/CIRCHEARTFAILURE.112.000200
29. Dhingra R et al (2014) Bnip3 mediates doxorubicin-induced cardiac myocyte necrosis and mortality through changes in mitochondrial signaling. Proc Natl Acad Sci 111(51):E5537–E5544. https://doi.org/10.1073/pnas.1414665111
30. Nagoor Meeran MF, Laham F, Azimullah S, Tariq S, Ojha S (2019) α-Bisabolol abrogates isoproterenol-induced myocardial infarction by inhibiting mitochondrial dysfunction and intrinsic pathway of apoptosis in rats. Mol. Cell. Biochem 453(1):89–102. https://doi.org/10.1007/s11010-018-3434-5
31. Scull CM, Tabas I (2011) Mechanisms of ER stress-induced apoptosis in atherosclerosis. Arterioscler Thromb Vasc Biol 31(12):2792–2797. https://doi.org/10.1161/ATVBAHA.111.224881
32. Gorman AM, Healy SJM, Jäger R, Samali A (2012) Stress management at the ER: regulators of ER stress-induced apoptosis. Pharmacol Ther 134(3):306–316. https://doi.org/10.1016/j.pharmthera.2012.02.003
33. Bae S, Yalamarti B, Kang PM (2008) Role of caspase-independent apoptosis in cardiovascular diseases. Front Biosci J Virtual Libr 13:2495–2503
34. Cheung ECC et al (2006) Dissociating the dual roles of apoptosis-inducing factor in maintaining mitochondrial structure and apoptosis. EMBO J 25(17):4061–4073. https://doi.org/10.1038/sj.emboj.7601276
35. Penaloza C, Lin L, Lockshin RA, Zakeri Z (2006) Cell death in development: shaping the embryo. Histochem Cell Biol 126(2):149. https://doi.org/10.1007/s00418-006-0214-1
36. Kunapuli S, Rosanio S, Schwarz ER (2006) 'How Do Cardiomyocytes Die?' apoptosis and autophagic cell death in cardiac myocytes. J Card Fail 12(5):381–391. https://doi.org/10.1016/j.cardfail.2006.02.002
37. Bento CF et al (2016) Mammalian autophagy: how does it work? Annu Rev Biochem 85(1):685–713. https://doi.org/10.1146/annurev-biochem-060815-014556

38. Li W, Li J, Bao J (2011) Microautophagy: lesser-known self-eating. Cell Mol Life Sci 69(7):1125–1136. https://doi.org/10.1007/s00018-011-0865-5
39. Noda NN, Inagaki F (2015) Mechanisms of autophagy. Annu Rev Biophys 44(1):101–122. https://doi.org/10.1146/annurev-biophys-060414-034248
40. Sciarretta S, Hariharan N, Monden Y, Zablocki D, Sadoshima J (2010) Is autophagy in response to ischemia and reperfusion protective or detrimental for the heart? Pediatr Cardiol 32(3):275–281. https://doi.org/10.1007/s00246-010-9855-x
41. Li M, Gao P, Zhang J (2016) Crosstalk between autophagy and apoptosis: potential and emerging therapeutic targets for cardiac diseases. Int J Mol Sci 17(3):332. https://doi.org/10.3390/ijms17030332
42. Mariño G, Niso-Santano M, Baehrecke EH, Kroemer G (2014) Self-consumption: the interplay of autophagy and apoptosis. Nat Rev Mol Cell Biol 15(2):81–94. https://doi.org/10.1038/nrm3735
43. Hamacher-Brady A, Brady NR, Gottlieb RA (2006) The interplay between pro-death and pro-survival signaling pathways in myocardial ischemia/reperfusion injury: apoptosis meets autophagy. Cardiovasc Drugs Ther 20(6):445. https://doi.org/10.1007/s10557-006-0583-7
44. Vacek TP, Vacek JC, Tyagi N, Tyagi SC (2011) Autophagy and heart failure: a possible role for homocysteine. Cell Biochem Biophys 62(1):1–11. https://doi.org/10.1007/s12013-011-9281-6
45. Wu D, Zhang K, Hu P (2019) The role of autophagy in acute myocardial infarction. Front Pharmacol 10. https://doi.org/10.3389/fphar.2019.00551
46. Lavandero S, Chiong M, Rothermel BA, Hill JA (2015) Autophagy in cardiovascular biology. J Clin Invest 125(1):55–64. https://doi.org/10.1172/JCI73943
47. Oka T et al (2012) Mitochondrial DNA that escapes from autophagy causes inflammation and heart failure. Nature 485(7397):251–255. https://doi.org/10.1038/nature10992
48. Taneike M et al (2010) Inhibition of autophagy in the heart induces age-related cardiomyopathy. Autophagy 6(5):600–606. https://doi.org/10.4161/auto.6.5.11947
49. Zhang J, Liu J, Liu L, McKeehan WL, Wang F (2012) The fibroblast growth factor signaling axis controls cardiac stem cell differentiation through regulating autophagy. Autophagy 8(4):690–691. https://doi.org/10.4161/auto.19290
50. Zhu H et al (2007) Cardiac autophagy is a maladaptive response to hemodynamic stress. J Clin Invest 117(7):1782–1793. https://doi.org/10.1172/JCI27523
51. Kostin S et al (2003) Myocytes die by multiple mechanisms in failing human hearts. Circ Res 92(7):715–724. https://doi.org/10.1161/01.RES.0000067471.95890.5C
52. Djavaheri-Mergny M, Maiuri MC, Kroemer G (2010) Cross talk between apoptosis and autophagy by caspase-mediated cleavage of Beclin 1. Oncogene 29(12):1717–1719. https://doi.org/10.1038/onc.2009.519
53. Galonek HL, Hardwick JM (2006) Upgrading the BCL-2 network. Nat Cell Biol 8(12):1317–1319. https://doi.org/10.1038/ncb1206-1317
54. Zalckvar E et al (2009) DAP-kinase-mediated phosphorylation on the BH3 domain of beclin 1 promotes dissociation of beclin 1 from Bcl-XL and induction of autophagy. EMBO Rep 10(3):285–292. https://doi.org/10.1038/embor.2008.246
55. Lee J-S et al (2009) FLIP-mediated autophagy regulation in cell death control. Nat Cell Biol 11(11):1355–1362. https://doi.org/10.1038/ncb1980
56. Fortunato F et al (2009) Impaired autolysosome formation correlates with Lamp-2 depletion: role of apoptosis, autophagy, and necrosis in pancreatitis. Gastroenterology 137(1):350-360.e5. https://doi.org/10.1053/j.gastro.2009.04.003
57. Norman JM, Cohen GM, Bampton ETW (2010) The in vitro cleavage of the hAtg proteins by cell death proteases. Autophagy 6(8):1042–1056. https://doi.org/10.4161/auto.6.8.13337
58. Pyo J-O et al (2005) Essential roles of Atg5 and FADD in autophagic cell death dissection of autophagic cell death into vacuole formation and cell death. J Biol Chem 280(21):20722–20729. https://doi.org/10.1074/jbc.M413934200
59. Shi M et al (2013) Calpain, Atg5 and Bak play important roles in the crosstalk between apoptosis and autophagy induced by influx of extracellular calcium. Apoptosis 18(4):435–451. https://doi.org/10.1007/s10495-012-0786-2

60. Rubinstein AD, Eisenstein M, Ber Y, Bialik S, Kimchi A (2011) The Autophagy protein Atg12 Associates with antiapoptotic Bcl-2 family members to promote mitochondrial apoptosis. Mol Cell 44(5):698–709. https://doi.org/10.1016/j.molcel.2011.10.014
61. Quinsay MN, Thomas RL, Lee Y, Gustafsson ÅB (2010) Bnip3-mediated mitochondrial autophagy is independent of the mitochondrial permeability transition pore. Autophagy 6(7):855–862. https://doi.org/10.4161/auto.6.7.13005
62. Hamacher-Brady A, Brady NR, Gottlieb RA, Gustafsson AB (2006) Autophagy as a protective response to Bnip3-mediated apoptotic signaling in the heart. Autophagy 2(4):307–309
63. You M et al (2013) TRAIL induces autophagic protein cleavage through caspase activation in melanoma cell lines under arginine deprivation. Mol Cell Biochem 374(1):181–190. https://doi.org/10.1007/s11010-012-1518-1

Chapter 7
Fibroblasts, Fibrosis and Autophagy

Sikta Chattopadhyaya and Michael P. Czubryt

Abstract The major distinguishing feature of fibrosis is significant deposition of collagen and other extracellular matrix (ECM) proteins, which can result in scarring if sufficiently excessive. Fibrosis affects many tissue types, and thus contributes to a broad group of diseases which, with few exceptions, continues to lack specific therapy. It has been estimated that nearly 45% of deaths in the developed world are caused by fibroproliferative diseases, which contribute to cardiovascular disease, pulmonary, renal, gut and liver fibrosis, and scleroderma (Bitterman and Henke in Chest 99:81S–84S, [1]). Fibroblasts are the most common stromal cell type of the connective tissues found in the body, and are the primary source of ECM in physiological conditions, i.e. in the absence of disease. The conversion of fibroblasts or similar stromal cells to myofibroblasts is a principal mediator of pathological fibrosis in many tissue types, and frequently occurs in response to ongoing tissue injury and chronic inflammation. While the fibrotic response can occur in response to existing disease, the phenotype conversion of fibroblasts to myofibroblasts due to transient stress or damage may lead to the initiation of long-term fibrotic disease (Bagchi et al. in BMC Biol 14:21, [2]). Inflammation has been found to be a critical inducer of fibrosis, with immune cells generating a variety of growth factors and cytokines that play critical roles in fibroblast activation and subsequent tissue remodelling and fibrosis. A common cellular response to stress stimuli such as inflammation is autophagy, and recent studies have tightly linked the activation or inhibition of autophagy with fibrotic diseases in myriad tissues. Here, we discuss the inter-relationship of these pathways to provide insight into their potential as therapeutic targets in fibrotic disease.

S. Chattopadhyaya · M. P. Czubryt
Department of Physiology and Pathophysiology, Rady Faculty of Health Sciences, University of Manitoba, Winnipeg, Canada

M. P. Czubryt (✉)
Institute of Cardiovascular Sciences, R4008 St. Boniface Hospital Albrechtsen Research Centre, 351 Tache Avenue, Winnipeg, MB R2H 2A6, Canada
e-mail: mczubryt@sbrc.ca

© Springer Nature Switzerland AG 2022

L. A. Kirshenbaum (ed.), *Biochemistry of Apoptosis and Autophagy*,
Advances in Biochemistry in Health and Disease 18,
https://doi.org/10.1007/978-3-030-78799-8_7

Keywords Inflammation · Fibrosis · Fibroblast · Myofibroblast · Apoptosis · Cytokine

Introduction

Fibrosis occurs due to the abnormal regulation of the synthesis and/or degradation of ECM, resulting in excessive extracellular deposition of fibrillar collagens (particularly type I and III) and other proteins and proteoglycans, and can alter the function of virtually every organ system in the body. While the precise impact of altered ECM production varies according to tissue type, fibrosis typically results in significant organ dysfunction, and frequently organ failure, contributing to patient morbidity and mortality. For example, idiopathic pulmonary fibrosis (IPF) progressively and severely compromises lung function, and shortens lifespan in affected individuals by many years [3]. Fibrosis frequently occurs secondary to other comorbidities—in the heart, hypertension and diabetes can both be significant drivers of fibrosis, as can smoking in the lung or acute or chronic toxic agent exposure in the liver [4–8]. Cardiac fibrosis can present in multiple patterns, but mid-wall fibrosis appears to be particularly dangerous, increasing the risk of death or hospitalization by up to 18-fold in dilated cardiomyopathy patients exhibiting this pattern compared to those without [9]. Thus, fibrosis is not only an outcome of pre-existing disease, it can also be a risk factor for further organ dysfunction, exacerbating adverse outcomes for patients. Clinical treatments for fibrosis in any organ remain scant, thus a better understanding of fibrosis pathogenesis is critically needed to enable the identification and development of novel therapeutics.

Fibroblasts are a polymorphic cell type which may arise from a variety of precursors such as endothelial or epithelial cells, depending on tissue type. Stromal fibroblasts play a variety of roles, including facilitating organ development, generating the supportive ECM of a tissue, and communicating with nearby parenchymal cells to maintain tissue homeostasis [10]. The ECM generated by fibroblasts consists of fibrous proteins such as collagen and elastin, gelatinous ground substance rich in glycosaminoglycans, and adhesive proteins such as laminin. This complex ECM provides overall tissue, organ and body integrity, and fibroblasts not only synthesize this material, they also play central roles in ECM maintenance and reabsorption. The specific composition of ECM can vary widely across tissues, depending on the specific mechanical stresses to which the tissue is subjected [11]. The critical importance of ECM, and thus of the fibroblasts that produce it, is reflected in the population of virtually all tissue types by fibroblasts or cells that fulfil a fibroblast function, such as hepatic stellate cells. Fibroblasts themselves are heterogeneous, even within tissues, and this is likely to help tune tissue integrity to local mechanical stresses [11, 12].

A general hypothesis that has gained significant consensus is that fibrosis represents a wound healing process that has somehow gone awry. In brief, the normal reparative process following tissue injury involves an initial inflammatory response, increased stromal tissue coupled with elevated ECM synthesis to support the healing injury, replacement of the dead or injured tissue with new parenchyma and a resolution phase in which the temporary ECM and stromal cells such as myofibroblasts are removed [13]. However, if the transition from the inflammatory phase to subsequent proliferation (of stroma and parenchyma) and wound remodelling (including ECM synthesis) fails to execute properly, fibrosis may occur.

A critical event during wound healing and the development of fibrosis is the activation of fibroblasts to a proliferative, migratory state, followed by a further phenotype transition to that of the myofibroblast [11, 14]. While fibroblasts synthesize and maintain ECM levels in homeostatic balance, myofibroblasts are the arbiters of excessive ECM production that occurs during tissue fibrosis. Not all myofibroblasts arise from fibroblasts, however they are largely responsible for disease progression and dysfunction in organs as varied as the heart, lungs, skin, kidneys, liver and gastrointestinal tract [12]. Myofibroblasts secrete high levels of ECM, and both secrete and are hypersensitive to a variety of cytokines, growth factors, and chemokines that promote and maintain the pro-fibrotic myofibroblast phenotype [15]. Their hallmark functional change is the acquisition of a contractile apparatus due to the induction of expression of α-smooth muscle actin which is incorporated into stress fibers, which may permit these cells to exert physical traction to help close or reduce the margins of wounds [12, 16, 17]. Cardiac myofibroblasts have been shown to also express the matricellular protein periostin, which itself has been implicated as a driver of fibrosis, and which is not expressed in non-activated fibroblasts [18, 19]. Fibroblast activation is unquestionably important in the wound healing process, but long-term maintenance of a myofibroblast-like phenotype contributes to pathological fibrosis [20, 21].

A number of cytokines and growth factors have been demonstrated to induce or facilitate the conversion of fibroblasts to myofibroblasts. While the most potent of these is TGFβ, a variety of other factors have also been implicated in fibroblast activation and fibrosis in multiple tissue types, including endothelin-1, Platelet-Derived Growth Factor (PDGF), angiotensin II, and Connective Tissue Growth Factor (CTGF/CCN2) [22–27]. Elevated TGFβ expression has long been associated with a host of fibrotic diseases including cystic fibrosis, scleroderma, and fibrosis of the lungs, heart, kidneys and liver [28–34]. TGFβ is produced and secreted to the extracellular matrix in a protein-bound, latent form by various cell types, including fibroblasts themselves as well as inflammatory cells [35, 36]. Various processes, including protease action or physical disruption, results in the release of active TGFβ, which acts via cell surface receptors to activate intracellular signalling cascades that can be Smad-dependent (canonical) or Smad-independent (non-canonical) [35, 37–39]. TGFβ-mediated expression of CTGF/CCN2, a matricellular protein, can further amplify pro-fibrotic processes, while upstream agents such as angiotensin II can up-regulate expression of TGFβ itself [40–43]. Antagonism of TGFβ signaling is effective in reducing evidence of fibrosis, and drugs that interfere with the renin–angiotensin–aldosterone system have been shown to exert anti-fibrotic effects [44–46]. It is unclear if these various factors act solely as a trigger to activate fibroblasts, or if their presence is required for ongoing maintenance of the myofibroblast phenotype. In either scenario, the transition of fibroblasts to myofibroblasts appears to be a critical step in the development and progression of fibrosis across tissue and organ types.

Immune Cells as Mediators of Fibrosis

Inflammation is a potent inducer of fibrosis across tissue types. While inflammation is an important initial step of the normal wound healing process, it is critical that inflammation resolves in a timely manner as a chronic inflammatory state leads to tissue remodelling and fibrosis [47]. Inflammation is mediated by a variety of immune cells including macrophages, mast cells, eosinophils, neutrophils, and CD4+ and CD8+ lymphocytes, and fibroblasts themselves can generate pro-inflammatory products such as growth factors and cytokines. Two primary types of immune responses can contribute to fibrosis. In type 1 immunity, Type 1 T helper (Th1) cells release factors such as interferon gamma (IFNγ) and interleukin-2 (IL-2). In turn, these factors stimulate phagocytic cells including mast cells and macrophages. In type 2 immunity, characterized by high antibody titers, Th2 cells secrete a variety of cytokines including IL-4, IL-5, IL-9, IL-10, and IL-13, resulting in eosinophil activation [48]. Fibrosis is initially characterized by the production of Th1 cytokines followed by the Th2 response, producing TGFβ and IL-13 and leading to activation of fibroblasts and their conversion to myofibroblasts to promote fibrosis [14, 49, 50]. Th17 cells have also been implicated in fibrosis of the skin and lung [51].

Mast Cells

Mast cells are involved in both innate and adaptive immunity, and play a pivotal role in inflammation as well as in tissue remodelling leading to fibrosis [52]. Mast cells produce, store and release various growth factors, inflammatory factors and cytokines which contribute to fibrosis, including TGFβ [53]. In addition, mast cells can produce proteoglycans such as hyaluronic acid which contribute to matrix composition directly, but which can also stimulate fibroblast activation [54]. Mast cells release a variety of proteases, including chymase, leukocyte elastase, and plasmin, that in turn release latent TGFβ from ECM niches to induce fibroblast activation [39]. Mice that were mast cell deficient were protected from bleomycin-induced pulmonary fibrosis, and this same study found that mast cell release of histamine and renin could activate fibroblasts either directly, or via the eventual generation of angiotensin II, respectively [55]. Thus, mast cells can trigger the initial activation of fibroblasts leading to their proliferation and ECM production via several complementary mechanisms, demonstrating their key role in the development of fibrosis.

Macrophages

Macrophages can be broadly classified into two phenotypes: the classical M1 phenotype that is pro-inflammatory and activated by cytokines such as IFNγ, and the alternative M2 phenotype that is involved in the resolution of inflammation and is activated by interleukins such as IL-4 and IL-13 [56, 57]. While M2 macrophages have been shown to specifically promote tissue remodelling and fibrosis, M1 macrophages can also drive fibrosis via the sustained maintenance of tissue inflammation, although the specific interplay of factors that determine whether M1 macrophages act in a pro- or anti-fibrosis fashion remain to be determined.

The pro-fibrotic role of M2 macrophages has been demonstrated in multiple tissue types [58]. When human monocyte THP-1 cells were polarized to M1 or M2 macrophages and exposed to human dermal fibroblasts, M2 macrophages, specifically, induced fibrotic responses including up-regulation of αSMA and expression of collagen [59]. However, the source of these macrophages appears to be important for their role in promoting fibrosis. Alveolar macrophages derived from monocytes were required for lung fibrosis in mice, whereas tissue-resident alveolar macrophages were not [60]. While M2 macrophages can serve as a source of TGFβ that subsequently acts directly on fibroblasts to induce their activation and promote fibrosis, TGFβ can instead act on bone marrow-derived macrophages to induce their conversion to myofibroblasts [61]. Macrophages thus exert a variety of effects on the induction of fibrosis across tissue types.

Interleukins

Interleukins are cytokines produced by white blood cells. They play a central role in the body's immune and inflammatory responses, and their action is modulated by the inflammasome—a cytosolic multiprotein complex produced by myeloid cells which can contribute to tissue fibrosis [62]. In response to stress or injury, inflammasome assembly results in the activation of inflammatory caspases that in turn activate inflammatory cytokines via cleavage of inactive precursors [63]. The inflammasome acts as a binding site for a variety of caspases responsible for activating different interleukins, including IL-1 family members and related cytokines that stimulate fibrosis via the inflammatory response [63, 64].

Inflammatory cytokines including IL-1, Tissue Necrosis Factor and IL-33, can promote fibroblast activation, proliferation and collagen synthesis, potentially by increasing TGFβ expression [65–67]. Activated cytokines IL-18 and IL-33, along with IL-1, enhance inflammation and through the involvement of TGFβ can lead to the production of other cytokines such as IL-4 and IL-13, which in the setting of the lung can activate fibroblasts to behave as inflammatory cells, releasing pro-inflammatory cytokines and chemokines [68]. IL-5 is secreted by several inflammatory cell types, including mast cells and eosinophils, and eosinophils in turn are activated by IL-5 and Granulocyte–Macrophage Colony Stimulating Factor to contribute to activation of fibroblasts or other stromal cells to promote lung and intestinal fibrosis [69, 70].

In contrast to these pro-fibrotic cytokines, some demonstrate anti-fibrotic activity. IL-37 is an IL-1 family member, but has been found to attenuate the production of pro-inflammatory cytokines and chemokines by mast cells, macrophages or other immune cells that contribute to inflammation, and has shown promise as an anti-fibrotic in the lung and liver [71, 72]. Kim et al. reported that IL-37 reduced extracellular matrix protein expression in primary human lung fibroblasts, attenuated their proliferation in response to TGFβ, and was more highly expressed in alveolar epithelial cells and macrophages compared to cells from IPF patients [73]. In this study, the beneficial effect of IL-37 was dependent upon its ability to induce autophagy by inhibition of mTOR. IL-37 may also induce the expression of IL-10, and over-expression of IL-10 in the lung was able to reduce pulmonary fibrosis induced by bleomycin [74]. IL-10 has been reported to promote wound healing via activation of fibroblast-specific STAT3 and down-stream hyaluronan synthesis without driving fibrosis [75]. IL-22 is an IL-10 family member that has been found to act as an anti-inflammatory and anti-fibrotic cytokine in liver injury [76].

Autophagy

Autophagy is a critical process for the removal of excessive or defective cellular components including proteins and organelles, which can be recycled both to remove their detrimental effects on cell function and to provide energy and resources back to the cell. While autophagy can help the body to cope with stress, damage, injury, or pathogen infection, excessive levels of autophagy can be detrimental, potentially leading to cell apoptosis and tissue dysfunction.

Recent studies have linked autophagy, apoptosis and fibrosis in a variety of tissue types (Fig. 7.1). Stimulation of human atrial myofibroblasts with TGFβ induced both collagen synthesis and autophagy, while blockade of autophagy attenuated the fibrotic effect of TGFβ and in a separate study, prevented the conversion of rat cardiac fibroblasts to myofibroblasts [77, 78]. Over-expression of the TGFβ/Smad repressor Ski induced apoptosis in rat cardiac myofibroblasts, and this effect was increased if autophagy was simultaneously inhibited, suggesting that autophagy provides energy required for cell survival [79]. In normal human lung parenchymal and airway fibroblasts, TGFβ induced collagen synthesis and autophagy in parallel, however while the effect of TGFβ on collagen synthesis was increased in cells derived from IPF patients, autophagy induction was reduced demonstrating that the relationship between autophagy and fibrosis is variable [80]. Instead, the unfolded protein response was induced in IPF cells. These pathways may thus represent mechanisms to meet the high energy demands of ECM synthesis when fibroblasts become myofibroblasts, including in the setting of fibrosis. Conversely, apoptosis may be a means by which myofibroblasts are removed when energy levels are insufficient to support their ECM synthesizing function. These mechanisms may be highly

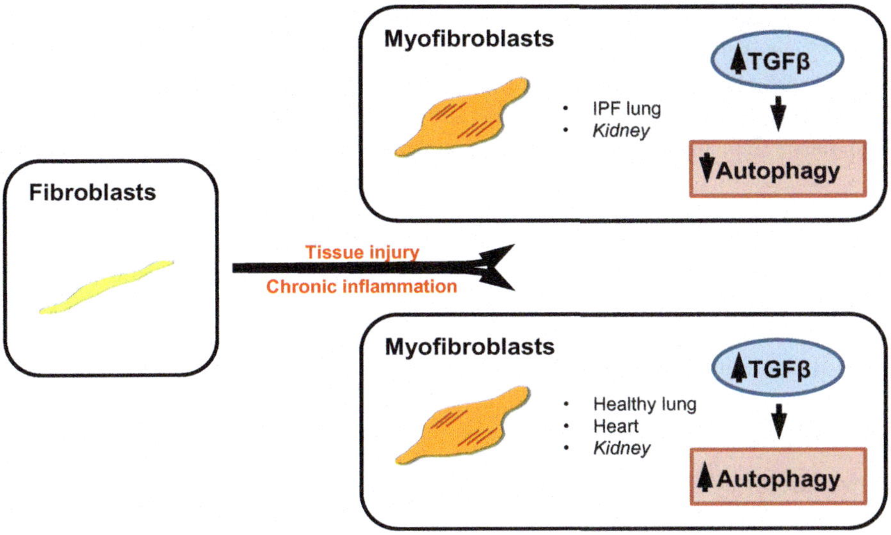

Fig. 7.1 The association of autophagy with TGFβ treatment is context-dependent. In response to tissue injury, chronic inflammation or impaired healing, fibroblasts undergo a phenotype conversion to myofibroblasts. The cytokine TGFβ can also independently induce this cellular change, however the induction of autophagy concomitantly with fibrosis is context-dependent. In a variety of cells including healthy lung fibroblasts, and atrial and ventricular cardiac fibroblasts, TGFβ induces both fibrotic gene expression and autophagy during conversion to myofibroblasts, and inhibition of autophagy typically attenuates fibrosis (lower panel). Conversely, in lung fibroblasts isolated from idiopathic pulmonary fibrosis (IPF) patients, autophagy decreases in response to TGFβ despite an induction in fibrotic gene expression during myofibroblast conversion. In kidney mesangial cells, TGFβ induced autophagy and collagen synthesis together, however reducing autophagy by decreasing Beclin-1 expression also stimulated collagen synthesis, further demonstrating that these pathways can be separated. Autophagy is positively correlated with fibrosis in other collagen-producing non-fibroblast cell types such as hepatocellular carcinoma cells, while in gut epithelium, attenuation or stimulation of autophagy increased or decreased fibrosis, respectively. The link between autophagy and fibrosis thus likely depends not only on cell type, but also on the specific environment and/or health of the cells

context- and signalling pathway-dependent, however: in kidney mesangial cells, TGFβ induced both collagen synthesis and autophagy, while knockout or knockdown of the autophagy protein Beclin 1 also increased collagen synthesis [81].

While autophagy may be required for fibroblast activation to myofibroblasts, the response of fibroblasts to the induction of autophagy may be distinct depending on the relative health status of the tissue, further supporting the idea that the specific cellular context is important in the relationship between autophagy and fibrosis. In normal lung fibroblasts, culturing on polymerized collagen can induce cell stress that results in apoptosis, but in contrast, IPF fibroblasts resist this stress and instead become proliferative and pro-fibrotic [82]. This was found to be due to alterations in PTEN/Akt/mTOR signalling in IPF fibroblasts compared to healthy lung fibroblasts, such that healthy cells undergo autophagy and subsequent apoptosis. Conversely, autophagy is down-regulated in IPF fibroblasts following culture on collagen resulting in increased cell survival [82]. A recent study suggests that IPF fibroblasts exhibit a senescent phenotype with reduced apoptosis and proliferation, rather than an activated fibroblast phenotype, with TGFβ inducing endoplasmic reticulum stress [83]. The disparity in these studies may represent differences in cell collection, nature or length of culture, biological variability across patients from which the cells are derived, or may simply be reflective of high levels of fibroblast heterogeneity. It is noteworthy, however, that both studies identified reduced apoptosis as a feature of IPF fibroblasts. It remains to be seen whether similar mechanisms linking fibrosis, autophagy and apoptosis occur in fibroblasts derived from other healthy or diseased tissues.

Regulation of Autophagy by TGFβ

TGFβ has been shown to regulate autophagy via both Smad-dependent and Smad-independent pathways, in a variety of disease contexts including cancer and fibrosis. In human hepatocellular carcinoma cells, for example, knockdown of Smad2/3 or Smad4 resulted in the inhibition of TGFβ-induced autophagy, although knockdown of c-Jun NH_2-terminal kinase had a similar effect, implicating both canonical and non-canonical TGFβ pathways [84]. In several cancer cell lines, TGFβ induced pRb/E2F1-mediated up-regulation of autophagy genes and induction of autophagosome formation [85]. Conversely, berberine administration in a rat bleomycin model of IPF attenuated Smad and PI3K/Akt signalling, but also inhibited mTOR to increase autophagy [86]. As noted above, TGFβ both induced or inhibited autophagy in lung fibroblasts depending on whether they were derived from healthy or IPF donors, respectively, despite inducing collagen expression in both cell types [80]. Thus, the variable induction of canonical and non-canonical signalling pathways downstream of TGFβ may account, in part, for the differential effects noted on fibrotic gene expression and the induction of autophagy. In turn, alterations in the level of autophagy may be beneficial or detrimental with respect to fibrosis in different contexts. Clearly, additional investigation in this area is required.

Potential Therapeutic Targets for Fibrosis

Given the central roles of inflammation and autophagy in the induction and/or progression of fibrosis in various tissue types, it is tempting to consider these areas for exploitation in the quest for novel anti-fibrotic treatments, particularly given the current and conspicuous lack of such medications. Given its central role as a product of inflammatory cells and inducer of fibrosis and autophagy, TGFβ presents a tempting target. However, with its myriad roles within individual cells and across cell and tissue types, some of which may oppose one another depending on context or disease state, direct therapeutic interference with TGFβ itself is complicated at best [87].

Other targets may be more tractable for development [11]. As noted, IL-37 is an anti-inflammatory cytokine which can act directly by downregulating pro-inflammatory mediators and interfering with TGFβ signalling, and indirectly by promoting the production of other anti-inflammatory cytokines such as IL-10 [73, 74]. IL-37 may be of particular use in the setting of IPF. IL-10 and its family member IL-22 may similarly be useful therapeutically via their anti-inflammatory and anti-fibrotic properties [74, 76].

The regulation of autophagy provides another opportunity to reduce fibrosis. Autophagy may be required for the activation of fibroblasts to myofibroblasts, and autophagy inhibition leads to cardiac myofibroblast apoptosis [79]. However, the link between autophagy and fibrosis appears to be variable across tissues, and may further be influenced by not only the presence of disease, but also by the specific nature of the disease. For example, autophagy inhibition may actually increase activation of lung fibroblasts, with consequences for IPF [88]. Stimulation of autophagy by rapamycin in a gut fibrosis model reduced fibrosis, while autophagy inhibition with 3-methyladenine exacerbated fibrosis [89]. Thus, the positive or negative manipulation of autophagy to treat fibrosis will be critically dependent on the specific pathology involved.

Conclusion

Fibrosis is an aberrant wound healing process in which fibroblasts, in response to various stimuli including growth factors like TGFβ, which in turn is secreted and activated through inflammation-activated immune cells, increase collagen and extracellular matrix protein deposition. The end result is tissue remodeling leading to organ failure, with increased risk of patient morbidity and mortality. The interplay of fibrosis development and autophagy is complex, variable across cell and tissue types, and dependent on a variety of intracellular signaling pathways including mTOR/AKT/PTEN, Smads and others, despite the superficial similarity of fibrosis in different tissues.

The identification of novel anti-fibrotic medications is arguably one of the most urgent clinical challenges at present, given the widespread occurrence of fibrosis and dearth of treatments. Targeting the relationships between inflammation, autophagy and fibrosis provides an exciting new frontier for therapeutic development. However, caution is required given the heterogeneity of fibrotic disease mechanisms, and further mechanism-focused research in this area is critically needed.

Acknowledgements This work was supported by a Project Grant from the Canadian Institutes of Health Research (PJT-162422 to MPC).

References

1. Bitterman PB, Henke CA (1991) Fibroproliferative disorders. Chest 99:81S-84S
2. Bagchi RA, Roche P, Aroutiounova N, Espira L, Abrenica B, Schweitzer R, Czubryt MP (2016) The transcription factor scleraxis is a critical regulator of cardiac fibroblast phenotype. BMC Biol 14:21
3. Guenther A, Krauss E, Tello S, Wagner J, Paul B, Kuhn S, Maurer O, Heinemann S, Costabel U, Barbero MAN et al (2018) The European IPF registry (eurIPFreg): baseline characteristics and survival of patients with idiopathic pulmonary fibrosis. Respir Res 19:141
4. Silvestre JS, Heymes C, Oubenaissa A, Robert V, Aupetit-Faisant B, Carayon A, Swynghedauw B, Delcayre C (1999) Activation of cardiac aldosterone production in rat myocardial infarction: effect of angiotensin II receptor blockade and role in cardiac fibrosis. Circulation 99:2694–2701
5. Lee JY, Ihm HS, Kim JS, Hwang HS, Jeong KH, Ihm CG (2019) Baseline high blood pressure is associated with clinico-pathologic findings and later renal progression in chronic glomerulonephritis. Electrolyte Blood Press 17:54–61
6. Bian H, Zhu X, Xia M, Yan H, Chang X, Hu X, Pan B, Guo W, Li X, Gao X (2020) Impact of type 2 diabetes on nonalcoholic steatohepatitis and advanced fibrosis in patients with nonalcoholic fatty liver disease. Endocr Pract
7. Katzenstein AL, Mukhopadhyay S, Zanardi C, Dexter E (2010) Clinically occult interstitial fibrosis in smokers: classification and significance of a surprisingly common finding in lobectomy specimens. Hum Pathol 41:316–325
8. Tabet E, Genet V, Tiaho F, Lucas-Clerc C, Gelu-Simeon M, Piquet-Pellorce C, Samson M (2016) Chlordecone potentiates hepatic fibrosis in chronic liver injury induced by carbon tetrachloride in mice. Toxicol Lett 255:1–10
9. Leyva F, Taylor RJ, Foley PW, Umar F, Mulligan LJ, Patel K, Stegemann B, Haddad T, Smith RE, Prasad SK (2012) Left ventricular midwall fibrosis as a predictor of mortality and morbidity after cardiac resynchronization therapy in patients with nonischemic cardiomyopathy. J Am Coll Cardiol 60:1659–1667
10. Espira L, Czubryt MP (2009) Emerging concepts in cardiac matrix biology. Can J Physiol Pharmacol 87:996–1008
11. Czubryt MP (2019) Cardiac fibroblast to myofibroblast phenotype conversion-an unexploited therapeutic target. J Cardiovasc Dev Dis 6
12. Nagalingam RS, Al-Hattab DS, Czubryt MP (2019) What's in a name? On fibroblast phenotype and nomenclature. Can J Physiol Pharmacol 97:493–497
13. Czubryt MP (2012) Common threads in cardiac fibrosis, infarct scar formation, and wound healing. Fibrogenesis Tissue Repair 5:19
14. Ma Y, Iyer RP, Jung M, Czubryt MP, Lindsey ML (2017) Cardiac fibroblast activation post-myocardial infarction: current knowledge gaps. Trends Pharmacol Sci 38:448–458
15. Blaauboer ME, Boeijen FR, Emson CL, Turner SM, Zandieh-Doulabi B, Hanemaaijer R, Smit TH, Stoop R, Everts V (2014) Extracellular matrix proteins: a positive feedback loop in lung fibrosis? Matrix Biol 34:170–178
16. Hinz B (2010) The myofibroblast: paradigm for a mechanically active cell. J Biomech 43:146–155
17. Tomasek JJ, Gabbiani G, Hinz B, Chaponnier C, Brown RA (2002) Myofibroblasts and mechano-regulation of connective tissue remodelling. Nat Rev Mol Cell Biol 3:349–363
18. Zhao S, Wu H, Xia W, Chen X, Zhu S, Zhang S, Shao Y, Ma W, Yang D, Zhang J (2014) Periostin expression is upregulated and associated with myocardial fibrosis in human failing hearts. J Cardiol 63:373–378
19. Snider P, Standley KN, Wang J, Azhar M, Doetschman T, Conway SJ (2009) Origin of cardiac fibroblasts and the role of periostin. Circ Res 105:934–947
20. Kendall RT, Feghali-Bostwick CA (2014) Fibroblasts in fibrosis: novel roles and mediators. Front Pharmacol 5:123

21. Kanisicak O, Khalil H, Ivey MJ, Karch J, Maliken BD, Correll RN, Brody MJ, SC, JL, Aronow BJ, Tallquist MD et al (2016) Genetic lineage tracing defines myofibroblast origin and function in the injured heart. Nat Commun 7:12260
22. Shi-Wen X, Chen Y, Denton CP, Eastwood M, Renzoni EA, Bou-Gharios G, Pearson JD, Dashwood M, du Bois RM, Black CM et al (2004) Endothelin-1 promotes myofibroblast induction through the ETA receptor via a rac/phosphoinositide 3-kinase/Akt-dependent pathway and is essential for the enhanced contractile phenotype of fibrotic fibroblasts. Mol Biol Cell 15:2707–2719
23. Mori T, Kawara S, Shinozaki M, Hayashi N, Kakinuma T, Igarashi A, Takigawa M, Nakanishi T, Takehara K (1999) Role and interaction of connective tissue growth factor with transforming growth factor-beta in persistent fibrosis: a mouse fibrosis model. J Cell Physiol 181:153–159
24. Shi-Wen X, Rodriguez-Pascual F, Lamas S, Holmes A, Howat S, Pearson JD, Dashwood MR, du Bois RM, Denton CP, Black CM et al (2006) Constitutive ALK5-independent c-Jun N-terminal kinase activation contributes to endothelin-1 overexpression in pulmonary fibrosis: evidence of an autocrine endothelin loop operating through the endothelin A and B receptors. Mol Cell Biol 26:5518–5527
25. Yu X, Xia Y, Zeng L, Zhang X, Chen L, Yan S, Zhang R, Zhao C, Zeng Z, Shu Y et al (2018) A blockade of PI3Kgamma signaling effectively mitigates angiotensin II-induced renal injury and fibrosis in a mouse model. Sci Rep 8:10988
26. Tanikawa AA, Grotto RM, Silva GF, Ferrasi AC, Sarnighausen VC, Pardini MI (2017) Platelet-derived growth factor A mRNA in platelets is associated with the degree of hepatic fibrosis in chronic hepatitis C. Rev Soc Bras Med Trop 50:113–116
27. Xu SW, Liu S, Eastwood M, Sonnylal S, Denton CP, Abraham DJ, Leask A (2009) Rac inhibition reverses the phenotype of fibrotic fibroblasts. PloS one 4:e7438
28. Wojnarowski C, Frischer T, Hofbauer E, Grabner C, Mosgoeller W, Eichler I, Ziesche R (1999) Cytokine expression in bronchial biopsies of cystic fibrosis patients with and without acute exacerbation. Eur Respir J: Off J Eur Soc Clin Respir Physiol 14:1136–1144
29. Ludwicka A, Ohba T, Trojanowska M, Yamakage A, Strange C, Smith EA, Leroy EC, Sutherland S, Silver RM (1995) Elevated levels of platelet derived growth factor and transforming growth factor-beta 1 in bronchoalveolar lavage fluid from patients with scleroderma. J Rheumatol 22:1876–1883
30. Rudnicka L, Varga J, Christiano AM, Iozzo RV, Jimenez SA, Uitto J (1994) Elevated expression of type VII collagen in the skin of patients with systemic sclerosis. Regulation by transforming growth factor-beta. J Clin Invest 93:1709–1715
31. Broekelmann TJ, Limper AH, Colby TV, McDonald JA (1991) Transforming growth factor beta 1 is present at sites of extracellular matrix gene expression in human pulmonary fibrosis. Proc Natl Acad Sci U S A 88:6642–6646
32. Espira L, Lamoureux L, Jones SC, Gerard RD, Dixon IM, Czubryt MP (2009) The basic helix-loop-helix transcription factor scleraxis regulates fibroblast collagen synthesis. J Mol Cell Cardiol 47:188–195
33. Runyan CE, Schnaper HW, Poncelet AC (2003) Smad3 and PKCdelta mediate TGF-beta1-induced collagen I expression in human mesangial cells. Am J Physiol Renal Physiol 285:F413-422
34. Gressner AM, Weiskirchen R, Breitkopf K, Dooley S (2002) Roles of TGF-beta in hepatic fibrosis. Front Biosci 7:d793-807
35. Buscemi L, Ramonet D, Klingberg F, Formey A, Smith-Clerc J, Meister JJ, Hinz B (2011) The single-molecule mechanics of the latent TGF-beta1 complex. Curr Biol 21:2046–2054
36. Hinz B (2009) Tissue stiffness, latent TGF-beta1 activation, and mechanical signal transduction: implications for the pathogenesis and treatment of fibrosis. Curr Rheumatol Rep 11:120–126
37. Zeglinski MR, Roche P, Hnatowich M, Jassal DS, Wigle JT, Czubryt MP, Dixon IM (2016) TGFbeta1 regulates Scleraxis expression in primary cardiac myofibroblasts by a Smad-independent mechanism. Am J Physiol Heart Circ Physiol 310:H239-249
38. Klingberg F, Chow ML, Koehler A, Boo S, Buscemi L, Quinn TM, Costell M, Alman BA, Genot E, Hinz B (2014) Prestress in the extracellular matrix sensitizes latent TGF-beta1 for activation. J Cell Biol 207:283–297

39. Taipale J, Lohi J, Saarinen J, Kovanen PT, Keski-Oja J (1995) Human mast cell chymase and leukocyte elastase release latent transforming growth factor-beta 1 from the extracellular matrix of cultured human epithelial and endothelial cells. J Biol Chem 270:4689–4696
40. Chen MM, Lam A, Abraham JA, Schreiner GF, Joly AH (2000) CTGF expression is induced by TGF- beta in cardiac fibroblasts and cardiac myocytes: a potential role in heart fibrosis. J Mol Cell Cardiol 32:1805–1819
41. Gu J, Liu X, Wang QX, Tan HW, Guo M, Jiang WF, Zhou L (2012) Angiotensin II increases CTGF expression via MAPKs/TGF-beta1/TRAF6 pathway in atrial fibroblasts. Exp Cell Res 318:2105–2115
42. Yang F, Chung AC, Huang XR, Lan HY (2009) Angiotensin II induces connective tissue growth factor and collagen I expression via transforming growth factor-beta-dependent and -independent Smad pathways: the role of Smad3. Hypertension 54:877–884
43. Campbell SE, Katwa LC (1997) Angiotensin II stimulated expression of transforming growth factor-beta1 in cardiac fibroblasts and myofibroblasts. J Mol Cell Cardiol 29:1947–1958
44. Leask A (2010) Potential therapeutic targets for cardiac fibrosis: TGFbeta, angiotensin, endothelin, CCN2, and PDGF, partners in fibroblast activation. Circ Res 106:1675–1680
45. Ma L, Hua J, He L, Li Q, Zhou J, Yu J (2012) Anti-fibrotic effect of Aliskiren in rats with deoxycorticosterone induced myocardial fibrosis and its potential mechanism. Bosn J Basic Med Sci/Udruzenje basicnih mediciniskih znanosti = Assoc Basic Med Sci 12:69–73
46. Dobaczewski M, Bujak M, Li N, Gonzalez-Quesada C, Mendoza LH, Wang XF, Frangogiannis NG (2010) Smad3 signaling critically regulates fibroblast phenotype and function in healing myocardial infarction. Circ Res 107:418–428
47. Borthwick LA, Wynn TA, Fisher AJ (2013) Cytokine mediated tissue fibrosis. Biochim Biophys Acta 1832:1049–1060
48. Spellberg B, Edwards JE Jr (2001) Type 1/type 2 immunity in infectious diseases. Clin Infect Dis 32:76–102
49. Sandler NG, Mentink-Kane MM, Cheever AW, Wynn TA (2003) Global gene expression profiles during acute pathogen-induced pulmonary inflammation reveal divergent roles for Th1 and Th2 responses in tissue repair. J Immunol 171:3655–3667
50. Hoffmann KF, McCarty TC, Segal DH, Chiaramonte M, Hesse M, Davis EM, Cheever AW, Meltzer PS, Morse HC 3rd, Wynn TA (2001) Disease fingerprinting with cDNA microarrays reveals distinct gene expression profiles in lethal type 1 and type 2 cytokine-mediated inflammatory reactions. FASEB J 15:2545–2547
51. Lei L, Zhao C, Qin F, He ZY, Wang X, Zhong XN (2016) Th17 cells and IL-17 promote the skin and lung inflammation and fibrosis process in a bleomycin-induced murine model of systemic sclerosis. Clin Exp Rheumatol 34(Suppl 100):14–22
52. Conti P, Caraffa A, Mastrangelo F, Tettamanti L, Ronconi G, Frydas I, Kritas SK, Theoharides TC (2018) Critical role of inflammatory mast cell in fibrosis: Potential therapeutic effect of IL-37. Cell Prolif 51:e12475
53. Hugle T (2014) Beyond allergy: the role of mast cells in fibrosis. Swiss Med Wkly 144:w13999
54. Eggli PS, Graber W (1993) Cytochemical localization of hyaluronan in rat and human skin mast cell granules. J Invest Dermatol 100:121–125
55. Veerappan A, O'Connor NJ, Brazin J, Reid AC, Jung A, McGee D, Summers B, Branch-Elliman D, Stiles B, Worgall S et al (2013) Mast cells: a pivotal role in pulmonary fibrosis. DNA Cell Biol 32:206–218
56. Dona A, Abera M, Alemu T, Hawaria D (2018) Timely initiation of postpartum contraceptive utilization and associated factors among women of child bearing age in Aroressa District, Southern Ethiopia: a community based cross-sectional study. BMC Public Health 18:1100
57. Pechkovsky DV, Prasse A, Kollert F, Engel KM, Dentler J, Luttmann W, Friedrich K, Muller-Quernheim J, Zissel G (2010) Alternatively activated alveolar macrophages in pulmonary fibrosis-mediator production and intracellular signal transduction. Clin Immunol 137:89–101
58. Gieseck RL 3rd, Wilson MS, Wynn TA (2018) Type 2 immunity in tissue repair and fibrosis. Nat Rev Immunol 18:62–76

59. Zhu Z, Ding J, Ma Z, Iwashina T, Tredget EE (2017) Alternatively activated macrophages derived from THP-1 cells promote the fibrogenic activities of human dermal fibroblasts. Wound Repair Regen 25:377–388
60. Misharin AV, Morales-Nebreda L, Reyfman PA, Cuda CM, Walter JM, McQuattie-Pimentel AC, Chen CI, Anekalla KR, Joshi N, Williams KJN et al (2017) Monocyte-derived alveolar macrophages drive lung fibrosis and persist in the lung over the life span. J Exp Med 214:2387–2404
61. Wang S, Meng XM, Ng YY, Ma FY, Zhou S, Zhang Y, Yang C, Huang XR, Xiao J, Wang YY et al (2016) TGF-beta/Smad3 signalling regulates the transition of bone marrow-derived macrophages into myofibroblasts during tissue fibrosis. Oncotarget 7:8809–8822
62. Suetomi T, Willeford A, Brand CS, Cho Y, Ross RS, Miyamoto S, Brown JH (2018) Inflammation and NLRP3 Inflammasome Activation Initiated in Response to Pressure Overload by Ca(2^+)/Calmodulin-Dependent Protein Kinase II delta Signaling in Cardiomyocytes Are Essential for Adverse Cardiac Remodeling. Circulation 138:2530–2544
63. Groslambert M, Py BF (2018) Spotlight on the NLRP3 inflammasome pathway. J Inflamm Res 11:359–374
64. Artlett CM (2018) The IL-1 family of cytokines. Do they have a role in scleroderma fibrosis? Immunol Lett 195:30–37
65. Lonnemann G, Shapiro L, Engler-Blum G, Muller GA, Koch KM, Dinarello CA (1995) Cytokines in human renal interstitial fibrosis. I. Interleukin-1 is a paracrine growth factor for cultured fibrosis-derived kidney fibroblasts. Kidney Int 47:837–844
66. Tan Z, Liu Q, Jiang R, Lv L, Shoto SS, Maillet I, Quesniaux V, Tang J, Zhang W, Sun B et al (2018) Interleukin-33 drives hepatic fibrosis through activation of hepatic stellate cells. Cell Mol Immunol 15:388–398
67. Sullivan DE, Ferris M, Nguyen H, Abboud E, Brody AR (2009) TNF-alpha induces TGF-beta1 expression in lung fibroblasts at the transcriptional level via AP-1 activation. J Cell Mol Med 13:1866–1876
68. Doucet C, Brouty-Boye D, Pottin-Clemenceau C, Canonica GW, Jasmin C, Azzarone B (1998) Interleukin (IL) 4 and IL-13 act on human lung fibroblasts. Implication in asthma. J Clin Invest 101:2129–2139
69. Gharaee-Kermani M, Phan SH (1997) Lung interleukin-5 expression in murine bleomycin-induced pulmonary fibrosis. Am J Respir Cell Mol Biol 16:438–447
70. Takemura N, Kurashima Y, Mori Y, Okada K, Ogino T, Osawa H, Matsuno H, Aayam L, Kaneto S, Park EJ et al (2018) Eosinophil depletion suppresses radiation-induced small intestinal fibrosis. Sci Transl Med 10
71. Li Y, Gao Q, Xu K, Peng X, Yuan X, Jiang W, Li M (2018) Interleukin-37 attenuates bleomycin-induced pulmonary inflammation and fibrosis in mice. Inflammation 41:1772–1779
72. Feng XX, Chi G, Wang H, Gao Y, Chen Q, Ru YX, Luo ZL, Yan W, Li PY, Liu M et al (2019) IL-37 suppresses the sustained hepatic IFN-gamma/TNF-alpha production and T cell-dependent liver injury. Int Immunopharmacol 69:184–193
73. Kim MS, Baek AR, Lee JH, Jang AS, Kim DJ, Chin SS, Park SW (2019) IL-37 Attenuates lung fibrosis by inducing autophagy and regulating TGF-beta1 production in mice. J Immunol 203:2265–2275
74. Kurosaki F, Uchibori R, Sehara Y, Saga Y, Urabe M, Mizukami H, Hagiwara K, Kume A (2018) AAV6-mediated IL-10 expression in the lung Ameliorates bleomycin-induced pulmonary fibrosis in mice. Hum Gene Ther 29:1242–1251
75. Balaji S, Wang X, King A, Le LD, Bhattacharya SS, Moles CM, Butte MJ, de Jesus Perez VA, Liechty KW, Wight TN et al (2017) Interleukin-10-mediated regenerative postnatal tissue repair is dependent on regulation of hyaluronan metabolism via fibroblast-specific STAT3 signaling. FASEB J 31:868–881
76. Khawar MB, Azam F, Sheikh N, Abdul Mujeeb K (2016) How does interleukin-22 mediate liver regeneration and prevent injury and fibrosis? J Immunol Res 2016:2148129
77. Ghavami S, Cunnington RH, Gupta S, Yeganeh B, Filomeno KL, Freed DH, Chen S, Klonisch T, Halayko AJ, Ambrose E et al (2015) Autophagy is a regulator of TGF-beta1-induced fibrogenesis in primary human atrial myofibroblasts. Cell Death Dis 6:e1696

78. Gupta SS, Zeglinski MR, Rattan SG, Landry NM, Ghavami S, Wigle JT, Klonisch T, Halayko AJ, Dixon IM (2016) Inhibition of autophagy inhibits the conversion of cardiac fibroblasts to cardiac myofibroblasts. Oncotarget 7:78516–78531
79. Zeglinski MR, Davies JJ, Ghavami S, Rattan SG, Halayko AJ, Dixon IM (2016) Chronic expression of Ski induces apoptosis and represses autophagy in cardiac myofibroblasts. Biochim Biophys Acta 1863:1261–1268
80. Ghavami S, Yeganeh B, Zeki AA, Shojaei S, Kenyon NJ, Ott S, Samali A, Patterson J, Alizadeh J, Moghadam AR et al (2018) Autophagy and the unfolded protein response promote profibrotic effects of TGF-beta1 in human lung fibroblasts. Am J Physiol Lung Cell Mol Physiol 314:L493–L504
81. Kim SI, Na HJ, Ding Y, Wang Z, Lee SJ, Choi ME (2012) Autophagy promotes intracellular degradation of type I collagen induced by transforming growth factor (TGF)-beta1. J Biol Chem 287:11677–11688
82. Nho RS, Hergert P (2014) IPF fibroblasts are desensitized to type I collagen matrix-induced cell death by suppressing low autophagy via aberrant Akt/mTOR kinases. PloS one 9:e94616
83. Alvarez D, Cardenes N, Sellares J, Bueno M, Corey C, Hanumanthu VS, Peng Y, D'Cunha H, Sembrat J, Nouraie M et al (2017) IPF lung fibroblasts have a senescent phenotype. Am J Physiol Lung Cell Mol Physiol 313:L1164–L1173
84. Kiyono K, Suzuki HI, Matsuyama H, Morishita Y, Komuro A, Kano MR, Sugimoto K, Miyazono K (2009) Autophagy is activated by TGF-beta and potentiates TGF-beta-mediated growth inhibition in human hepatocellular carcinoma cells. Cancer Res 69:8844–8852
85. Korah J, Canaff L, Lebrun JJ (2016) The retinoblastoma tumor suppressor protein (pRb)/E2 promoter binding factor 1 (E2F1) pathway as a novel mediator of TGFbeta-induced autophagy. J Biol Chem 291:2043–2054
86. Chitra P, Saiprasad G, Manikandan R, Sudhandiran G (2015) Berberine inhibits Smad and non-Smad signaling cascades and enhances autophagy against pulmonary fibrosis. J Mol Med (Berl) 93:1015–1031
87. Al Hattab D, Czubryt MP (2017) A primer on current progress in cardiac fibrosis. Can J Physiol Pharmacol 95:1091–1099
88. Araya J, Kojima J, Takasaka N, Ito S, Fujii S, Hara H, Yanagisawa H, Kobayashi K, Tsurushige C, Kawaishi M et al (2013) Insufficient autophagy in idiopathic pulmonary fibrosis. Am J Physiol Lung Cell Mol Physiol 304:L56-69
89. Cosin-Roger J, Canet F, Macias-Ceja DC, Gisbert-Ferrandiz L, Ortiz-Masia D, Esplugues JV, Alos R, Navarro F, Barrachina MD, Calatayud S (2019) Autophagy stimulation as a potential strategy against intestinal fibrosis. Cells 8

Chapter 8
Gene Therapy and Its Application in Cardiac Diseases

Sikta Chattopadhyaya and Michael P. Czubryt

Abstract Cardiovascular diseases are responsible for high mortality rates throughout the world, and diseases specifically affecting the myocardium including fibrosis, myocardial infarction, arrhythmias, and cardiomyopathies leading to cardiac dysfunction and heart failure remain significant contributors to this problem. For decades, therapeutic intervention has relied almost exclusively on surgical or pharmacologic approaches, and while these have improved survival and quality of life, side effects can be significant and cardiac death remains a major problem, leading to extensive efforts to identify novel means of treatment. The potential for cardiac gene therapy, in which disease is treated via the introduction of therapeutic genetic material to correct defective genes, improve cellular function, and restore cardiac health, has been recognized for many years, but early challenges and adverse outcomes severely limited adoption. Improvements in gene therapeutic approaches have resulted in safer and more precisely-targeted means of treating heart diseases, with considerable advancement in the field in recent years. In this chapter, we review the viral and non-viral vectors which have been utilized in treating cardiac diseases such as ischemic cardiomyopathy, myocardial infarction, and fibrosis. We also consider the use of these approaches to facilitate gene editing using CRISPR/Cas9-based methods, provide a brief overview of stem cell therapy including the use of engineered stem cells, and discuss how combinational strategies are gaining in popularity due to therapeutic advantages over individual strategies.

Keywords Gene therapy · Transgenesis · Heart disease · Stem cells · Viral vectors · Transgene

---

S. Chattopadhyaya · M. P. Czubryt
Department of Physiology and Pathophysiology, Rady Faculty of Health Sciences, University of Manitoba, Winnipeg, Canada

M. P. Czubryt (✉)
Institute of Cardiovascular Sciences, R4008 St. Boniface Hospital Albrechtsen Research Centre, 351 Tache Avenue, Winnipeg, MB R2H 2A6, Canada
e-mail: mczubryt@sbrc.ca

© Springer Nature Switzerland AG 2022
L. A. Kirshenbaum (ed.), *Biochemistry of Apoptosis and Autophagy*,
Advances in Biochemistry in Health and Disease 18,
https://doi.org/10.1007/978-3-030-78799-8_8

Introduction

Cardiovascular diseases which lead to heart failure—in brief, an impaired ability of the ventricles of the heart to fill with or eject blood—are responsible for high mortality rates worldwide. Standard treatments for heart failure (HF) are primarily pharmacological, targeting critical signaling pathways such as the renin–angiotensin–aldosterone or adrenergic receptor-mediated systems, and include angiotensin converting enzyme inhibitors, angiotensin receptor blockers, β-adrenergic receptor antagonists, aldosterone antagonists, and diuretics [1]. Surgical approaches may also be taken, including heart transplantation or implantation of a ventricular assist device [2]. Despite improvements in patient care and treatment, mortality rates remain high, and heart failure treatments themselves may cause unwanted side effects arising from pharmacotherapy or required anti-rejection medications. Moreover, many therapies improve symptomology but may not correct underlying defects at the cellular or subcellular level, such as derangements in cardiomyocyte metabolism. Recently, significant attention has shifted to the exciting potential of gene therapy, which can be used to effectively target specific signaling pathways in cardiac myocytes as an approach to treat heart failure [2]. The heart, due to its anatomical compartmentalization and relatively easy accessibility by surgical and percutaneous approaches, presents an excellent target for gene therapy [3].

Gene Therapy

Gene therapy refers to a type of genetic modification involving the introduction of exogenous genetic material (a transgene) to a target cell with the aim of providing new functions inside the cell, thus correcting genotypic or phenotypic abnormalities [4]. This exogenous genetic material can encode a protein, peptide, minipeptide or small interfering RNA, which affects and corrects altered or defective intracellular signaling pathways in the targeted cardiac cells [5, 6]. For example, key proteins such as the Ca^{2+}-sensor S100A1 or the sarcoplasmic reticulum (SR) Ca^{2+} ATPase (SERCA2a) are downregulated in the failing heart, therefore upregulation of these proteins through gene therapy may provide novel treatment options [7]. Conversely, heart failure may be induced in part via the upregulation of proteins or enzymes causing altered and detrimental cell responses, such as the G-protein-coupled receptor kinase 2 (GRK2). In this situation, delivery of inhibitors, dominant negative variants, short interfering RNA, short hairpin RNA, microRNA, truncated proteins for expressing enzyme auto-inhibitory domains, or hindering of protein–protein interactions can offer potential therapies [8] (Fig. 8.1).

In cardiovascular gene therapy, there are a number of key points that must be considered for the procedure to be safe and effective: first, the correct transgene must be selected for achieving the specific clinical aim, such as inhibition of fibrosis, improving cardiac contractility, reversal or improvement of remodelling, increased

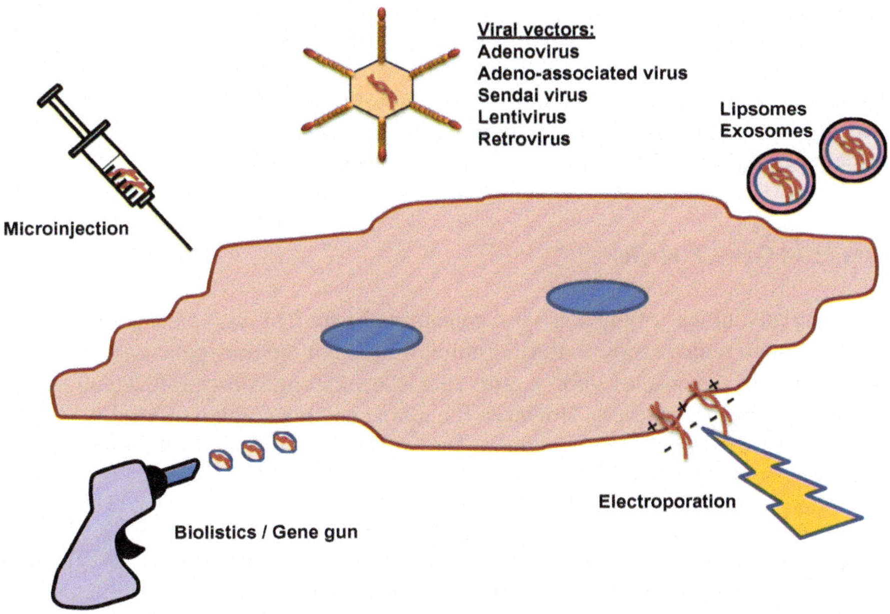

Fig. 8.1 Vectors for cardiac gene therapy. Vectors used for delivery of transgenic material for cardiac therapy comprise a variety of non-viral options such as liposomes, exosomes, electroporation, direct microinjection to the myocardium, and biological ballistics (gene gun), as well as viral vectors including adenovirus, adeno-associated virus, Sendai virus, retrovirus, and lentivirus. The non-viral approaches of direct injection, electroporation, and biological ballistics are those most commonly used for the delivery of bacterial plasmids (pDNA) encoding therapeutic transgenes. Gene editing, employing CRISPR/Cas9 delivery, is typically accomplished via microinjection, electroporation or hybrid exosome-liposome nanoparticles. As the technology of viral vectors improves, along with a better understanding of potential complications such as immune or inflammatory responses and their avoidance, the use of such vectors is anticipated to sharply increase

angiogenesis, regeneration of myocytes, correction of a specific genetic effect, or inhibiting apoptosis. Second, the correct delivery option must be used such that the efficiency of transfer of the transgene, its stability and its expression in the target cell is high, and results in achievement of the desired therapeutic effect. Third, the selected delivery route should be clinically safe. Fourth, all of the above three considerations must be used at the minimum effective dose to avoid the risk of increased immunogenicity, to reduce adverse effects at the time of delivery, and to obtain efficient therapeutic effect [9].

Different Options for Transgene Delivery

Transgene delivery methods for gene therapy are subdivided into two categories: non-viral and viral delivery vectors. Both delivery mechanisms have their own associated advantages and disadvantages.

Non-viral Gene Transfer

The main advantages of the non-viral gene transfer are that it is safe, and it provokes a low immunogenic response. Furthermore, the size of the transgene to be delivered is restricted in the case of viral vectors due to packaging constraints which are not present in non-viral delivery. However, the main challenge with non-viral delivery is that the transfection efficiency remains low, regardless of the specific method used for transgenesis [3].

i. ***Liposomes***
Liposomes act as vehicles for transferring transgenes to target cells. They are usually prepared by the addition of lipid bilayer constituents to an aqueous solution of DNA molecules. By a self-organization process they form lipid bilayer spheres containing the aqueous solution along with the transgene-bearing DNA. The transgene is in turn delivered by fusing of the bilayer spheres with the recipient cell membrane, or by endocytosis [10, 11]. There are several challenges with liposome-mediated transgene delivery. Liposomes may be phagocytosed and cleared mostly by liver and spleen macrophages, reducing their transfection efficiency [12]. While liposomes are less toxic and less immunogenic than viral delivery methods, and while the transgenic cargo is delivered to the cytoplasm with reasonable efficiency, only a small fraction further traverse into the nucleus, making this process less efficient in terms of overall transgene stability and expression [4, 13]. Recently, liposomal magnetofection, i.e. the application of a magnetic field to liposome-mediated transgene delivery, has been found to increase transfection efficiency in vivo [14]. Liposomal magnetofection has been used for the efficient delivery of shIGF1R for downregulating IGF1R (insulin-like growth factor 1 receptor), an oncogene overexpressed in the hypertrophic heart and responsible in part for its pathology. Liposomal magnetofection mediated site-specific delivery of shIGF1R was able to attenuate cardiac hypertrophy [14].

ii. ***Exosomes***
Exosomes are secreted from virtually all types of cells, and are extracellular vesicles of endocytic origin [15]. They are a cell-free naturally-derived system that exhibits significant advantages as compared to other vectors for transgene delivery. Exosomes efficiently transport exogenous RNA (siRNA and miRNA) to target cells primarily via receptor-mediated fusion with the recipient cell membrane [15, 16]. The tough exosome membrane protects the RNA/transgene from degradation, and exosomes are very efficiently taken up by target cells

[15]. Another advantage of this transgene delivery approach is that the size of the exosomes prevents them from undergoing phagocytosis. They can be derived from the patient, thus are recognised as self, lowering immunogenicity and cytotoxicity [16]. Recently, exosomes secreted from stem cells have been successfully used for transferring miRNA to recipient cells in the heart to facilitate cardiac repair [15].

iii. ***Electroporation***
Cell membranes exposed to high intensity electrical impulses become temporarily destabilized and thus highly permeable. Electroporation harnesses this phenomenon to transfer transgenes across the lipid bilayer membrane of the cell [4, 17]. Although this method is fast and inexpensive, the disadvantages faced are that broad empirical optimisation is required for every cell type, and the efficiency of genomic integration of the transgene is very low [4]. Nonetheless, recent research indicates that electroporation of pDNA (bacterial plasmids containing the transgene; see below) can be an effective measure in treating heart diseases in vivo via the safe and effective transfer of pDNA into beating porcine hearts [18].

iv. ***Microinjection***
This process of transgene delivery permits germline targeting of the genome by direct injection of naked DNA to zygotic cell nuclei. This approach may be effective in treating genetic cardiovascular diseases via delivery of transgenes, but has been limited to animal models to date—and in fact has been a critical technique for the production of novel transgenic model organisms, particularly mice. The manipulation of the germline carries significant ethical implications, thus its use in humans remains a more distant possibility. This method is one of the most efficient physical or non-viral methods of transporting gene editing cargos of the CRISPR/Cas9 system [19, 20].
DNA injection can also be used for delivery of naked DNA into specific tissues, although this has been more often performed in liver and muscle cells [21, 22]. It has not been used for direct DNA injection into heart cells to date as it has been associated with disadvantages including acute inflammatory responses, post needle injury and poor uniformity of expression of the transgene, remaining primarily at the site of injection and thus being insufficient for correction of autosomal and X-linked recessive cardiomyopathies [23, 24]. However, this approach may provide an option for regionally-restricted therapy, for example inducing border zone angiogenesis following myocardial infarction [9].

v. ***Biological ballistics***
Biolistic transfection refers to the use of micrometer-sized heavy metal particles (including gold or tungsten) coated with exogenous DNA and accelerated to high velocity using helium gas under pressure. At this velocity, these particles bombard cells and pass through the lipid bilayer membrane and the exogenous DNA is transferred to the cytoplasm, where the transgene is solubilized and expressed. The cell membrane rupture is transient and self-healing [25]. This process was first established in plants, and recently has been used for transgene delivery to animal cells in vivo [25, 26]. Biolistics are highly efficient compared

to liposome-mediated transfection, but less efficient than adenovirus-mediated transgene delivery [25]. Disadvantages associated with this technique include limited penetration into the myocardium, as only the surface layer of cardiomyocytes are transfected, and low levels of transgene expression [27]. Gene guns have been used to successfully transfer plasmid DNA (pDNA—see below) into heart cells using Epstein-Barr virus-based episomal vectors, achieving long-lasing expression of the transgene [27]. A handheld gene gun induced efficient delivery of transgenes into the beating heart of adult Wistar rats [28]. Biological ballistics thus show significant potential for gene therapy of cardiovascular diseases.

vi. ***pDNA***

While not a delivery method per se, bacterial plasmids (pDNA) encoding transgenes represent the most basic method of gene delivery with significant advantages over naked DNA fragments. Following transfection by electroporation, biological ballistics or direct injection, cells take up the transgene DNA, which is transported to the nucleus and expressed. The advantages of using pDNA include the fact that it can be obtained at higher concentrations and purity by means of HPLC or other purification methods, plasmids remain stable and can be stored for longer periods of time without loss of potency, they can be easily manipulated, they have much lower packaging restrictions compared to viral methods thus the size of the transgene is less problematic, and they provoke relatively low immunogenic responses. The primary disadvantage is that transfection efficiency and persistence after introduction can be very low, as when pDNA was directly injected into mouse skeletal muscle in vivo [29, 30]. However, as noted above, transfection efficiency can be significantly boosted via electroporation or biological ballistics procedures, with recent studies showing pDNA can be used effectively for cardiac gene therapy [18, 27].

Viral Gene Transfer

A primary advantage of using viruses as transgene delivery vectors is their typically much higher transfection efficiency compared to non-viral methods. Conversely, viral transgene delivery is also accompanied by a number of risks or caveats that must be considered in any therapy to be developed, notably the potential to give rise to an immunogenic response, packaging constraints that limit the size of the transgenes to be delivered, and the possibility of targeting the wrong cells. The most common vectors used for cardiovascular gene transfer are adenovirus, adeno-associated virus (AAV), Sendai virus, and retrovirus. AAV, lentivirus and retrovirus vectors are capable of integrating the transgene into the host genome, which increases the risk of oncogenic transformation and is not critically required in non-dividing cardiomyocytes, but may be important for stem cell-mediated therapies in which the transgene needs to be replicated and maintained in dividing cells [3].

i. ***Adenovirus***

Adenoviruses are very commonly used in gene therapy, second only to retroviruses. Adenoviruses are DNA viruses encoding early promoter regions (E1 to E4) which initiate viral replication, followed by late promoter regions responsible for transcription of the viral genes. The adenovirus contains genes in both cis and trans, with cis genes intrinsic to the virus and responsible for its origin of replication and packaging of the DNA, while trans genes can be replaced by the exogenous transgene payload which is to be delivered to recipient cells [31]. There are several advantages of adenoviruses, including the ability to efficiently carry large transgenes, a lack of host specificity enabling them to be used for treating different types of tissues and organs, and ease of manipulation. Adenoviruses may be used to provide stable transgene expression in both dividing and quiescent cells, such as fibroblasts and cardiomyocytes, respectively. Since they do not integrate into the host genome, the chances of transgene mutation are reduced [4, 31, 32]. In the first generation of therapeutically-used adenoviruses, the E1 region was replaced by the transgene to inhibit viral replication, but still gave rise to high immunogenic responses—a primary disadvantage of adenoviral gene therapy. This high immunogenic response moreover resulted in the loss of transgene expression [33]. The second and third generation adenoviruses have been created by deletion of combinations of the E1, E2, and E4 genes as they were found to be associated with induction of the immune response. Although this approach reduced the toxicity and stabilized gene expression, the efficiency of gene transfer was lowered [34]. In addition to high immunogenicity, another issue with adenoviruses is transient expression of the transgene due to a lack of integration of the transgene into the host genome, thereby limiting their usage in gene therapy [31]. The main problem which remains with the use of adenoviruses is their ability to elicit a potent immune response. This issue was brought into sharp focus following the tragic death of Jesse Gelsinger, which occurred due to systemic response inflammatory syndrome when adenovirus was used for transfer of ornithine transcarbamylase cDNA to treat an enzyme deficiency [35]. Although the use of adenovirus for gene therapy in the heart has not given rise to a similar event, the potential for a dangerous inflammatory response remained a significant concern which limited the clinical use of adenovirus for gene transfer in cardiac gene therapy to date [3]. This concern has only recently started to decrease due to improved approaches and a better understanding of the body's response.

ii. ***AAV (Adeno-associated virus)***

The parvovirus AAV is nonpathogenic—a key beneficial feature of this vector—was discovered first as a contaminant in laboratory stocks of adenovirus, and was referred to as a "defective satellite virus" [36, 37]. It is a single-stranded DNA virus which generates a double-stranded DNA template upon infection of a host cell. It lacks replication potential, unless in the presence of other helper viruses (mainly adenovirus or herpesvirus), thus provoking a much lower immune response compared to adenoviruses, and thus is much safer [37–39]. In addition to having a number of advantages in common with adenoviruses, including low host cell specificity and the ability to transfect both dividing and quiescent cells, AAV is capable of site-specific genome integration into human chromosome 19 using Rep protein, and thus can maintain the stable expression of transgenes for longer periods of time, making AAV one of the more efficient viral vectors in gene therapy [40]. The primary disadvantage of AAV is its low packaging capacity, limiting payload size [41]. With respect to gene therapy for treatment of cardiomyocytes, AAV (primarily AAV9) exhibits greater efficiency of transfection than lentiviruses [42, 43]. In the early years of gene therapy, AAV was found to be effective in the treatment of dilated cardiomyopathy in hamsters, where AAV-mediated recombinant δ-sarcoglycan gene transduction improved sarcolemmal integrity and function, resulting in improved cardiac wall thickness, demonstrating the potential of this approach for clinical cardiac gene therapy [44]. Similarly, another study involved MYBPC3, a gene encoding cardiac myosin-binding protein C (cMyBP-C); frameshift mutation of MYBPC3 causes neonatal cardiomyopathy, which was successfully treated by using MYBPC3 gene therapy with AAV as the vector [45]. Patients with non-ischemic cardiomyopathy or chronic systolic heart failure were treated with AAV1-encoded sarcoplasmic reticulum calcium ATPase 2a (SERCA2a) in the CUPID2 trial. However, while previous smaller trials had given positive results, this was not the case in CUPID2, owing to low gene transfer efficiency [46].

iii. ***Sendai virus***

Sendai virus vectors are non-pathogenic, single-stranded RNA viruses that can infect both dividing and non-dividing cells [47]. It does not integrate into the host chromosome, resulting in transient gene expression and limiting its use in cardiovascular research or treatment [3]. However, in a recent study, integration-free induced pluripotent stem cells were generated from urine samples (UiPSCs) of a ventricular septal defect patient using Sendai virus which carried the Yamanaka factors responsible for generation of induced pluripotent cells. Through further modification of the canonical Wnt signalling pathway with small molecules, the UiPSCs were induced to convert into functional cardiomyocytes, providing a useful and abundant model for studying the genetic signature, pathology and treatment of ventricular septal defect [48]. A similar study was done in which iPSCs were generated from skin fibroblasts of a short QT syndrome patient using Sendai virus. This approach provided a model for elucidating the mechanisms associated with the disease, which results in life-threatening cardiac arrhythmias leading to heart failure, and for discovering potential therapeutics [49].

iv. ***Retrovirus***

The most commonly used therapeutic vectors are the retroviruses. A key characteristic of retrovirus is how it alternately uses both DNA and RNA as its genetic material during the infectious life cycle. Normally retrovirus carries its genetic material inside individual virions in the form of RNA, which is reverse transcribed into DNA after infection, i.e. after entering the host cell and integrating into the host genome (the provirus). The site of integration is usually random, but can also occur at preferred integration sites that have been identified in specific cell types [50]. Retroviruses can be manipulated to carry exogenous RNA for eventual integration into the host genome, and the resulting efficiency of transgene uptake is high. These advantages make retrovirus an efficient vector for somatic gene therapy [4, 51]. The primary disadvantage of retrovirus is that it is limited to infecting dividing cells undergoing mitosis, thus they are not effective for gene therapy in cardiomyocytes, although cardiac fibroblasts, vascular smooth muscle and endothelial cells can be targeted [52]. Recently retrovirus has been used for transgene delivery of cardiac reprogramming factors to convert cardiac fibroblasts into induced cardiac-like myocytes [53].

v. ***Lentivirus***

One of the most efficient viral vectors currently used for gene therapy of cardiovascular diseases is lentivirus, a genus of retroviruses. The primary difference between lentivirus and retrovirus is that lentivirus can infect both dividing and non-dividing cells, broadening its efficacy and applicability. Lentivirus can be used for both ex vivo and in vivo gene transfer with high efficiency, including for the generation of induced pluripotent stem cells or transfer of transgenes into hematopoietic stem cells, maintaining consistent transgene expression when the recipient cell divides [54–57]. Safety concerns exist due to the potential for insertional mutagenesis, however while oncogenic transformation is possible, lentiviruses are considered safe since transgenes typically integrate far from transcription start sites [58]. The self-inactivating (SIN) lentiviral vectors have further improved lentivirus safety, enabling their use in clinical trials [59]. SIN lentiviral vectors were found to maintain stable and long-lasting transgene expression in cardiomyocytes, and the efficiency of transfection increased upon forced diffusion into the myocardium, thus they can act as an effective tool for gene therapy in cardiovascular diseases [60, 61]. Although lentiviruses have been used for efficient gene transfer into cardiomyocytes, the efficiency of gene transfer was much lower compared to that obtained from AAV, particularly AAV9 [42, 43].

Emerging Therapeutic Approaches

CRISPR/Cas9

One of the most powerful and promising gene editing techniques is the clustered regularly interspaced short palindromic repeats (CRISPR)/CRISPR-associated protein 9 (Cas9) system which may be used for gene deletion, precise homology-directed repair or mutation by error-prone non-homologous end-joining [62]. Germline CRISPR/Cas9-directed genome editing offers the exciting potential to cure monogenic cardiovascular diseases not only in the affected offspring, but also in future generations [63]. A critical challenge, however, is how to deliver this therapy. Recently it has been found that, although liposomes and exosomes cannot deliver the CRISPR/Cas9 system individually, a hybrid exosome-liposome nanoparticle created though simple incubation of exosomes and liposomes together could efficiently deliver the CRISPR/Cas9 system to its specific target cell, offering an effective strategy in gene therapy for the treatment of cardiovascular disorders [62]. CRISPR/Cas9-directed germline alteration has also been achieved using electroporation and microinjection for delivery [20, 64]. In addition to these non-viral vector methods, gold nanoparticles can also be used for CRISPR/Cas9 delivery [65, 66].

Although the non-viral delivery options for CRISPR/Cas9 have advantages, viral delivery through AAV, adenovirus and lentivirus was initially predicted to be a better approach, and was successfully used for in vivo gene therapy [67, 68]. Recent cardiovascular studies indicate that non-viral delivery of CRISPR/Cas9 is gaining popularity [69]. For permanent correction of disease-causing mutations in the adult cardiomyocyte genome, although CRISPR/Cas9 appears to be an effective tool, it has not yet been as successful in clinical trials since a mosaic pattern of gene disruption was observed, however additional research will undoubtedly lead to greater adoption of this promising approach in time [70].

Stem Cell Therapy

Stem cells exhibit the key capabilities of self-renewal and the capacity to differentiate into a wide variety of mature cell types [71]. In cardiac stem cell therapy, i.e. the delivery of stem cells to effect tissue repair or to treat disease, a key feature is that, rather than involving the transfer of isolated genes, it comprises the adoptive transfer of healthy cells which either directly replace the unhealthy cells, or operate indirectly through secretion of growth factors, paracrine factors, exosomes or even miRNAs to modulate immune response mechanisms and stimulate cardiac regeneration [72].

The main criteria that must be considered in developing a stem cell-based therapy include selection of the best population of cells, identification of the optimum mode of administration, and determination of the correct dose and timing of the treatment [72–74]. Stem cell therapy typically involves the use of pluripotent stem cells, or cells taken from tissues such as bone marrow, heart, or skeletal muscle [75–81]. Stem cells may be autologous, such as the mesenchymal stem cells and mononuclear bone marrow cells used in the TAC-HFT randomized clinical trial, which showed that these cells were safe for patients with ischemic cardiomyopathy and improved myocardial function [82]. The stem cells can also be allogeneic, such as the mesenchymal stem cells used in the POSEIDON randomized clinical trial, which showed that injection of these cells reduced scar size, improving the myocardial function in patients with chronic ischemic cardiomyopathy [83]. Allogeneic induced pluripotent stem cells (iPSCs) have also been used for cardiac repair [84]. The mode of delivery of stem cells is also important. In myocardial infarction, transendocardial stem cell injection reduced infarct size compared to other modes of delivery including intracoronary infusion, intramyocardial injection and intravenous infusion [74].

Several clinical trials have successfully used stem cells for treating heart disease, including the CADUCEUS trial, which showed that stem cell therapy decreased infarct size, and the SCIPIO trial, which was the first human phase 1 clinical trial employing autologous cardiac stem cells, showing local and global improvement in left ventricular function, reduction in scar size and an increase in viable tissue that persisted for one year with evidence of cardiac regeneration in patients with ischemic cardiomyopathy [80, 85]. Several studies have shown that bone marrow mesenchymal stem cells can be successfully used for the treatment of myocardial infarction [77, 86, 87]. Recently, human umbilical cord blood-derived mesenchymal stem cells were modified to carry the transgene lymphoid enhancer-binding factor 1 (LEF-1) integrated into the AAV integration site 1 locus by CRISPR/Cas9. Because of the AAV safe and stable integration site on chromosome 19, this approach resulted in stable transgene expression of LEF-1, resulting in cardioprotective effects in the case of ischemic heart diseases such as myocardial infarction, and decreasing fibrosis and infarct size [88].

Other strategies that have been explored include combining tissue engineering with iPSCs to improve heart function post-myocardial infarction. Polyethylene glycol hydrogel, iPSCs and erythropoietin in combination were introduced by injection into rat hearts after myocardial infarction, and after 10 weeks there was a decrease in ventricular remodeling, increased scar thickness and improved cardiac function, with this combination providing greater improvement than iPSCs alone [89]. Hydrogel injection did not induce arrythmias or negative effects on cardiac function. This tissue engineering strategy resulted in significantly lower loss of the grafted cells within 24 h as compared to the direct injection of cells suspended in liquid solution to the infarcted heart [90].

The main caveats associated with stem cell therapy include minimal evidence of differentiation of stem cells into cardiomyocytes, and a lack of long-term engraftment of the cells due to immune clearance. This problem of poor engraftment is associated with all varieties of cardiomyogenic cells, including cardiac precursors and cardiomyocytes derived from pluripotent stem cells, reprogrammed cells, and mesenchymal stem cells [91, 92]. In spite of these issues, stem cells have been found to promote cardiac precursor differentiation, endogenous cardiac repair, and immune modulation, and the grafted cells have been found to secrete pro-survival and pro-angiogenic signals through a probable paracrine mechanism which still remains poorly elucidated [93, 94]. Additional study is required to determine these unknown mechanisms, however the demonstrated beneficial effects ensure that stem cell therapy will remain a key emerging strategy for the treatment of myocardial infarction and other cardiac diseases.

The survival and retention of stem cells in the infarcted heart represent significant problems facing the adoption of stem cell treatment techniques. Recent studies suggest that the sequential delivery of mesenchymal stem cell-derived exosomes and bone marrow-derived stem cells has the potential to solve this problem. This approach improved cardiac function after myocardial infarction, reduced infarct size and inflammatory factors, promoted stem cell survival and increased angiogenesis and neovascularization significantly better than exosomes or bone marrow-derived stem cells applied separately [87].

IL-10 was reported to have anti-inflammatory, anti-apoptotic, and cardioprotective effects, increasing angiogenesis and resulting in a reduction in infarct size thereby ameliorating left ventricular dysfunction [95–97]. Improvement in cardiac function post-MI was observed when IL-10 was overexpressed in mesenchymal stem cells using adenovirus followed by their transplantation into the heart [95]. A more recent study has shown that IL-10-overexpressing bone marrow-derived stem cells, created using CRISPR/Cas9 methods and transplanted into the heart for the treatment of myocardial infarction in diabetic mice, exhibited greater sustained survival, enhanced angiogenesis and attenuated inflammation [98]. This study is particularly noteworthy since the diabetic environment has been found to decrease the functional efficiency of mesenchymal stem cells, affecting their reparative potency, including their ability to proliferate and induce angiogenesis [98, 99].

Another recent study, although not involving stem cells, is noteworthy as an alternative form of cardiac cell therapy. An editorial paper suggested that gene therapy can be used for reducing cardiac arrhythmias [100]. This suggestion was made based on a study in which human dermal fibroblasts underwent myogenic differentiation using a lentivirus vector to express the skeletal muscle myogenic determination factor, MyoD, and gap junction protein connexin 43, resulting in cellular electrical coupling [101]. This study showed that fibroblasts could be genetically modified to generate excitable cells, which can electrically couple and be used to repair defects in cardiac conduction.

Use of Gene Therapy Techniques for Treating Cardiac Fibrosis

Several studies have employed gene therapy techniques to treat cardiac fibrosis. Since raising high density lipoprotein (HDL) helps protect against heart failure, Apo-AI, the major apolipoprotein in HDL, was over-expressed via gene transfer using adeno-associated virus serotype 8 in mice subjected to transverse aortic constriction (TAC). Compared to control TAC mice, the TAC mice that received AAV8-mediated Apo-AI gene therapy showed significant improvement in cardiac function, reduced cardiac fibrosis and a reduction in hypertrophy [102].

Cardiac fibroblasts activated to the myofibroblast phenotype are the major players responsible for excessive deposition of extracellular matrix proteins in cardiac fibrosis. iPSCs were used to generate quiescent cardiac fibroblasts which could be used for testing drugs that could induce or inhibit fibrogenesis [103]. Using this approach, it was found that crosstalk between iPSC-derived cardiomyocytes and iPSC-derived cardiac fibroblasts via the atrial/brain natriuretic peptide receptor 1 pathway plays a major role in suppressing fibrosis, and thus can be targeted as a novel anti-fibrotic therapy [103].

Conclusion

Gene therapy shows remarkable promise for the effective treatment of cardiovascular diseases including heart failure. Recent gene therapy studies suggest that, rather than using a single strategy, a combination of strategies may yield better results via more efficient treatment, and additional studies will undoubtedly improve the effectiveness of these approaches. As noted above, both non-viral and viral delivery vectors for transgene delivery have their own associated advantages and disadvantages, and current studies will better elucidate which combination strategies involving non-viral, viral and stem cell-based therapies will be most effective against specific forms of cardiac disease. Clinical trials have already demonstrated the tremendous promise of such approaches, and new research in the field will help bring cardiac gene therapy to the frontline of care for patients.

References

1. Vinge LE, Raake PW, Koch WJ (2008) Gene therapy in heart failure. Circ Res 102:1458–1470
2. Raake PW, Tscheschner H, Reinkober J, Ritterhoff J, Katus HA, Koch WJ, Most P (2011) Gene therapy targets in heart failure: the path to translation. Clin Pharmacol Ther 90:542–553
3. Tilemann L, Ishikawa K, Weber T, Hajjar RJ (2012) Gene therapy for heart failure. Circ Res 110:777–793
4. Smith KR (2003) Gene therapy: theoretical and bioethical concepts. Arch Med Res 34:247–268
5. Kaye DM, Preovolos A, Marshall T, Byrne M, Hoshijima M, Hajjar R, Mariani JA, Pepe S, Chien KR, Power JM (2007) Percutaneous cardiac recirculation-mediated gene transfer of an inhibitory phospholamban peptide reverses advanced heart failure in large animals. J Am Coll Cardiol 50:253–260
6. del Monte F, Harding SE, Dec GW, Gwathmey JK, Hajjar RJ (2002) Targeting phospholamban by gene transfer in human heart failure. Circulation 105:904–907
7. Jiang MT, Narayanan N (1990) Effects of aging on phospholamban phosphorylation and calcium transport in rat cardiac sarcoplasmic reticulum. Mech Ageing Dev 54:87–101
8. Raake PW, Vinge LE, Gao E, Boucher M, Rengo G, Chen X, DeGeorge BR Jr, Matkovich S, Houser SR, Most P et al (2008) G protein-coupled receptor kinase 2 ablation in cardiac myocytes before or after myocardial infarction prevents heart failure. Circ Res 103:413–422
9. Katz MG, Fargnoli AS, Williams RD, Bridges CR (2013) Gene therapy delivery systems for enhancing viral and nonviral vectors for cardiac diseases: current concepts and future applications. Hum Gene Ther 24:914–927
10. Kolasinac R, Jaksch S, Dreissen G, Braeutigam A, Merkel R, Csiszar A (2019) Influence of environmental conditions on the fusion of cationic liposomes with living mammalian cells. Nanomaterials (Basel) 9
11. Connor J, Yatvin MB, Huang L (1984) pH-sensitive liposomes: acid-induced liposome fusion. Proc Natl Acad Sci U S A 81:1715–1718
12. Gabizon A, Papahadjopoulos D (1988) Liposome formulations with prolonged circulation time in blood and enhanced uptake by tumors. Proc Natl Acad Sci U S A 85:6949–6953
13. Duzgunes, Nir S (1999) Mechanisms and kinetics of liposome-cell interactions. Adv Drug Deliv Rev 40:3–18
14. Xu Y, Li X, Kong M, Jiang D, Dong A, Shen Z, Duan Q (2014) Cardiac-targeting magnetic lipoplex delivery of SH-IGF1R plasmid attenuate norepinephrine-induced cardiac hypertrophy in murine heart. Biosci Rep 34

15. Mathiyalagan P, Sahoo S (2017) Exosomes-based gene therapy for MicroRNA delivery. Methods Mol Biol 1521:139–152
16. Zhao L, Gu C, Gan Y, Shao L, Chen H, Zhu H (2020) Exosome-mediated siRNA delivery to suppress postoperative breast cancer metastasis. J Control Release 318:1–15
17. Chang, D.C. (1992). Guide to electroporation and electrofusion. Academic Press, San Diego
18. Hargrave B, Downey H, Strange R Jr, Murray L, Cinnamond C, Lundberg C, Israel A, Chen YJ, Marshall W Jr, Heller R (2013) Electroporation-mediated gene transfer directly to the swine heart. Gene Ther 20:151–157
19. Yang H, Wang H, Shivalila CS, Cheng AW, Shi L, Jaenisch R (2013) One-step generation of mice carrying reporter and conditional alleles by CRISPR/Cas-mediated genome engineering. Cell 154:1370–1379
20. Horii T, Arai Y, Yamazaki M, Morita S, Kimura M, Itoh M, Abe Y, Hatada I (2014) Validation of microinjection methods for generating knockout mice by CRISPR/Cas-mediated genome engineering. Sci Rep 4:4513
21. Hickman MA, Malone RW, Lehmann-Bruinsma K, Sih TR, Knoell D, Szoka FC, Walzem R, Carlson DM, Powell JS (1994) Gene expression following direct injection of DNA into liver. Hum Gene Ther 5:1477–1483
22. Andre FM, Cournil-Henrionnet C, Vernerey D, Opolon P, Mir LM (2006) Variability of naked DNA expression after direct local injection: the influence of the injection speed. Gene Ther 13:1619–1627
23. Lin H, Parmacek MS, Morle G, Bolling S, Leiden JM (1990) Expression of recombinant genes in myocardium in vivo after direct injection of DNA. Circulation 82:2217–2221
24. Buttrick PM, Kass A, Kitsis RN, Kaplan ML, Leinwand LA (1992) Behavior of genes directly injected into the rat heart in vivo. Circ Res 70:193–198
25. Gainer AL, Korbutt GS, Rajotte RV, Warnock GL, Elliott JF (1996) Successful biolistic transformation of mouse pancreatic islets while preserving cellular function. Transplantation 61:1567–1571
26. Sanford JC, Klein TM, Wolf ED, Allen N (1987) Delivery of substances into cells and tissues using a particle bombardment process. Part Sci Technol 5:27–37
27. Nishizaki K, Mazda O, Dohi Y, Kawata T, Mizuguchi K, Kitamura S, Taniguchi S (2000) in vivo gene gun-mediated transduction into rat heart with Epstein-Barr virus-based episomal vectors. Ann Thorac Surg 70:1332–1337
28. Matsuno Y, Iwata H, Umeda Y, Takagi H, Mori Y, Miyazaki J, Kosugi A, Hirose H (2003) Nonviral gene gun mediated transfer into the beating heart. ASAIO J 49:641–644
29. Davis J, Westfall MV, Townsend D, Blankinship M, Herron TJ, Guerrero-Serna G, Wang W, Devaney E, Metzger JM (2008) Designing heart performance by gene transfer. Physiol Rev 88:1567–1651
30. Wolff JA, Malone RW, Williams P, Chong W, Acsadi G, Jani A, Felgner PL (1990) Direct gene transfer into mouse muscle in vivo. Science 247:1465–1468
31. Vorburger SA, Hunt KK (2002) Adenoviral gene therapy. Oncologist 7:46–59
32. Stratford-Perricaudet LD, Makeh I, Perricaudet M, Briand P (1992) Widespread long-term gene transfer to mouse skeletal muscles and heart. J Clin Invest 90:626–630
33. Zoltick PW, Chirmule N, Schnell MA, Gao GP, Hughes JV, Wilson JM (2001) Biology of E1-deleted adenovirus vectors in nonhuman primate muscle. J Virol 75:5222–5229
34. Schiedner G, Morral N, Parks RJ, Wu Y, Koopmans SC, Langston C, Graham FL, Beaudet AL, Kochanek S (1998) Genomic DNA transfer with a high-capacity adenovirus vector results in improved in vivo gene expression and decreased toxicity. Nat Genet 18:180–183
35. Raper SE, Chirmule N, Lee FS, Wivel NA, Bagg A, Gao GP, Wilson JM, Batshaw ML (2003) Fatal systemic inflammatory response syndrome in a ornithine transcarbamylase deficient patient following adenoviral gene transfer. Mol Genet Metab 80:148–158
36. Atchison RW, Casto BC, Hammon WM (1965) Adenovirus-associated defective virus particles. Science 149:754–756
37. Xiao X, Li J, Samulski RJ (1996) Efficient long-term gene transfer into muscle tissue of immunocompetent mice by adeno-associated virus vector. J Virol 70:8098–8108

38. Afione SA, Conrad CK, Kearns WG, Chunduru S, Adams R, Reynolds TC, Guggino WB, Cutting GR, Carter BJ, Flotte TR (1996) In vivo model of adeno-associated virus vector persistence and rescue. J Virol 70:3235–3241
39. Fisher KJ, Jooss K, Alston J, Yang Y, Haecker SE, High K, Pathak R, Raper SE, Wilson JM (1997) Recombinant adeno-associated virus for muscle directed gene therapy. Nat Med 3:306–312
40. Samulski RJ, Zhu X, Xiao X, Brook JD, Housman DE, Epstein N, Hunter LA (1991) Targeted integration of adeno-associated virus (AAV) into human chromosome 19. EMBO J 10:3941–3950
41. Wu Z, Yang H, Colosi P (2010) Effect of genome size on AAV vector packaging. Mol Ther 18:80–86
42. Gregorevic P, Blankinship MJ, Allen JM, Crawford RW, Meuse L, Miller DG, Russell DW, Chamberlain JS (2004) Systemic delivery of genes to striated muscles using adeno-associated viral vectors. Nat Med 10:828–834
43. Vandendriessche T, Thorrez L, Acosta-Sanchez A, Petrus I, Wang L, Ma L, L, DEW, Iwasaki Y, Gillijns V, Wilson JM et al (2007) Efficacy and safety of adeno-associated viral vectors based on serotype 8 and 9 vs. lentiviral vectors for hemophilia B gene therapy. J Thromb Haemost 5:16–24
44. Kawada T, Nakazawa M, Nakauchi S, Yamazaki K, Shimamoto R, Urabe M, Nakata J, Hemmi C, Masui F, Nakajima T et al (2002) Rescue of hereditary form of dilated cardiomyopathy by rAAV-mediated somatic gene therapy: amelioration of morphological findings, sarcolemmal permeability, cardiac performances, and the prognosis of TO-2 hamsters. Proc Natl Acad Sci U S A 99:901–906
45. Mearini G, Stimpel D, Geertz B, Weinberger F, Kramer E, Schlossarek S, Mourot-Filiatre J, Stoehr A, Dutsch A, Wijnker PJ et al (2014) Mybpc3 gene therapy for neonatal cardiomyopathy enables long-term disease prevention in mice. Nat Commun 5:5515
46. Greenberg B, Butler J, Felker GM, Ponikowski P, Voors AA, Desai AS, Barnard D, Bouchard A, Jaski B, Lyon AR et al (2016) Calcium upregulation by percutaneous administration of gene therapy in patients with cardiac disease (CUPID 2): a randomised, multinational, double-blind, placebo-controlled, phase 2b trial. Lancet 387:1178–1186
47. Armeanu S, Ungerechts G, Bernloehr C, Bossow S, Gregor M, Neubert WJ, Lauer UM, Bitzer M (2003) Cell cycle independent infection and gene transfer by recombinant Sendai viruses. J Virol Methods 108:229–233
48. Cao Y, Xu J, Wen J, Ma X, Liu F, Li Y, Chen W, Sun L, Wu Y, Li S et al (2018) Generation of a urine-derived Ips cell line from a patient with a ventricular septal defect and heart failure and the robust differentiation of these cells to cardiomyocytes via small molecules. Cell Physiol Biochem 50:538–551
49. Guo F, Sun Y, Wang X, Wang H, Wang J, Gong T, Chen X, Zhang P, Su L, Fu G et al (2019) Patient-specific and gene-corrected induced pluripotent stem cell-derived cardiomyocytes elucidate single-cell phenotype of short QT syndrome. Circ Res 124:66–78
50. Laufs S, Gentner B, Nagy KZ, Jauch A, Benner A, Naundorf S, Kuehlcke K, Schiedlmeier B, Ho AD, Zeller WJ et al (2003) Retroviral vector integration occurs in preferred genomic targets of human bone marrow-repopulating cells. Blood 101:2191–2198
51. Morgan RA, Couture L, Elroy-Stein O, Ragheb J, Moss B, Anderson WF (1992) Retroviral vectors containing putative internal ribosome entry sites: development of a polycistronic gene transfer system and applications to human gene therapy. Nucleic Acids Res 20:1293–1299
52. Miller DG, Adam MA, Miller AD (1990) Gene transfer by retrovirus vectors occurs only in cells that are actively replicating at the time of infection. Mol Cell Biol 10:4239–4242
53. Hashimoto H, Wang Z, Garry GA, Malladi VS, Botten GA, Ye W, Zhou H, Osterwalder M, Dickel DE, Visel A et al (2019) Cardiac reprogramming factors synergistically activate genome-wide cardiogenic stage-specific enhancers. Cell Stem Cell 25:69–86 e65
54. Somers A, Jean JC, Sommer CA, Omari A, Ford CC, Mills JA, Ying L, Sommer AG, Jean JM, Smith BW et al (2010) Generation of transgene-free lung disease-specific human induced pluripotent stem cells using a single excisable lentiviral stem cell cassette. Stem Cells 28:1728–1740

55. Sommer CA, Stadtfeld M, Murphy GJ, Hochedlinger K, Kotton DN, Mostoslavsky G (2009) Induced pluripotent stem cell generation using a single lentiviral stem cell cassette. Stem Cells 27:543–549
56. Nightingale SJ, Hollis RP, Pepper KA, Petersen D, Yu XJ, Yang C, Bahner I, Kohn DB (2006) Transient gene expression by nonintegrating lentiviral vectors. Mol Ther 13:1121–1132
57. Biffi A, De Palma M, Quattrini A, Del Carro U, Amadio S, Visigalli I, Sessa M, Fasano S, Brambilla R, Marchesini S et al (2004) Correction of metachromatic leukodystrophy in the mouse model by transplantation of genetically modified hematopoietic stem cells. J Clin Invest 113:1118–1129
58. Montini E, Cesana D, Schmidt M, Sanvito F, Ponzoni M, Bartholomae C, Sergi Sergi L, Benedicenti F, Ambrosi A, Di Serio C et al (2006) Hematopoietic stem cell gene transfer in a tumor-prone mouse model uncovers low genotoxicity of lentiviral vector integration. Nat Biotechnol 24:687–696
59. Throm RE, Ouma AA, Zhou S, Chandrasekaran A, Lockey T, Greene M, De Ravin SS, Moayeri M, Malech HL, Sorrentino BP et al (2009) Efficient construction of producer cell lines for a SIN lentiviral vector for SCID-X1 gene therapy by concatemeric array transfection. Blood 113:5104–5110
60. Fleury S, Simeoni E, Zuppinger C, Deglon N, von Segesser LK, Kappenberger L, Vassalli G (2003) Multiply attenuated, self-inactivating lentiviral vectors efficiently deliver and express genes for extended periods of time in adult rat cardiomyocytes in vivo. Circulation 107:2375–2382
61. Bonci D, Cittadini A, Latronico MV, Borello U, Aycock JK, Drusco A, Innocenzi A, Follenzi A, Lavitrano M, Monti MG et al (2003) Advanced' generation lentiviruses as efficient vectors for cardiomyocyte gene transduction in vitro and in vivo. Gene Ther 10:630–636
62. Lin Y, Wu J, Gu W, Huang Y, Tong Z, Huang L, Tan J (2018) Exosome-liposome hybrid nanoparticles deliver CRISPR/Cas9 system in MSCs. Adv Sci (Weinh) 5:1700611
63. Ma H, Marti-Gutierrez N, Park SW, Wu J, Lee Y, Suzuki K, Koski A, Ji D, Hayama T, Ahmed R et al (2017) Correction of a pathogenic gene mutation in human embryos. Nature 548:413–419
64. Troder SE, Ebert LK, Butt L, Assenmacher S, Schermer B, Zevnik B (2018) An optimized electroporation approach for efficient CRISPR/Cas9 genome editing in murine zygotes. PLoS One 13:e0196891
65. Lee K, Conboy M, Park HM, Jiang F, Kim HJ, Dewitt MA, Mackley VA, Chang K, Rao A, Skinner C et al (2017) Nanoparticle delivery of Cas9 ribonucleoprotein and donor DNA in vivo induces homology-directed DNA repair. Nat Biomed Eng 1:889–901
66. Mout R, Ray M, Yesilbag Tonga G, Lee YW, Tay T, Sasaki K, Rotello VM (2017) Direct cytosolic delivery of CRISPR/Cas9-ribonucleoprotein for efficient gene editing. ACS Nano 11:2452–2458
67. Long C, Amoasii L, Mireault AA, McAnally JR, Li H, Sanchez-Ortiz E, Bhattacharyya S, Shelton JM, Bassel-Duby R, Olson EN (2016) Postnatal genome editing partially restores dystrophin expression in a mouse model of muscular dystrophy. Science 351:400–403
68. Carroll KJ, Makarewich CA, McAnally J, Anderson DM, Zentilin L, Liu N, Giacca M, Bassel-Duby R, Olson EN (2016) A mouse model for adult cardiac-specific gene deletion with CRISPR/Cas9. Proc Natl Acad Sci U S A 113:338–343
69. Schuh RS, Gonzalez EA, Tavares AMV, Seolin BG, Elias LS, Vera LNP, Kubaski F, Poletto E, Giugliani R, Teixeira HF et al (2020) Neonatal nonviral gene editing with the CRISPR/Cas9 system improves some cardiovascular, respiratory, and bone disease features of the mucopolysaccharidosis I phenotype in mice. Gene Ther 27:74–84
70. Johansen AK, Molenaar B, Versteeg D, Leitoguinho AR, Demkes C, Spanjaard B, de Ruiter H, Akbari Moqadam F, Kooijman L, Zentilin L et al (2017) Postnatal cardiac gene editing using CRISPR/Cas9 With AAV9-mediated delivery of short guide RNAs results in mosaic gene disruption. Circ Res 121:1168–1181
71. Malaver-Ortega LF, Sumer H, Liu J, Verma PJ (2012) The state of the art for pluripotent stem cells derivation in domestic ungulates. Theriogenology 78:1749–1762

72. Bagno L, Hatzistergos KE, Balkan W, Hare JM (2018) Mesenchymal stem cell-based therapy for cardiovascular disease: progress and challenges. Mol Ther 26:1610–1623
73. Florea V, Rieger AC, DiFede DL, El-Khorazaty J, Natsumeda M, Banerjee MN, Tompkins BA, Khan A, Schulman IH, Landin AM et al (2017) Dose comparison study of allogeneic mesenchymal stem cells in patients with ischemic cardiomyopathy (The TRIDENT Study). Circ Res 121:1279–1290
74. Kanelidis AJ, Premer C, Lopez J, Balkan W, Hare JM (2017) Route of delivery modulates the efficacy of mesenchymal stem cell therapy for myocardial infarction: a meta-analysis of preclinical studies and clinical trials. Circ Res 120:1139–1150
75. Menasche P, Vanneaux V, Hagege A, Bel A, Cholley B, Cacciapuoti I, Parouchev A, Benhamouda N, Tachdjian G, Tosca L et al (2015) Human embryonic stem cell-derived cardiac progenitors for severe heart failure treatment: first clinical case report. Eur Heart J 36:2011–2017
76. Menasche P, Vanneaux V, Fabreguettes JR, Bel A, Tosca L, Garcia S, Bellamy V, Farouz Y, Pouly J, Damour O et al (2015) Towards a clinical use of human embryonic stem cell-derived cardiac progenitors: a translational experience. Eur Heart J 36:743–750
77. Stamm C, Westphal B, Kleine HD, Petzsch M, Kittner C, Klinge H, Schumichen C, Nienaber CA, Freund M, Steinhoff G (2003) Autologous bone-marrow stem-cell transplantation for myocardial regeneration. Lancet 361:45–46
78. Lunde K, Solheim S, Aakhus S, Arnesen H, Abdelnoor M, Egeland T, Endresen K, Ilebekk A, Mangschau A, Fjeld JG et al (2006) Intracoronary injection of mononuclear bone marrow cells in acute myocardial infarction. N Engl J Med 355:1199–1209
79. Malliaras K, Makkar RR, Smith RR, Cheng K, Wu E, Bonow RO, Marban L, Mendizabal A, Cingolani E, Johnston PV et al (2014) Intracoronary cardiosphere-derived cells after myocardial infarction: evidence of therapeutic regeneration in the final 1-year results of the CADUCEUS trial (CArdiosphere-Derived aUtologous stem CElls to reverse ventricUlar dySfunction). J Am Coll Cardiol 63:110–122
80. Chugh AR, Beache GM, Loughran JH, Mewton N, Elmore JB, Kajstura J, Pappas P, Tatooles A, Stoddard MF, Lima JA et al (2012) Administration of cardiac stem cells in patients with ischemic cardiomyopathy: the SCIPIO trial: surgical aspects and interim analysis of myocardial function and viability by magnetic resonance. Circulation 126:S54-64
81. Sawa Y, Yoshikawa Y, Toda K, Fukushima S, Yamazaki K, Ono M, Sakata Y, Hagiwara N, Kinugawa K, Miyagawa S (2015) Safety and efficacy of autologous skeletal myoblast sheets (TCD-51073) for the treatment of severe chronic heart failure due to ischemic heart disease. Circ J 79:991–999
82. Heldman AW, DiFede DL, Fishman JE, Zambrano JP, Trachtenberg BH, Karantalis V, Mushtaq M, Williams AR, Suncion VY, McNiece IK et al (2014) Transendocardial mesenchymal stem cells and mononuclear bone marrow cells for ischemic cardiomyopathy: the TAC-HFT randomized trial. JAMA 311:62–73
83. Suncion VY, Ghersin E, Fishman JE, Zambrano JP, Karantalis V, Mandel N, Nelson KH, Gerstenblith G, DiFede Velazquez DL, Breton E et al (2014) Does transendocardial injection of mesenchymal stem cells improve myocardial function locally or globally?: an analysis from the percutaneous stem cell injection delivery effects on neomyogenesis (POSEIDON) randomized trial. Circ Res 114:1292–1301
84. Shiba Y, Gomibuchi T, Seto T, Wada Y, Ichimura H, Tanaka Y, Ogasawara T, Okada K, Shiba N, Sakamoto K et al (2016) Allogeneic transplantation of iPS cell-derived cardiomyocytes regenerates primate hearts. Nature 538:388–391
85. Kisiel AH, McDuffee LA, Masaoud E, Bailey TR, Esparza Gonzalez BP, Nino-Fong R (2012) Isolation, characterization, and in vitro proliferation of canine mesenchymal stem cells derived from bone marrow, adipose tissue, muscle, and periosteum. Am J Vet Res 73:1305–1317
86. Yoon YS, Wecker A, Heyd L, Park JS, Tkebuchava T, Kusano K, Hanley A, Scadova H, Qin G, Cha DH et al (2005) Clonally expanded novel multipotent stem cells from human bone marrow regenerate myocardium after myocardial infarction. J Clin Invest 115:326–338

87. Huang P, Wang L, Li Q, Xu J, Xu J, Xiong Y, Chen G, Qian H, Jin C, Yu Y et al (2019) Combinatorial treatment of acute myocardial infarction using stem cells and their derived exosomes resulted in improved heart performance. Stem Cell Res Ther 10:300
88. Cho HM, Lee KH, Shen YM, Shin TJ, Ryu PD, Choi MC, Kang KS, Cho JY (2020) Transplantation of hMSCs genome edited with LEF1 improves cardio-protective effects in myocardial infarction. Mol Ther Nucleic Acids 19:1186–1197
89. Chow A, Stuckey DJ, Kidher E, Rocco M, Jabbour RJ, Mansfield CA, Darzi A, Harding SE, Stevens MM, Athanasiou T (2017) Human induced pluripotent stem cell-derived cardiomyocyte encapsulating bioactive hydrogels improve rat heart function post myocardial infarction. Stem Cell Reports 9:1415–1422
90. Sheikh AY, Lin SA, Cao F, Cao Y, van der Bogt KE, Chu P, Chang CP, Contag CH, Robbins RC, Wu JC (2007) Molecular imaging of bone marrow mononuclear cell homing and engraftment in ischemic myocardium. Stem Cells 25:2677–2684
91. Fernandes S, Chong JJH, Paige SL, Iwata M, Torok-Storb B, Keller G, Reinecke H, Murry CE (2015) Comparison of human embryonic stem cell-derived cardiomyocytes, cardiovascular progenitors, and bone marrow mononuclear cells for cardiac repair. Stem Cell Reports 5:753–762
92. Lalit PA, Salick MR, Nelson DO, Squirrell JM, Shafer CM, Patel NG, Saeed I, Schmuck EG, Markandeya YS, Wong R et al (2016) Lineage reprogramming of fibroblasts into proliferative induced cardiac progenitor cells by defined factors. Cell Stem Cell 18:354–367
93. Ong SG, Huber BC, Lee WH, Kodo K, Ebert AD, Ma Y, Nguyen PK, Diecke S, Chen WY, Wu JC (2015) Microfluidic single-cell analysis of transplanted human induced pluripotent stem cell-derived cardiomyocytes after acute myocardial infarction. Circulation 132:762–771
94. Duran JM, Makarewich CA, Sharp TE, Starosta T, Zhu F, Hoffman NE, Chiba Y, Madesh M, Berretta RM, Kubo H et al (2013) Bone-derived stem cells repair the heart after myocardial infarction through transdifferentiation and paracrine signaling mechanisms. Circ Res 113:539–552
95. Meng X, Li J, Yu M, Yang J, Zheng M, Zhang J, Sun C, Liang H, Liu L (2018) Transplantation of mesenchymal stem cells overexpressing IL10 attenuates cardiac impairments in rats with myocardial infarction. J Cell Physiol 233:587–595
96. Kaur K, Sharma AK, Singal PK (2006) Significance of changes in TNF-alpha and IL-10 levels in the progression of heart failure subsequent to myocardial infarction. Am J Physiol Heart Circ Physiol 291:H106-113
97. Krishnamurthy P, Rajasingh J, Lambers E, Qin G, Losordo DW, Kishore R (2009) IL-10 inhibits inflammation and attenuates left ventricular remodeling after myocardial infarction via activation of STAT3 and suppression of HuR. Circ Res 104:e9-18
98. Meng X, Zheng M, Yu M, Bai W, Zuo L, Bu X, Liu Y, Xia L, Hu J, Liu L et al (2019) Transplantation of CRISPRa system engineered IL10-overexpressing bone marrow-derived mesenchymal stem cells for the treatment of myocardial infarction in diabetic mice. J Biol Eng 13:49
99. Kim H, Han JW, Lee JY, Choi YJ, Sohn YD, Song M, Yoon YS (2015) Diabetic mesenchymal stem cells are ineffective for improving limb ischemia due to their impaired angiogenic capability. Cell Transplant 24:1571–1584
100. Allessie M, Schotten U, Verheule S, Harks E (2005) Gene therapy for repair of cardiac fibrosis: a long way to Tipperary. Circulation 111:391–393
101. Kizana E, Ginn SL, Allen DG, Ross DL, Alexander IE (2005) Fibroblasts can be genetically modified to produce excitable cells capable of electrical coupling. Circulation 111:394–398
102. Amin R, Muthuramu I, Aboumsallem JP, Mishra M, Jacobs F, De Geest B (2017) Selective HDL-raising human Apo A-I gene therapy counteracts cardiac hypertrophy, reduces myocardial fibrosis, and improves cardiac function in mice with chronic pressure overload. Int J Mol Sci 18
103. Zhang H, Tian L, Shen M, Tu C, Wu H, Gu M, Paik DT, Wu JC (2019) Generation of quiescent cardiac fibroblasts from human induced pluripotent stem cells for in vitro modeling of cardiac fibrosis. Circ Res 125:552–566

Chapter 9
Circadian Regulation of Autophagy in the Heart Via the mTOR Pathway

Matthew Love, Inna Rabinovich-Nikitin, and Lorrie A. Kirshenbaum

Abstract Disruption of the normal circadian clock has been associated with greater incidence of cardiovascular disease in shift workers. While the underlying mechanisms for this phenomenon are poorly understood, recent evidence from our laboratory has identified a novel signalling axis that functionally connects the mechanistic target of rapamycin (mTOR) to circadian biology in cardiac myocytes. The mTOR pathway regulates several processes such as cell growth, metabolism, and homeostasis throughout the body. mTOR has been linked to the cardiovascular systems through its regulation of both physiological and pathological processes, making it a desirable suspect for the investigating of its role in cardiovascular disease emanating from circadian dysfunction. Herein, we review the relevant literature highlighting the interworking of circadian interaction with metabolism including nutrient stress and autophagy within the cardiovascular system. We hope to use this information to spark interest in the potential for circadian intervened therapies designed for improving cardiovascular metabolism.

Keywords Circadian · Clock · Cardiac myocytes · Autophagy · Mitochondria · mTOR

M. Love · I. Rabinovich-Nikitin · L. A. Kirshenbaum (✉)
Department of Physiology and Pathophysiology, The Institute of Cardiovascular Sciences, St. Boniface Hospital Albrechtsen Research Centre, Winnipeg, MB R2H2A6, Canada
e-mail: Lorrie@sbrc.ca

L. A. Kirshenbaum
Department of Pharmacology and Therapeutics, Rady College of Medicine, Max Rady Faculty of Health Sciences, University of Manitoba, Winnipeg, MB R2H2A6, Canada

© Springer Nature Switzerland AG 2022

L. A. Kirshenbaum (ed.), *Biochemistry of Apoptosis and Autophagy*, Advances in Biochemistry in Health and Disease 18,
https://doi.org/10.1007/978-3-030-78799-8_9

Introduction

Circadian and the Cardiovascular System

The circadian clock is responsible for regulating various cellular, metabolic, and physiological processes associated with transitioning between sleep/wake periods. Stimuli such as light, nutrient availability, and other environmental cues can lead to changes in circadian driven proteins that are found in virtually every cell type in the body [8, 68, 74, 93, 114]. In mammals, the circadian clock is regulated centrally by the hypothalamic pacemaker found in the suprachiasmatic nucleus (SCN) and coordinates circadian oscillations peripherally to other systems, [68]. A key component of the molecular circadian machinery is the heterodimerization of two basic helix-loop-helix domain proteins that include the Circadian locomotor output cycles kaput (Clock) and brain and muscle Arnt-like protein 1 (BMAL1). The Clock:BMAL1 dimer allows the complex to bind to E- box elements on certain promotors and transcriptionally activate circadian output genes [45, 68]. This drive a transcriptional core-clock network (CCN) comprised of other circadian regulators that include Period (Per) and Cryptochrome (Cry) proteins which serve as a negative feedback loop on the Clock:BMAL1 dimer, supressing their activity. The Clock:BMAL1 complex also activates other circadian genes such as ROR and REV-ERB and similarly feedback to regulate the expression of BMAL1. While the binding of REV-ERB to BMAL1 represses its function ROR activates Bmal [45, 68]. Disruption of this intricate feedback cycle such as during jet lag [6], shift work (Knutsson et al. 1986), or sleep disorders [13, 54] can be detrimental and contribute to the pathophysiology of cancer, cell metabolism, and the cardiovascular disease.

It is well known that components of the cardiovascular system including vascular smooth muscle, aorta and endothelial cells possess internal clocks that in addition to receiving input from the SCN. Previous research has pointed towards this strong association between circadian clock and the cardiovascular system. Most notably, their relationship is highlighted by the circadian influence on blood pressure (BP) and heart rate (HR); where both parameters peak in the morning and progressively fall throughout the day [29, 115]. This association is important physiologically because individuals who deviate from this diurnal oscillation have reported increased risk of cardiovascular diseases (CVD's) [52, 115, 118]. Circadian genes such as BMAL1 and Cry were shown to cause BP levels to fluctuate and the deletion of PPAR γ, a known activator of BMAL1 resulted in reduced diurnal variation in heart rate [26, 122]. Additionally, myocardial infarction (MI) and other CVD's such as stroke, arrhythmias, and heart failure are more likely to occur in the morning partly due to the circadian controlled genetic and immune responses at that time of day [9]. MI in murine models strengthen the understanding of the circadian machinery in the heart by showing that short term disruption of diurnal rhythms worsens cardiac outcomes resulting in increased scarring and left ventricular dysfunction, ultimately leading to decline in ejection fraction when compared to normal mice [3]. Appropriate inflammatory responses allow for removal of dead tissue and remodelling of myocardium, however, the diurnal disruption leads to an altered inflammatory responses and wound healing which triggers maladaptive cardiac remodelling [36, 74]. Therefore, the importance of a normal intact circadian rhythm is necessary for prevention of CVD's.

Another CVD of interest to circadian rhythm is Ischemia reperfusion (IR) injury, a condition where coronary artery flow is compromised resulting in decreased oxygen delivery to the heart muscle. This causes tissue hypoxia that ultimately leads to cardiac dysfunction from increased cardiac cell death [20, 85]. Despite promising therapies for treating IR injury such as revascularization and cardiac catherization, the risk for adverse cardiac remodelling is extremely high and may lead to heart failure [104]. One of the major underlying causes of cardiac dysfunction is cardiomyocyte cell death. While there are many types of programmed cell death, several involve biological cues from the mitochondria—a highly dynamic organelle which constantly undergoes morphological changes by highly coordinated fusion and fission [105]. Mitochondria change their structural morphology depending on nutrient and ATP availability, with long elongated mitochondria associated with abundant ATP from oxidative metabolism, fragmented mitochondria exhibited a reduced pool of cellular ATP and impaired mitochondrial bioenergetics [37]. Under normal physiological conditions mitochondria oxidize glucose and fatty acids to produce ATP essential for continuous contraction of cardiac myocytes, however CVD's such as IR disrupt calcium homeostasis resulting in the production of reactive oxygen species (ROS), mitochondrial dysfunction and mitochondria mediated cell death [10, 113].

The mTOR Pathway and the Cardiovascular System

A key player that links mitochondrial function and circadian rhythm is the mechanistic Target Of Rapamycin (mTOR) protein which is an evolutionary conserved serine/threonine kinase involved in cellular metabolism and growth [101, 106]. mTOR forms two multiprotein complexes: mTOR complex 1 (mTORC1) and mTOR complex 2 (mTORC2) [102]. These two complexes contain mTOR, mammalian lethal with sec13 protein 8 (mLST8) [57], the inhibitory DEP domain containing mTOR interacting protein (DEPTOR) [82] and Tel two interacting protein 1 (Tel 2) [53]. mTORC1 contains the regulator associated protein of mammalian target of rapamycin (Raptor) [42, 57] and Proline-rich AKT substrate of 40 kDa (PRAS 40 [98] while mTORC2 has the rapamycin insensitive companion of mTOR (Rictor) [100], mammalian stress activated MAP kinase-interacting protein 1 (mSIN1), and Proteins observed with rictor 1 and 2 (PROTOR1&2) [102]. These complexes both regulate anabolic processes such as protein, nucleotide, and lipid synthesis while inhibition of the mTOR complexes leads to catabolic pathways such as autophagy [101, 125].

There are several proteins that interact with the mTOR complex including protein kinase B (Akt), AMP protein activated kinase (AMPK), tuberous sclerosis protein

1/2 (TSC-1/2), and Ras homolog enriched in brain (Rheb). As seen in Fig. 9.1, both AMPK and TSC-1/2 regulate mTOR activity in response to changes in nutrient availability, hypoxia, DNA damage or cytokines. Specifically, TSC2 was shown to inhibit the small GTPase lysosomal membrane protein Rheb, from activating mTORC1 in cardiomyocytes [48, 66, 73]. When AMPK is active, it not only inactivates mTOR by phosphorylating TSC-2, but it also inhibits RAPTOR—a critical positive regular of mTOR [40, 50]. Conversely, both phosphorylation of Akt and presence of growth factors dissociate TSC1/2 from the lysosomal membrane therefore inhibiting TSC1/2 and allowing Rheb to activate mTOR and cell growth [63, 64, 73]. Below, we highlight the cellular dysfunction upon manipulation of up and downstream mTOR targets Fig. 9.1.

Important downstream targets of mTOR are S6-kinase1 (S6K1) and elF4E binding protein 1 (4E-BP1) which promote growth when activated [125]). In the heart, mTORC1 phosphorylates 4E-BP1, thus inhibiting interaction with initiation factor 4E (elF4E) which results in reduced protein synthesis along with less 4E-BP1 accumulation [62, 131]. In contrast, when mTOR phosphorylates S6K1, cell growth is promoted by activating eIF4E, leading to protein translation while programmed

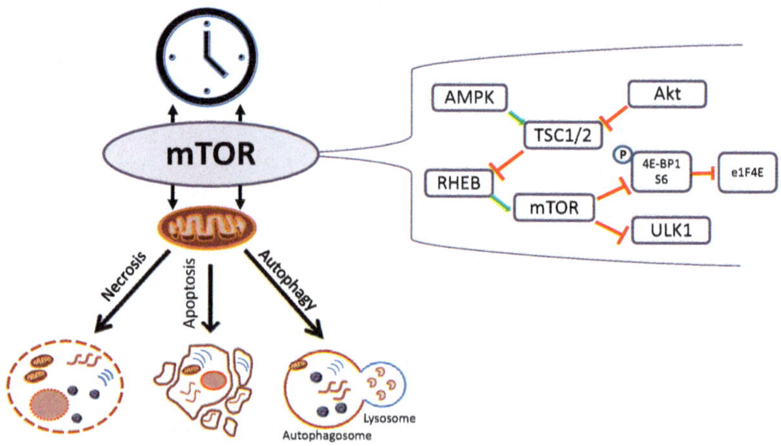

Fig. 9.1 Hypothesized mechanism linking circadian rhythm to cardiovascular disease mediated through the mTOR pathway. Circadian mechanisms manipulate the mitochondrial metabolism through the mTOR pathway by interacting with upstream targets AMPK or AKT, leading to cell death by necrosis, apoptosis or autophagy. AMPK activates TSC1/2 which inhibits RHEB from activating mTOR promoting autophagy through ULK1 while reducing translation by phosphorylating 4E-BP1 and S6 activity. This interferes with the association of 4E-BP1 and e1F4E. AKT can inhibit TSC1/2 preventing it from inhibiting RHEB which activates mTOR leading to increased protein translation from triggering of 4E-BP1 and S6. Mechanistic Target Of Rapamycin (mTOR), AMP Activated Protein Kinase (AMPK), Protein Kinase B (AKT), Tuberous Sclerosis Protein – 1/2 (TSC-1/2), Ras Homolog Enriched In Brain (RHEB), Eukaryotic Translation Initiation Factor 4E Binding Protein 1 (4E-BP1),Ribosomal S6 Kinase (S6), Eukaryotic translation initiation factor 4E (e1F4E), Unc- 51Like Autophagy Activating Kinase (ULK1)

cell death 4 (PDCD4) is inhibited [28]. It is also noteworthy, that cardiomyocytes undergoing cellular stress from hypoxia or nutrient deprivation experience reduced mTOR expression. It was shown that nutrient stress increased AMPK levels which was reported to interact with mTORC1 directly through RAPTOR or indirectly via TSC-1/2, thereby inhibiting mTOR activity [40, 49]. During oxidative stress in cardiomyocytes it was recently shown that Thioredoxin-1 bindingto mTOR reduces its activity, while simultaneously maintaining the complex for cell survival [83]. Due to the important role for mTOR and its effector proteins in regulating cell growth and autophagy, it is not surprisingthat cardiomyocytes heavily rely on this complex for normal cardiac development and homeostatic function.

Both mTORC1 and mTORC2 are identified as vital contributors of cardiac and vascular growth starting at embryonic development. Notably, complete genetic ablation of mTOR in the mouse resulted in cardiovascular developmental abnormalities and metabolic defects in utero [118]. Cardio-specific deletion of mTOR and Raptor resulted in similar cardiac defects in adult mouse hearts that included sarcomere disorganization, apoptosis, and autophagy [131]. Cardiomyopathies were also found when Rheb was deleted from these mice resulting in mortality after 10 days, because of loss of fundamental protein synthesis production in cardiomyocytes [131]. Simultaneous deletion of 4E-BP1 rescued the catastrophic phenotype seen with deletion of mTOR, RAPTOR, and Rheb demonstrating that the mTOR pathway was significantly required for normal cardiac development [116, 131]. Postnatal mice with cardiac mTOR deletion also displayed cardiac dilation, fibrosis, and apoptosis leading to heart failure with the mice dying within three weeks [71]. These data show that genetic disruption of the mTOR pathway during development or in post-natal heart triggers metabolic and aberrant defects in cardiac cell growth.

Although complete genetic ablation of the mTOR complex is incompatible with both normal cardiac development and cardiac function in adulthood, partial ablation of mTOR has been suggested to be beneficial to the cardiovascular system [123, 124]. This is because the partial inhibition of mTOR allows for less hypertrophy inducing proteins to be synthesized , leading to a decrease in energy expenditure, misfolded proteins and activation of autophagy [102]. These factors have been linked to a decrease in cardiac aging by lowering heart inflammation and fibroblast accumulation [35]. The beneficial effects of partial ablation could be reversed when both mTOR and Rheb were overexpressed due to their effects on on the activation of growth dependent gene expression and sarcomere assembly proteins in cardiomyocytes [133]. This implies that partial inhibition of the mTOR pathway is an adaptive response in cardiomyocytes by upholding key mechanisms such as energy preservation, and proper protein folding leading to cardiac cell survival, and autophagy.

Circadian Mediated Mitochondrial Autophagy in the Cardiovascular System

In the heart, autophagy is necessary for maintaining cardiac homeostasis, function, and cell survival. [107]. Autophagy is a catabolic process that permits cells to re-cycle and/or discard damaged organelles, and macromolecular structures during metabolic stress imposed by CVDs such as IR injury, MI, and heart failure [33, 47, 69]. Autophagy is activated in response to nutrient stress or when cellular ATP falls below a threshold resulting in AMP/ADP accumulation. The catabolism of damaged macromolecules through this highly specialized lysosomal regulated process allows for amino acids, fatty acids, and carbohydrates to be re-cycled as an energy source to maintain cellular ATP levels [81]. This is especially important to adult cardiomyocytes when taking into account their lost replicative ability and their role in cardiac contraction. Cardiomyocyte use ATP for contraction and therefore, during starvation autophagy becomes a vital process to determine whether cells have sufficient energy to survive nutrient deprivation [46, 94]. Although autophagy is viewed as a protective mechanism, excessive or de-regulated autophagy can be detrimental and exacerbate injury resulting in cell death [134].

Autophagy is a highly regulated process that involves the lysosomal digestion of autophagosomes and their cargo. Autophagosomes are formed with the activation and recruitment of unc-51-like-autophagy activating kinase (ULK) to cellular cargo destined for removal, that creates the phagophore structure—the outer membrane of the eventual autophagosome. ULK1 in particular phosphorylates Beclin 1 activating the Vps34 kinase complex leading to formation of the developing autophagosome. Autophagy-related genes (ATG) allow the autophagosome to mature and subsequently fuse with the lysosome associated N-ethylmaleimide-sensitive fusion proteins attachment protein receptor (SNARE) to form the autolysosome.

Notably, mTOR inhibits autophagy by phosphorylating ULK1, thereby inhibiting the downstream activation of Vsp34 complex and autophagosome formation. Conversely, when mTOR is inactive, such as during nutrient stress conditions, autophagosomes readily form and fuse with lysosomes with help from autophagic machinery, including the Transport of transcript factor EB (TFEB) which drives lysosome biogenesis [83]. TFEB's role in autophagic clearance has been highlighted by its importance as a therapeutic target for lysosomal storage disorders, accumulation of cell toxins, and undigested autophagosomes [113]. In the presence of nutrient abundance, mTOR phosphorylates TFEB and retains it in the cytoplasm, thereby inhibiting lysosomal synthesis and autophagy [109]. In contrast, starvation inactivates mTOR allowing nuclear translocation of TFEB and lysosomal synthesis to occur which promote autophagy [83]. In this way, mTOR plays a central role in autophagy by regulating not only the formation of the autophagosomes but also lysosomal biogenesis.

In mammalian tissue, light/dark cycles provide physiological cues that help regulate energy metabolism such as gluconeogenesis [88], lipogenesis [44], and cholesterol biosynthesis [31]. Autophagy is another circadian controlled mechanism, through its ability to degrade protein aggregates, remove damaged organelles, and suppress the effects of invading pathogens through immune responses [75]. This connection recieved tremendous attention in the 1970s, when electron microscopy was used to discover how autophagic vacuoles varied throughout the day [86]. In a further study, the diurnal cycle of vacuole formation and density, was found to peak in the heart during the late phase of the light period and fell towards the early phase of the dark period [87]. These observations were further supported by recent data from our lab showing that the Clock gene transcriptionally coordinates mitochondrial autophagy during myocardial ischemia by directly controlling transcription of genes required for mitochondrial fission, fusion and quality control, demonstrating that Clock regulates an adaptive stress response critical for cardiac cell survival [91]. The energy sensing kinase AMPK is another mechanism that functionally links Clock and cell metabolism. As previously mentioned, AMPK is an mTOR inhibitor, meaning that it promotes catabolic mechanisms such as autophagy [56]. In hepatocytes, AMPK was shown to follow a cyclic oscillation that inversely mimicked that of the circadian protein Cry1 highlighting, the importance of catabolic inducers in destabilizing Cry1 proteins [60]. Limited function of AMPK resulted in significant peripheral clock impairment [60]. This directly connects to autophagy, because AMPK can interact with ULK1 and abrogate the inhibitory effects of mTOR on ULK1 [56]. Since AMPK provides crosstalk between circadian and metabolism via mTOR and ULK1, there appears to be a close relationship between mitochondrial fitness and circadian rhythm suggesting a mechanism for metabolism synchronizing with time of a day cycles, [91].

Interaction of Circadian Machinery and Cell Metabolism

As previously mentioned, the circadian rhythm is important for ensuring metabolic homeostasis . During sleep, the body is nutrient deprived and does not require anabolic complexes such as the mTOR pathway to be activated. This would suggest an evolutionarily advantageforcell metabolism to be in communication with circadian rhythms. For example, the insulin signalling pathway has become one of the most investigated areas for circadian and metabolic interaction. Studies highlighted the importance of circadian regulated oscillations of insulin-like growth factors (IGFs) for activating PIK3-Akt-mTOR pathway for cell growth [132]. While these studies provided exciting advancements for the of CCN proteins on metabolism, there is recent literature that highlights a bi-directional relationship between circadian clock and metabolism from mTOR signalling, and resynchronizing of the SCN.

Cao et al. found that light pulses at night activated the mTORC1/S6K pathway within the SCN causing photic entrainment and later found that mTORC1 mediated activation of 4E-BP1 resulted in increased mRNA transcripts of vasoactive intestinal polypeptide (VIP) [18, 19]. Additionally, it was found that drosophila with elevated AKT and mTOR expression levels induced longer circadian periods, conversely, reducing these genes shortenes circadian periods [55]. In mice, in which mTOR was knocked down showed impaired circadian behaviour and decreased synchronization in SCN cells [19].

In hepatocytes, lentiviral shRNA mediated mTOR knockdown, increased the circadian period length in mice, that was reversed when gain of function mutations of Rheb was mutations increased mTOR. This in turn could be reversed by inhibition of mTOR via rapamycin, torin1 and PP242 which act as negative regulators of mTORC1&2 [92]. Additionally, CCN proteins Clock, BMAL1, and Cry1 were elevated when TSC-2 was knocked down compared to its activation, however the effects of TSC-2 knockdown overexpression could be inhibited by rapamycin [92]. Studies have investigated the affects of circadian rhythm on other autophagic regulators such as TFEB and have shown, its nuclear localization to follow a a cyclic oscillation. Additionally, overexpression of TFEB leads to increased REV-ERB which inhibits autophagic flux, suggesting that TFEB works together with circadian regulated genes to create its own autophagic rhythm [84]. The following results point to a complex interaction between autophagy related proteins such as mTOR and TFEB with circadian genes, however, more can be uncovered by exploring if these proteins are still interacting with circadian proteins in cardiomyocytes.

SIRT1 as a Potential Mediator Between Circadian Rhythm, mTOR, and the Cardiovascular System

Potential mediators for the circadian/mTOR interaction in the cardiovascular system come from the Sirtuin (SIRT) protein family, which are a family of seven proteins encoded by the Silent Information Regulatory gene [25, 41]. SIRT proteins can be found in the cytoplasm or nucleus and deacetylate a wide range of proteins promoting survival during extreme conditions [25, 41, 70, 127]. Certain SIRT proteins such as SIRT3 can play an important role in mitochondria biogenesis, metabolizing both lipids and glucose, and insulin sensitivity [11, 89, 127, 133]. SIRT3 along with SIRT5 and SIRT6, have protective roles in IR injury [12, 59] while SIRT2 and SIRT3 have similar positive effects on atherosclerosis and heart failure respectively [39, 130].

In both the brain and peripheral clocks, SIRT1 alters the expression of many CCN proteins. In particular, SIRT1 directly deacetylates BMAL1 on E-Box containing promotors which antagonize the heterodimerization of Clock and BMAL1 [79]. This deacetylation also seems to be circadian controlled, implying that SIRT1 also regulates circadian rhythm [7, 79]. Additionally, mice deficient in SIRT1 had impaired regulation of CCN proteins such as Per1, Per2, cry1, and cry2 with Per2 specifically repressing SIRT1 activity [97, 112]. Concordant with CCN genes, SIRT1 is another regulator of metabolic health status. SIRT1 can also function as an energy sensor, consuming nicotinamide adenine dinucleotide (NAD+) which is a metabolite of nicotinamide phosphoribosyltranferase (NAMPT) [70]. SIRT1 was shown to be activated by NAD+ in caloric restricted yeast leading to extended lifespan [5, 4].

Another study found that NAD+ depletion from nutrient stress is a major cause of cell death highlighting the importance of SIRT1 and cell survival [129]. Caloric restriction also increases SIRT1 expression in a plethora of tissues [22] while high fat diets and obesity decrease SIRT1 [21]. Within prostate cancer cells, SIRT1 was upregulated leading to resistance towards oxidative stress increasing cell proliferation [51, 121].

With this body of information regarding the underlying relationship between SIRT1 and metabolism, we can begin to appreciate an association between mTOR and SIRT1. The previously mentioned energy currency produced by the mitochondria such as ATP and NADH need to be tightly regulated to ensure depletion does not trigger aforementioned mitochondrial injury and cell death. The mTOR pathway detects the metabolic end products through upstream targets such as AMPK, which becomes activated by low cellular levels of ATP. [80]. SIRT1 utilizes NAD+ for enzymatic activity and sensing NAD+/NADH levels, therefore supporting a role for SIRT1 in mitochondrial quality control and thus mTOR complex activity [4, 80]. AMPK increases SIRT1 by increasing levels of NAD+, while at the same time activating fatty acid oxidation, thereby indirectlyproviding a feedback loop for activating AMPK [17, 34, 117] The polyphenol resveratrol was found to be dependent on SIRT1 for mediating resverastrol's activation of AMPK and NAD+ for improving mitochondrial function.

SIRT6 is found to associate with Forkhead box protein O (FOXO), however SIRT1 also has a relevant relationship with FOXO. SIRT1 has been shown to catalyze the deacetylation of FOXO during oxidative stress preventing FOXO induced cell death [15, 38, 77]. While intriguing on its own, this circles back to mTOR because the upstream mTOR regulator Akt along with its upstream partner phosphoinositide kinase 3 (PI3K), phosphorylates FOXO transcription factors in the presence of growth factors such as IGF which promote cell survival through the Akt-mTOR pathway [14, 27, 38, 90]. Additionally, PI3K and Akt contribute to the activation of mTORC1 to promote physiological cardiac hypertrophy [72]. Notably, FOXO1 overexpression resulted in Akt-mTOR mediated cell growth while deficiency of FOXO1 led to the loss of Akt-mTOR feedback loop [27]. We see here that SIRT1 has the potential to be a major site for interaction between key upstream mTOR targets such as AMPK, Akt and circadian rhythm.

SIRT1 in the heart functions similarly to its role in circadian and metabolic physiology by deacetylating histone and non-histone proteins [30]. SIRT1 can be induced by phytoalexin which has been shown to improve cardiac function after CVDs such as IR-injury, atheroscelorsis, and cardiac aging [24, 61, 111]. Similarly, to the mTOR pathway, SIRT1 expression is also increased during embryonic cardiac development in mice suggesting a role in both maintenance and development of the heart [99]. It was found that low to moderate overexpression of SIRT1 leads to antioxidant enzyme activation which inhibits apoptotic pathways mentioned previously. Higher levels of SIRT1 were found to have negative effects by activating Akt thus leading to Akt mediated cardiac hypertrophy likely through the mTOR pathway [1]. Despite this, it has been shown that cardiac aging reduces levels of SIRT1 showing that SIRT1 may be linked to autophagy because of the diminished regulation of autophagy throughout cardiac aging.

MTOR and Starvation

Human metabolism was evolutionary influenced through periods of nutrient stress, and thus our bodies evolved ways to adapt to periods of starvation. Glucose is often considered the fuel for organismal life, however during times when glucose was sparse, specific mechanisms were completed to ensure enough glucose was present such as gluconeogenesis and glycogenolysis [65, 96]. As nutrient availability depletes, triglycerides release from lipid stores and allow fats to become the primary energy source and thus allows for the preservation of proteins. This highlights the importance of glucose sensing for starvation, and it should come as no surprise that mTOR acts as a glucose sensor. When glucose is present, mTOR is activated and can be used towards anabolic processes while glucose depletion leads to catabolic processes including autophagy as mentioned above [67]. The synthesis of proteins, lipids, and nucleotides are extremely ATP demanding, thus the ability to control mTOR when glucose levels are low is critical to avoid ATP depletion [65, 16]. Notably, dysregulated mTOR has been observed in a number of metabolic related diseases such as diabetes, obesity, and metabolic syndrome [101].

Depleted glucose levels cause upstream negative mTOR regulators AMPK and TSC-1/2 to activate by detecting low levels of ATP rather than AMP and ADP [43, 65]. It was found that the observed pathway of AMPK phosphorylating TSC-1/2 to inhibit mTOR allowed energy depleted cells to control cell size, reduce protein synthesis, and prevent apoptosis [48, 50]. TSC knockout cells were unable to adapt to glucose deprivation resulting in the increase of a known apoptosis inducer: p53. This led to cell death, however when mTOR was inhibited the cells had their ATP levels restored and were protected against apoptosis via TSC1/2 and AMPK [23, 63, 64]. When nutrients are present, the tight association between the pro-autophagic protein ULK1 and AMPK is disrupted by mTOR phosphorylating ULK1, keeping autophagy inactivated once again suggesting the importance of autophagy for cell longevity [32, 56]. Many studies have used amino acid deficient Hanks balanced buffer solution (HBSS) as a way to investigate the effect starvation has on this particular pathway [108]. While it is well understood that starvation causes the dissociation between mTOR and ULK1 leading to autophagic activity, this mechanism is thought to be independent of AMPK [2]. Since circadian is heavily tied to nutrient availability, which is linked to the mTOR complex, understanding the role of circadian rhythm in this nutrient sensitive protein network could provide new insight in treating starved cardiomyocytes. Future studies are warranted to investigate this important connection between mTOR, cell metabolism and circadian regulation and how disruption of these highly interconnected feedback pathways underlie human disease.

Acknowledgements This work was supported by a Foundation grant to L.A.K from the Canadian Institute for Health Research (CIHR) and Heart and Stroke Foundation of Canada, L.A.K. holds a Canada Research Chair in Molecular Cardiology. I.R.N received CIHR fellowship, M.L. Holds a University of Manitoba Graduate Studentship Award.

References

1. Alcendor RR, Gao S, Zhai P, Zablocki D, Holle E, Yu X, Tian B, Wagner T, Vatner S, Sadoshima J (2007) Sirt1 regulates aging and resistance to oxidative stress in the heart. Circ Res 100(10):1512–1521
2. Alers S, Loffler AS, Wesselborg S, Stork B (2012) Role of AMPK-mTOR-Ulk1/2 in the regulation of autophagy: cross talk, shortcuts, and feedbacks. Mol Cell Biol 32(1):2–11
3. Alibhai FJ, Tsimakouridze EV, Chinnappareddy N, Wright DC, Billia F, O'Sullivan ML, Pyle WG, Sole MJ, Martino TA (2014) Short-term disruption of diurnal rhythms after murine myocardial infarction adversely affects long term myocardial structure and function. Circ res 114(11):1713–1722
4. Anderson RM, Bitterman KJ, Wood JG, Medvedik O, Sinclair DA (2003) Nicotinamide and PNC1 govern lifespan extension by caloric restriction in Saccharomyces cerevisiae. Nature 423(6936):181–185
5. Anderson KA, Maden AS, Olsen CA, Hirschey MD (2017) Metabolic control by sirtuins and other enzymes that sense NAD+, NADH, or their ratio. Biochim Biophys Acta 1858:991–998
6. Arendt J, Skene DJ, Middleton B, Lockley SW, Deacon S (1997) Efficacy of melatonin treatment in jet lag, shift work, and blindness. J Biol Rhythms 12:604–617
7. Asher G, Gatfield D, Stratmann M, Reinke H, Dibner C, Kreppel F, Mostoslavsky R, Alt FW, Schibler U (2008) SIRT1 regulates circadian clock gene expression through PER2 Deacetylation. Cell 134(2):134–328
8. Bell-Pedersen D, Cassone VM, Earnest DJ, Golden SS, Hardin PE, Thomas TL, Zoran MJ (2005) Cardiac rhythms from multiple oscillators: lessons from diverse organisms. Nat Rev Genet 6:544–566
9. Bennardo M, Alibhai F, Tsimakouridze E, Chinnappareddy N, Podobed P, Reitz C, Pyle WG, Simpson J, Martimo TA (2016) Day-night dependence of gene expression and inflammatory responses in the remodelling murine heart post-myocardial infarction. Am J Physiol Regul Integr Comp Physiol 311(6):1243–1254
10. Biala AK, Kirshenbaum LA (2014) The interplay between cell death signalling pathways in the heart. Trends Cardiovasc Med 24:325–331
11. Bindu S, Pillai VB, Gupta MP (2016) Role of Sirtuins in regulating pathophysiology of the heart. Trends Endocrinol Metab 27:563–573
12. Boylston JA, Sun J, Chen Y, Gucek M, Sack MN, Murphy E (2015) Characterization of the cardiac succinylome and its role in ischemia-reperfusion injury. J Mol Cell Cardiol 88:73–81
13. Bradley TD, Floras JS (2009) Obstructive sleep apnea and its cardiovascular consequences. Lancet 373:82–93
14. Brunet A, Bonni A, Zigmond MJ, Lin MZ, Juo P, Hu LS, Anderson MJ, Arden KC, Blenis J, Greenberg ME (1999) AKT promotes cell survival by phosphorylating and inhibiting a Forkhead transcription factor. Cell 96(6):857–868
15. Brunet A, Sweeney LB, Sturgill JF, Chua KF, Greer PL, Lin Y, Tran H, Ross SE, Mostoslavsky R, Cohen HY, Hu LS, Cheng HL, Jedrychowski MP, Gygi SP, Sinclair DA, Alt FW, Greenberg ME (2004) Stress-dependent regulation of FOXO transcription factors by the SIRT1 deacetylase. Sci 303(5666):2011–2015
16. Buttgereit F, Brand MD (1995) A hierarchy of ATP-consuming processes in mammalian cells. Biochem J 312 (Pt 1):163–167

17. Canto C, Gerhart-Hines Z, Feige JN, Lagouge M, Milne JC, Elliot PJ, Puigserver P, Auwerx J (2009) AMPK regulated energy expenditure by modulating NAD+ metabolism and SIRT1 activity. Nature 458:1056–1060
18. Cao R, Li A, Cho HY, Lee B, Obrietan K (2010) Mammalian target of signalling modulates photic entrainment of the suprachiasmatic circadian clock. J Neurosci 30(8):6302–6314
19. Cao R, Robinson B, Xu H, Gkogkas C, Khoutorsky A, Alain T, Yanagiya A, Nevarko T, Liu AC, Amir S, Sonenberg N (2013) Translation control of entrainment and synchrony of the suprachiasmatic circadian clock by mTOR/4E-BP1 signalling. Neuron 79(4):712–724
20. Carden DL, Granger DN (2000) Pathophysiology of ischemia-reperfusion injury. J Pathol 190:255–266
21. Chalkiadaki A, Guarente L (2012) High fat diet triggers inflammation induced cleavage of SIRT1 in adipose tissue to promote metabolic dysfunction. Cell metab 16:180–188
22. Chang H-CC, Guarente L (2014) SIRT1 and other sirtuins in metabolism. Trends Endocrinol Metab 25(3):138–145
23. Choo AY, Kim SG, Heiden MV, Mahoney SJ, Hieu V, Yoon SO, Cantley LC, Blenis J (2010) Glucose addiction of TSC null cells is caused by failed mTORC1 dependent balancing of metabolic demand with supply. Mol Cell 38(4):487–499
24. Cote CD, Rasmussen BA, Duca FA, Zadeh-Tahmasebi M, Baur JA, Daljeet M, Breen DM, Filippi BM, Lam TK (2015) Resveratrol activates duodenal Sirt1 to reverse insulin resistance in rats through a neuronal network. Nat Med 21(5):498–505
25. Covington JD, Bajpeyi S (2016) The sirtuins: markers of metabolic health. Mol nutr 60:79–91
26. Curtis AM, Cheng Y, Kapoor S, Reilly D, Price TS, Fitzgerald GA (2007) Circadian variation of blood pressure and the vascular response to asynchronous stress. Proc Natl Acad Sci USA 104:3450–3455
27. Dharaneeswaran H, Abid R, Yuan L, Dupuis D, Beeler D, Spokes KC, Janes L, Sciuto T, Kang PM, Jaminet SS, Dvorak A, Grant MA, Regan ER, Aird WC (2014) Foxo1-mediated activation of AKT plays a critical role in vascular homeostasis. Circ Res 115(2):238–251
28. Dorello NV, Peschiaroli A, Guardavaccaro D, Colburn NH, Sherman NE, Pagano M (2006) Science 314(5798):467–471
29. Durgan DJ, Young ME (2010) The cardiomyocyte circadian clock: emerging roles in heath and disease. Circ Res 106:647–658
30. D'Onofrio N, Servillo L, Balestrieri ML (2018) SIRT1 and SIRT6 signalling pathways in cardiovascular disease protection. Antioxidant Redox Signal. 28:711–732
31. Edwards PA, Muroya H, Gould RG (1972). In vivo demonstration of the circadian thythm of cholesterol biosynthesis in the liver and intestine of the rat. J Lipid Res 13(3):396–401
32. Egan D, Kim J, Shaw RJ, Guan KL (2011) The autophagy initiating kinase Ulk1 is regulated via opposing phosphorylation by AMPK and mTOR. Autophagy 7(6):643–644
33. Eisenberg T, Abdellatif M, Schroeder S, Primessnig U, Stekovic S, Pendl T, Harger A, Schipke J, Zimmermann A, Schmidt A, Tong M, Ruckenstuhl C, Dammbrueck C, Gross AS, Herbst V, Magnes C, Trausinger G, Narath S, Meinitzer A, Hu Z, Kirsch A, Eller K, Carmona-Gutierrez D, Büttner S, Pietrocola F, Knittelfelder O, Schrepfer E, Rockenfeller P, Simonini C, Rahn A, Horsch M, Moreth K, Beckers J, Fuchs H, Gailus-Durner V, Neff F, Janik D, Rathkolb B, Rozman J, de Angelis MH, Moustafa T, Haemmerle G, Mayr M, Willeit P, von Frieling-Salewsky M, Pieske B, Scorrano L, Pieber T, Pechlaner R, Willeit J, Sigrist SJ, Linke WA, Mühlfeld C, Sadoshima J, Dengjel J, Kiechl S, Kroemer G, Sedej S, Madeo F (2016) Cardioprotection and lifespan extension by the natural polyamine spermidine. Nat Med 22(12):1428–1438
34. Feige JN, Logouge M, Canto C, Strehle A, Houten SM, Milne JC, Lambert PD, Mataki C, Elliot PJ, Auwerx J (2008) Specific SIRT1 activation mimics low energy levels and protects against diet inducing metabolic disorders by enhancing fat oxidation. Cell metab 8:347–358
35. Flynn JM, O'Leary MN, Zambataro CA, Academia EC, Presley MP, Garret BJ, Zykovich A, Mooney SD, Strong R, Rosen CJ, Kapahi P, Nelson MD, Kennedy BK, Melov S (2013) Late life rapamycin treatment reverses age-related heart dysfunction. Aging Cell 12:851–862

36. Frangogiannis NG (2012) Regulation of the inflammatory response in cardiac repair. Cir Res. 110(1):159–173
37. Gomes LC, Benedetto GD, Scorrano L (2011) During autophagy mitochondria elongate, are spared from degradation and sustain cell viability. Nat Cell Biol 13(5):589–598
38. Greer AL, Brunet A (2005) Foxo transcription factors at the interface between longevity and tumor suppression. Oncogene 24:7410–7425
39. Grillon JM, Johnson KR, Kotlo K, Danziger RS (2012) Non-histone lysine acetylated proteins in heart failure. Biochim Biophys Acta 1822:607–614
40. Gwinn DM, Shackleford DB, Egan DF, Mihaylova MM, Mery A, Vasquez DS, Turk BE, Shaw RJ (2008) AMPK phosphorylation of raptor mediates a metabolic checkpoint. Mol Cell 30:214–226
41. Haigis MC, Sinclair DA (2010) Mammalian sirtuins: biological insights and disease relevance. Anu rev pathol mech dis. 5:253–295
42. Hara K, Maruki Y, Long X, Yoshino K, Oshiro N, Hidayat S, Tokunaga C, Avruch J, Yonezawa K (2002) Raptor, a binding partner of target of rapamycin (TOR), mediates TOR action. Cell 26;110(2):177–189
43. Hardie DG, Ross FA, Hawley SA (2012) AMPK: a nutrient energy sensor that maintains homeostasis. Nat Rev Mol Cell Biol 13:251–262
44. Hems DA, Rath EA, Verrinder TR (1975). Fatty acid synthesis in liver and adipose tissue of normal and genetically obese (ob/ob) mice during the 24-hour cycle. Biochem J 150(2):167–173
45. Huang N, Chelliah Y, Shan Y, Taylor C, Yoo SH, Partch C, Green CB, Zhang H, Takahashi JS (2012) Crystal structure of heterodimeric clock: BMAL1 transcriptional activator complex. Science 337:189–194
46. Häseli S, Deubel S, Jung T, Grune T, Ott C (2020) Cardiomyocyte Contractility and Autophagy in a Premature Senescence Model of Cardiac Aging. Oxid Med Cell Longev
47. Ikeda Y, Shirakabe A, Maejima Y, Zhai P, Sciarretta S, Toli J, Nomura M, Mihara K, Egashira K, Ohishi M, Abdellatif M, Sadoshima J (2015) Endogenous Drp1 mediates mitochondrial autophagy and protects the heart against energy stress. Circ Res 116(2):264–278
48. Inoki K, Li Y, Xu T, Guan KL (2003) Rheb GTPase is a direct target of TCS2 GAP activity and regulates mTOR signalling. Genes Dev 17:1829–1834
49. Inoki K, Ouyang H, Zhu T, Lindvall C, Wang Y, Zhang X, Yang Q, Bennett C, Harada Y, Stankunas K, Wang CY, He X, MacDougald OA, You M, Williams BO, Guan KL (2006) TCS2 integrates WNT and energy signals via coordinated phosphorylation by AMPK and GSK3 to regulate cell growth. Cell 126:995–968
50. Inoki K, Zhu T, Guan KL (2003) TSC-2 mediates cellular energy response to control cell growth and survival. Cell 115:577–590
51. Jung-Hynes B, Schmit TL, Reagan-Shaw SR, Siddiqui IA, Mukhtar H, Ahmad N (2011) Melatonin, a novel Sirt1 inhibitor, imparts antiproliferative effect against prostate cancer in vitro in culture and in vivo in TRAMP model. J pineal res 50(2):140–149
52. Kairo K, Pickering TG, Matsuo T, Hoshide S, Schwartz JE, Shimada K (2001) Stroke prognosis and abnormal nocturnal blood pressure falls in older hypertensives. Hypertension 38:852–857
53. Kaizuka T, Hara T, Oshiro N, Kikkawa U, Yonezawa K, Takehana K, Iemura S, Natsune T, Mizushima N (2010) Tti1 and tel2 are critical factors in mammalian target of rapamycin complex assembly. J Biol Chem 285:20109–20116
54. Kasai T, Bradley TD (2011) Obstructive sleep apnea and heart failure: pathophysiologic and therapeutic implications. J am Coll Cardiol 57:119–127
55. Kijak E, Pyza E (2017) Tor signalling pathway and autophagy are involved in the regulation of circadian rhythms in behavior and plasticity of L2 interneurons in the brain of drosophila melanogaster. PLos One 12(2):e0171848
56. Kim J, Kundu M, Violett B, Guan KL (2011) AMPK and mTOR regulate autophagy through direct phosphorylation of Ulk1. Nature Cell Biol 13:132–141

57. Kim DH, Sarbassov DD, Ali SM, King JE, Latek RR, Erdjument-Bromage H, Tempst P, Sabatini DM (2002) mTOR interacts with raptor to form a nutrient-sensitive complex that signals to the cell growth machinery. Cell 110(2):163–175
58. Knutsson A, Akerstedt T, Jonsson BG, Orth-Gomer K (1986) Increased risk of ischaemic heart disease in shift workers. Lancet 12;2(8498):89–92
59. Koentges C, Pfeil K, Meyer-Steenbuck M, Lother A, Hoffman MM, Odening KE, Hein L, Bode C, Bugger H (2016) Preserved recovery of cardiac functioning following ischemia reperfusion in mice lacking SIRT3. Can J Physiol Pharmacol 94(1):72–80
60. Lamia KA, Sachdeva UM, DiTacchio L, Williams EC, Alvarez JG, Egan DF, Vasquez DS, Juguilon H, Panda S, Shaw RJ, Thompson CB, Evans RM (2009) AMPK regulates the circadian clock by cryptochrome phosphorylation and degradation Sci 326(5951):437–440
61. Lancake P, Price RJ (1977) A new class of phytoalexins from grapevines. Experientia 33:151–152
62. Laplante M, Sabatini DM (2012) mTOR signaling in growth control and disease. Cell 149(2):274–293
63. Lee CH, Inoki K, Karbowniczek M, Petroulakis E, Sonenberg N, Henske EP, Guan KL (2007) Constitutive mTOR activation in TSC mutants sensitizes cells to energy starvation and genomic damage via p53. EMBO J 26(23):4812–4823
64. Lee DF, Kuo HP, Chen CT, Hsu JM, Chou CK, Wei Y, Sun HL, Li LY, Ping B, Huang WC, He X, Hung JY, Lai CC, Ding Q, Su JL, Yang JY, Sahin AA, Hortobagyi GN, Tsai FJ, Tsai CH, Hung MC (2007) IKK beta suppression of TSC1 links inflammation and tumor angiogenesis via the mTOR pathway. Cell 130(3):440–455
65. Lepriver G, Rotblat B (2020) How does mTOR sense glucose starvation? AMPK is the usual suspect. Cell Death Discov 6:27
66. Lim JA, Li L, Shirihai OS, Trudeau KM, Puertollano R, Raben N (2017) Modulation of mTOR signalling as a strategy for the treatment of Pompe disease. EMBO Mol Med 9:353–370
67. Liu GY, Sabatini DM (2020) mTOR at the nexus of nutrition, growth, aging, and disease. Mol Cell Biol 21:183–293
68. Lowery PL, Takahashi JS (2011) Genetics of circadian rhythms in mammalian model organisms. Adv Genet 74:175–230
69. Matsui Y, Takagi H, Qu X, Abdellatif M, Sakoda H, Asano T, Levine B, Sadoshima J (2007) Distinct roles of autophagy in the heart during ischemia and reperfusion: roles of AMP-activated protein kinase and Beclin 1 in mediating autophagy. Circ Res 100(6):914–922
70. Mayo JC, Sainz RM, Menendez PG, Cepas V, Tan DX, Reiter RJ (2017) Melatonin and sirtuins: a not-so unexpected relationship. J Pineal Res 62(2)
71. Mayo JC, Sainz RM, Menendez PG, Cepas V, Tan DX, Reiter RJ (2017) Melatonin and sirtuins: a not-so unexpected relationship. Pineal Res 62(2)
72. Mazelin L, Panthu B, Nicot AS, Belotti E, Tingtignac L, Teixeira G, Zhang Q, Risson V, Bass D, Delaune E, Derumeaux G, Taillandier D, Ohlmann T, Ovize M, Gangloff YG, Schaeffer L (2016) J Mol Cell Cardiol 97:213–225
73. Mcmullen JR, Shioi T, Zhang L, Tarnavski O, Sherwood MC, Kang PM, Izumo S (2003) Phosphoinositide 3-kinase(p110alpha) plays a critical role for the induction of physiological, but not pathological, cardiac hypertrophy. Proc Natl Acad Sci 100(21):12355–12360
74. Mistry P, Duong A, Kirshenbaum L, Martino T (2017) Cardiac clock and preclinical translation. Heart Failure Clin 13(4):657–672
75. Mizushima N, Komatsu M (2011) Autophagy: enovation of cells and tissues. Cell 147(4):728–741
76. Morales CR, Li DL, Pedrozo Z, May HI, Jiang N, Kyrychenko V, Cho GW, Kim SY, Wang ZV, Rotter D, Rothermal BA, Schneider JW, Lavandero S, Gillette TG, Hill JA (2016) Inhibition of class I histone deacetylases blunts cardiac hypertrophy through TSC2-dependent mTOR repression. Sci Signal 9(422):ra34
77. Motta MC, Divecha N, Lemieux M, Kamel C, Chen D, Gu W, Bultsma Y, McBurney M, Guarente L (2004) Mammalian SIRT1 represses forkhead transcription factors. Cell 116(4):551–563

78. Mughal W, Kirshenbaum LA (2011) Cell death signalling mechanisms in heart failure. Exp Clin Cardiol 16:102–108
79. Nakahata Y, Kaluzova M, Grimaldi B, Sahar S, Hirayama J, Chen D, Guarente LP, Sassone-Corsi P (2008) The NAD+-dependent deacetylase SIRT1 modulates CLOCK-Mediated chromatin remodeling and circadian control. Cell 1134(2):329–340
80. Nakahata Y, Kaluzova M, Grimaldi B, Sahar S, Hirayama J, Chen D, Guarente LP, Sassone-Corsi P (2008) The NAD+-dependent deacetylase SIRT1 modulates CLOCK-Mediated chromatin remodeling and circadian control. Cell 134(2):329–340
81. Nakai A, Yamaguchi O, Takeda T, Higuchi Y, Hikoso S, Taniike M, Omiya S, Mizote I, Matsumura Y, Asahi M, Nishida K, Hori M, Mizushima N, Otsu K (2007) The role of autophagy in cardiomyocytes in the basal state and in response to hemodynamic stress. Nat Med 13(5):619–624
82. Nogueiras R, Habegger KM, Chaudhary N, Finan B, Banks AS, Dietrich MO, Horvath TL, Sinclair DA, Pfluger P, Tschop MH (2012) Sirtuin 1 and sirtuin 3: physiological modulator of metabolism. Physiol Rev 92:1479–1514
83. Oka SI, Hirata T, Suzuki W, Naito D, Chen Y, Chin A, Yaginuma H, Saito T, Nagarajan N, Zhai P, Bhat S, Schesing K, Shao D, Hirabayashi Y, Yodoi J, Scriarretta S, Sadoshima J (2017) Thioredoxin-1 maintains mechanistic target of rapamycin (mTOR) function during oxidative stress in cardiomyocytes. J Biol Chem 292(46):18988–19000
84. Pastore N, Vainshtein A, Herz NJ, Huynh T, Brunetti L, Klisch TJ, Mutarelli M, Annunziata P, Kinouchi K, Brunetti-Pierri N, Sassone-Corsi P, Ballabio A (2019) Nutrient-sensitive transcription factors TFEB and TFE3 couple autophagy and metabolism to the peripheral clock. EMBO J 38(12)
85. Peterson TR, Laplante M, Thoreen CC, Sancak Y, Kang SA, Kuehl WM, Gray NS, Sabatini DM (2009) DEPTOR is an mTOR inhibitor frequently overexpressed in multiple myeloma cells required for survival. Cell 137(5):873–886
86. Pfeifer U, Scheller H (1975) A morphometric study of cellular autophagy including diurnal variations in kidney tubules of normal rats. J Cell Biol 64(3):608–621
87. Pfeifer U, Strauss P (1981) Autophagic vacuoles in heart muscle and liver. A comparative morphometric study including circadian variations in meal-fed rats. J Mol Cell Cardiol 13(1):37–49
88. Phillips LJ, Berry LJ (1970) Circadian rhythm of mouse liver phosphoenolpyruvate carboxykinase. Am J Physiol 218(5):1440–1444
89. Poulose N, Raju R (2015) Sirtuin regulation in aging and injury. Biochim Biophys Acta 1852(11):2442–2455
90. Rabinovich-Nikitin I, Lieberman B, Martino TA, Kirshenbaum LA (2019) Circadian-regulated cell death in cardiovascular diseases. Circulation 139:965–980
91. Rabinovich-Nikitin I, Rasouli M, Reitz CJ, Posen I, Margulets V, Dhingra R, Khatua TN, Thliveris JA, Martino TA, Kirshenbaum LA (2021) Mitochondrial autophagy and cell survival is regulated by the circadian Clock gene in cardiac myocytes during ischemic stress. Autophagy 1–19
92. Ramanathan C, Kathale ND, Liu D, Lee C, Freeman DA, Hogenesch JB, Cao R, Liu AC (2018) mTOR signalling regulates central and peripheral circadian clock function. PLos Genet 14(5):e1007369
93. Ramanathan C, Kathale ND, Liu D, Lee C, Freeman DA, Hogenesch JB, Cao R, Liu AC (2018) mTOR signalling regulates central and peripheral circadian clock function. PLos Genet 14(5)
94. Ravikumar B, Sarkar S, Davies JE, Futter M, Garcia-Arencibia M, Green-Thompson ZW, Jimenez-Sanchez M, Korolchuk VI, Lichtenberg M, Luo S, Massey DC, Menzies FM, Moreau K, Narayanan U, Renna M, Siddiqi FH, Underwood BR, Winslow AR, Rubinsztein DC (2010) Regulation of mammalian autophagy in physiology and pathophysiology. Physiol Rev (4):1383–1435
95. Rena G, Guo S, Chichy S, Unterman TG, Cohen P (1999) Phosphorylation of the transcription factor forkhead family member FKHR by protein kinase B. J Biol Chem 274(24):17179–17183

96. Richards J, Gumz ML (2013) Mechanisms of the circadian clock in physiology. Am J Physiol Regul Integr Comp Physiol 304:1053–1064
97. Russell LK, Finick BN, Kelly DP (2005) Mouse models of mitochondrial dysfunction and heart failure. J Mol Cell Cardiol 38(1):81–91
98. Russell LK, Finick BN, Kelly DP (2005) Mouse models of mitochondrial dysfunction and heart failure. J Mol Cell Cardiol 38:81–91
99. Sakamoto J, Miura T, Shimamoto K, Horio Y (2004) Predominant expression of Sir2α, an NAD-dependant histone deacetylase, in the embryonic mouse heart and brain. FEBS Lett 556(1-3):281–286
100. Sakamoto J, Miura T, Shimamoto K, Horio Y (2004) Predominant expression of Sir2α, an NAD-dependant histone deacetylase, in the embryonic mouse heart and brain. FEBS Lett 556:281–286
101. Sancak Y, Thoreen CC, Peterson TR, Lindquist RA, Kang SA, Spooner E, Carr SA, Sabatini DM (2007) PRAS40 is an insulin inhibitor of the mTORC1 protein kinase. Mol Cell 25(6):903–915
102. Saxon RA, Sabatini DM (2017) mTOR signalling in growth, metabolism, and disease. Cell 169:361–371
103. Schirone L, Forte M, Palmerio S, Yee D, Norcella C, Angelini F, Pagano F, Schiavon S, Bordin A, Carrizzo A, Vecchione C, Valenti V, Chimenti I, De Falco E, Sciarretta S, Frati G (2017) A review of the molecular mechanisms under lying the development and progression of cardiac remodelling. Oxid Med Cell Longev 2017:3920195
104. Schirone L, Forte M, Palmerio S, Yee D, Norcella C, Angelini F, Pagano F, Schiavon S, Bordin A, Carrizzo A, Vecchione C, Valenti V, Chimenti I, De Falco E, Sciarretta S, Frati G (2017) A review of the molecular mechanisms under lying the development and progression of cardiac remodelling. Oxid Med Cell Longev 2017:3920195
105. Schirone L, Forte M, Palmerio S, Yee D, Norcella C, Angelini F, Pagano F, Schiavon S, Bordin A, Carrizzo A, Vecchione C, Valenti V, Chimenti I, De Falco E, Sciarretta S, Frati G (2017) A review of the molecular mechanisms under lying the development and progression of cardiac remodelling. Oxid Med Cell Longev
106. Sciarretta S, Forte M, Frati G, Sadoshima J (2019) New insights into the role of mTOR signalling in the cardiovascular system. Circ Res 122(3):489–505
107. Sciarretta S, Maejima Y, Zablocki D, Sadoshima J (2018) The role of autophagy in the heart. Annu Rev Physiol 80:1–26
108. Sciarretta S, Volpe N, Sadoshima J (2014) Mammalian target of rapamycin signalling in cardiac physiology and disease. Circ Res 114:549–564
109. Settembre C, Zoncu R, Medina DL, Vetrini F, Erdin S, Erdin S, Huynh T, Ferron M, Karsenty G, Vellard MC, Facchinetti V, Sabatini DM, Ballabio A (2012) A lysosome-to-nucleus signalling mechanism senses and regulates the lysosome via mTOR and TFEB. EMBO J 31(5):1095–1108
110. Shang L, Wang X (2011) AMPK and mTOR coordinate the regulation of ULK1 and mammalian autophagy initiation. Autophagy 7(8):924–926
111. Shende P, Xu L, Morandi C, Pentassuglia L, Heim P, Lebboukh S, Berthonneche C, Pedrazzini T, Kaufmann BA, Hall MN, Ruegg MA, Brink M (2016) Cardiac mTOR complex 2 preserves ventricular function in pressure-overload hypertrophy. Cardiovasc Res 109:103–114
112. Soni SK, Basu P, Singaravel M, Sharma R, Pandi-Perumal SR, Cardinali DP, Reiter RJ (2021) Sirtuins and the circadian clock interplay in cardioprotection: focus on sirtuin 1. Cell Mol Life Sci 78(6):2503–2515
113. Spampanato C, Feeney E, Li L, Cardone M, Lim JA, Annunziata F, Zare H, Polishchuk R, Puertollano R, Parenti G, Ballabio A, Raben N (2013) Transcription factor EB (TFEB) is a new therapeutic target for Pompe disease. EMBO Mol Med 5(5):691–706
114. Stein S, Matter CM (2011) Protective roles of SIRT1 in atherosclerosis. Cell Cycle 10:640–647
115. Szydlowska K, Tymianski M (2010) Calcium, ischemia, and excitotoxicity. Cell Calcium 47:122–129

116. Takahashi JS, Hong HK, Ko CH, Mcdearmon EL (2008) The genetics of mammalian circadian order and disorder: implications for physiology and disease. Nat rev genet 9:764–775
117. Takeda N, Maemura K (2011) Circadian clock and cardiovascular disease. J Cardiol 57:249–256
118. Tamai T, Yamaguchi O, Hikoso S, Takeda T, Taneike M, Oka T, Oyabu J, Murakawa T, Nakayama H, Uno Y, Horie K, Nishida K, Sonenberg N, Shah AM, Takeda J, Komuro I, Otsu K (2013) Rhen (ras homologue enriched in brain)-dependent mammalian target of rapamycin complex 1 (mTORC1) activation becomes indispensable for cardiac hypertrophic growth after early postnatal period. J Biol Chem 288:10176–10187
119. Tamai T, Yamaguchi O, Hikoso S, Takeda T, Taneike M, Oka T, Oyabu J, Murakawa T, Nakayama H, Uno Y, Horie K, Nishida K, Sonenberg N, Shah AM, Takeda J, Komuro I, Otsu K (2013) Rheb (Ras homologue enriched in brain)-dependent mammalian target of rapamycin complex 1 (mTORC1) activation becomes indispensable for cardiac hypertrophic growth after early postnatal period. J Biol Chem 288(14):10176–10187
120. Um JH, Park SJ, Kang H, Yang S, Foretz M, McBurney MW, Kim MK, Viollet B, Chung JH (2010) AMP-activated protein kinase-deficient mice are resistant to the metabolic effects of resveratrol. Diabetes 59:554–563
121. Verdecchia P, Schillaci G, Guerrieri M, Gatteschi C, Benemio G, Boldrini F, Porcellati C (1990) Circadian blood pressure changes and left ventricular hypertrophy in essential hypertension. Circulation 81(2):528–536
122. Völkers M, Toko H, Doroudgar S, Din S, Quijada P, Joyo AY, Ornelas L, Joyo E, Thuerauf DJ, Konstandin MH, Gude N, Glembotski CC, Sussman MA (2013) Pathological hypertrophy amelioration by PRAS40-mediated inhibition of mTORC1. Proc Natl Acad Sci U S A 110:12661–12666
123. Wang B, Hasan MK, Alvarado E, Yuan H, Wu H, Chen WY (2011) NAMPT overexpression in prostate cancer and its contribution to tumor cell survival and stress response. Oncogene 30:907–921
124. Wang N, Yang G, Jia Z, Zhang H, Aoyagi T, Soodvilai S, Symons JD, Schnermann JB, Gonzalez FJ, Litwin SE, Yang T (2008) Vascular PPAR γ controls circadian variation in blood pressure and heart rate through BMAL1. Cell Metab 8(6):482–491
125. Wang XX, Wang XL, Tong MM, Gan L, Chen H, Wu S, Chen J, Ru-Li L, Wu Y, Zhang H, Zhu Y, Li Y, He J, Wang M, Jiang W (2016) SIRT6 protects cardiomyocytes against ischemia reperfusion injury by augmenting FOXO3 α-dependent antioxidant defense mechanisms. Basic Res Cardiol 111(2):1–19
126. Wang XX, Wang XL, Tong MM, Gan L, Chen H, Wu S, Chen J, Ru-Li L, Wu Y, Zhang H, Zhu Y, Li Y, He J, Wang M, Jiang W (2016) SIRT6 protects cardiomyocytes against ischemia reperfusion injury by augmenting FOXO3 α-dependent antioxidant defense mechanisms. Basic Res Cardiol 1111(2):13
127. Wu X, Coa Y, Nie J, Liu H, Lu S, Hu X, Zhu J, Zhoa X, Chen J, Chen X, Yang Z, Li X (2013) Genetic and pharmacological inhibition of Rheb1-mTORC1 signaling exerts cardioprotection against adverse cardiac remodeling in mice. AM J Pathol 182(6):2005–2014
128. Wu JJ, Liu J, Chen EB, Wang JJ, Cao L, Narayan N, Fergusson MM, Rovira II, Allen M, Springer DA, Lago CU, Zhang S, DuBois W, Ward T, deCabo R, Gavrilova O, Mock B, Finkel T (2013) Increased mammalian lifespan and a segmental and tissue specific slowing of aging after genetic reduction of mTOR expression. Cell Rep 4(5):913–920
129. Wullschleger S, Loewith R, Hall MN (2006) TOR signalling in growth and metabolism. Cell 124(3):471–484
130. Xie M, Kong Y, Tan W, May H, Battiprolu PK, Pedrozo Z, Wang ZV, Morales C, Luo X, Cho G, Jiang N, Jessen ME, Warner JJ, Lavandero S, Gillette TG, Turer AT, Hill JA (2014) Histone deacetylase inhibition blunts ischemia/reperfusion injury by inducing cardiomyocyte autophagy. Circulation 129:1139–1151
131. Yamamota H, Schoonjans K, Auwerx (2007) Sirtuin functions in health and disease. Mol Endocrinol 21(8):1745–1755

132. Yang H, Yang T, Baur J, Perez E, Matsui T, Carmona JJ, Lamming DW, Souza-Pinto NC, Bohr VA, Rosenzweig A, Cabo RD, Sauce AA, Sinclair DA (2007) Nutrient-sensitive mitochondrial NAD+ levels dictate cell survival. Cell 130(6):1095–1107
133. Yang T, Fu M, Pestell R, Sauve AA (2006) SIRT1 and endocrine signalling. Trends Endocrinol Metab 17:186–191
134. Yu L, Wan F, Dutta S, Welsh S, Liu Z, Freundt E, Baehrecke EH, Lenardo M (2007) Autophagic programmed cell death by selective catalase degradation. Proc Natl Acad Sci USA 103(13):4952–4957
135. Zhang D, Contu R, Latronico MV, Zhang J, Zhang JL, Rizzi R, Catalucci D, Miyamoto S, Huang K, Ceci M, Gu Y, Dalton ND, Peterson KL, Guan KL, Brown JH, Chen J, Sonenberg N, Condorelli G (2010) mTORC1 regulates cardiac function and myocyte survival through 4EBP1 inhibition in mice. J Clin Invest 120:2805–2816
136. Zhang R, Lahens NF, Balance HI, Hughes ME, Hogenesch JB (2014) A circadian gene expression atlas in mammals: implications for biology and medicine. Proc Natl Acad Scie USA 111(45):16219–16224
137. Zhang Y, Long Z, Xu J, Tan S, Zhang N, Li A, Wang L, Wang T (2017) Hydrogen inhibits isoproterenol induced autophagy in cardiomyocytes in vitro and in vivo. Mol Med Rep 16:8253–8258
138. Zhang BC, Ma YF, Xiang CH (2018) SIRT2 decreases atherosclerotic plaque formation in low-density lipoprotein receptor-deficient mice by modulating macrophage polarization. Biomed pharmacother 97:1238–1242
139. Zhou J, Freeman TA, Ahmed F, Shang X, Mangano E, Gao E, Farber J, Wang Y, Ma XL, Woodgett J, Vagnozzi RJ, Lal H, Force T (2013) GSK-3 α is a central regulator of age-related pathologies in mice. J Clin Invest 123:1821–1832
140. Zhu Y, Pires KM, Whitehead KJ, Olsen CD, Wayment B, Zhang YC, Bugger H, Ilkun O, Litwin SE, Thomas G, Kozma SC, Abel ED (2013) Mechanistic target of rapamycin (mtor) is essential for murine embryonic health development and growth. PLoS one 8(1)

Chapter 10
Mitochondria and Their Cell Hosts: Best of Frenemies

Allen M. Andres, Somayeh Pourpirali, and Roberta A. Gottlieb

Abstract Mitochondria originated as an endosymbiont in the first eukaryotic cells, leading to a successful partnership that is at times an uneasy truce. In this review we discuss the interplay between mitochondria and the cells they occupy. Cells regulate the potentially lethal aspects of mitochondria by inserting both pro-apoptotic and anti-apoptotic machinery into mitochondria. In this way mitochondria can trigger apoptosis, ferroptosis, necrosis, and energy insufficiency; they also serve as a nexus for inflammatory signaling. Cellular mitigation of mitochondrial mischief includes anti-apoptotic Bcl-2 family members, mitophagy and other protein quality control mechanisms.

Keywords Apoptosis · Mitochondrial permeability transition pore · Programmed cell death · Inflammation · Mitophagy

Perspective: Since the dawn of eukaryotic life, when protobacteria established intracellular existence within eukaryotic host cells, there has been an uneasy truce that included pacts of mutually assured destruction. Cell death is a common and easily-triggered response to disruption of the homeostatic balance and proceeds by an evolutionarily conserved mechanism. In this chapter we will discuss various mitochondrial disruptions that trigger cell death, and will also examine corresponding responses to ensure cell survival.

Attacking the Host: Apoptosis and MPTP

Apoptosis. Programmed cell death is mediated by proteolytic destruction of multiple intracellular targets and activation of oligonucleosomal DNA cleavage. Initiation

---

A. M. Andres · S. Pourpirali · R. A. Gottlieb (✉)
Smidt Heart Institute, Cedars-Sinai Medical Center, 127 S. San Vicente Blvd, Los Angeles, CA 90048, USA
e-mail: Roberta.Gottlieb@cshs.org

© Springer Nature Switzerland AG 2022

L. A. Kirshenbaum (ed.), *Biochemistry of Apoptosis and Autophagy*,
Advances in Biochemistry in Health and Disease 18,
https://doi.org/10.1007/978-3-030-78799-8_10

of cell death may be through the extrinsic pathway involving death receptors (e.g., Fas, TRAIL receptors), or through the intrinsic pathway mediated by pro-apoptotic Bcl-2 family members, which activate mitochondrial outer membrane permeabilization and release of cytochrome *c*. Cytochrome *c*, essential for shuttling electrons between OXPHOS complex III and IV in the mitochondrial inner membrane, plays a more sinister role in the cytosol, where it is a cofactor in assembly of the apoptosome, a multiprotein complex made up of APAF-1, cytochrome *c*, and caspase-9. Apoptosome assembly leads to proteolytic activation of caspase 9, which in turn processes procaspase 3 to its active form as the prime death-effector protease of the cell. Release of cytochrome *c* from mitochondria is governed by interactions of pro-apoptotic and anti-apoptotic Bcl-2 family members, which have up to 4 conserved Bcl-2 homology (BH) domains. Proapoptotic Bax and Bak have BH1, BH2, and BH3 domains, whereas antiapoptotic Bcl-2 and Bcl-xL have, in addition to BH1, BH2, BH3, a BH4 domain that is considered to be a hallmark of the antiapoptotic Bcl-2 family members. A third branch of the Bcl-2 family is represented by the BH3-only proteins such as Bid and Bad. These cytosolic proteins respond to cellular derangements such as abnormal kinase signaling (Bad), activation of proteases (Bid) or ER stress (Bim). Many more Bcl-2 family members exist, particularly BH3-only proteins. Interestingly, apoptosis and autophagy are linked as Beclin 1, Ambra-1 and Atg12 contain a BH3 domain and have been shown to participate in both pathways [1–3]. The balance of pro-apoptotic and anti-apoptotic Bcl-2 family members is generally thought to determine whether mitochondrial outer membrane permeabilization and release of cytochrome *c* occurs, but it is more nuanced. Bak is constitutively present but inactive in the mitochondrial outer membrane, whereas Bax is often located in the cytosol and must undergo translocation and conformational change to associate with the mitochondrial outer membrane and mediate permeabilization. Phosphorylation of Bax by Akt (Ser184) stabilizes the inactive cytosolic conformation, preventing its mitochondrial association [4]. Once associated with the mitochondrial outer membrane, Bax or Bak can oligomerize to cause permeabilization of the mitochondrial outer membrane and release of cytochrome *c*. Cytochrome *c* must dissociate from the inner membrane and electron transfer complexes, and major reorganization of crista architecture must take place before egress to the cytosol is possible [5, 6]. Outer membrane permeabilization also permits the egress of additional proapoptotic factors including apoptosis-inducing factor (AIF), Smac/Diablo, HtrA2/OMI, and endonuclease G (reviewed in [7]).

The existence of this arcane and mutually sadistic arrangement between cytosolic factors that can disrupt the mitochondrial outer membrane and the consequent release of proapoptotic molecules that ensure death of the cell appears, at first glance, to be a case of mutually-assured destruction. But when one considers that all of the proteins involved are nuclear-encoded and which lack bacterial homologs, it appears that evolution has seen fit to ensure that mitochondrial damage results in death of the cell. An important stopgap measure is mitophagy (discussed in a later section) which is a mechanism to eliminate damaged mitochondria without triggering cell death. One can speculate that these measures are necessary to prevent release of mitochondrial components that would be recognized as damage-associated molecular patterns (DAMPs), including cardiolipin (ligand for NLRP3 inflammasome) [8], hsp60 (TLR4 ligand) [9], and mitochondrial DNA (mtDNA), which lacks CpG methylation and is a ligand for TLR9 [10]. As inflammation can affect many neighboring cells, apoptotic cell death, in the event of mitochondrial damage, may be preferable to permitting complete mitochondrial disintegration and wholesale activation of inflammation. This will be discussed in greater detail in a later section.

Mitochondrial permeability transition pore (MPTP). The MPTP is a large-conductance pore spanning the inner mitochondrial membrane that allows exchange of solutes and water up to—1500 daltons in size. As the matrix has a very high oncotic pressure, sustained opening of the pore will result in the influx of water and expansion of the matrix compartment, which normally is quite condensed and surrounded by a heavily infolded and serpiginous inner membrane. As the matrix swells with water, the outer membrane may rupture, releasing intermembrane space components (cytochrome *c*, endonuclease G, AIF, Smac/Diablo, etc.). At the same time, the matrix Ca^{2+} may re-equilibrate with cytosolic Ca^{2+}, abruptly raising cytosolic levels and activating calpains and other destructive enzymes. Disruption of the proton gradient across the inner membrane may result in the ATP synthase operating in reverse, hydrolyzing ATP in a futile attempt to restore membrane potential. This dysregulation can spread rapidly from one mitochondrion to the next and can culminate in death of the cell. The pattern of cell death may resemble apoptosis if release of cytochrome *c* leads to caspase activation before ATP (a necessary cofactor of the apoptosome) is depleted; however, if ATP is depleted rapidly, the cell ruptures due to failure of the Na^+/K^+ ATPase at the plasma membrane. The molecular constituents of the MPTP have been the target of investigation for decades. Cyclophilin D was implicated based on the ability of the inhibitor cyclosporin A to prevent MPTP opening [11], and efforts to identify interacting proteins led to partial purification of the most abundant and hydrophobic proteins of the outer and inner membranes, namely VDAC and ANT [12]. Despite thorough substantiation, the model was rapidly adopted and still appears in textbooks and slides. However, gene knockout studies disproved a role for VDAC and ANT in the MPTP [13, 14]. Cyclophilin D, however, was proven to be a critical regulator of the pore [15]. Recently Paolo Bernardi published compelling evidence representing more than a decade of work, showing the MPTP to comprise dimers of ATP synthase in a conformation that is consistent with the known features of the pore [16]. Cyclophilin D interacts with the ATP synthase, somehow triggering its reorganization to the MPTP state. Cell death mediated by the MPTP is likely to trigger more inflammation due to the release of mitochondrial DAMPs.

MPTP: With Friends Like This, Who Needs Enemies?

Benefits of inhibiting the MPTP. Many studies have documented the benefits of inhibiting the MPTP in catastrophic injury scenarios such as ischemia/reperfusion injury, anthracycline toxicity, and hypoglycemic coma [17–19]. Cyclosporine A is commonly used as an inhibitor of the MPTP, but it also inhibits calcineurin which also mediates very important signaling in cells. Therefore investigators often confirm their findings by one of 3 approaches: (a) inhibiting calcineurin but not the MPTP with FK506 [20]; (b) inhibiting the MPTP but not calcineurin with NIM811 [21]; or (c) using cyclophilin D knockout mice [15]. Nearly every instance of cell injury involving MPTP opening can be rescued by inhibiting cyclophilin D. However, cyclosporin A is toxic to some cell types such as renal tubular cells [22] although this may be due to its effects on calcineurin; not surprisingly, nephrotoxicity is a complication of long-term treatment with CsA [23]. At present, it is difficult to identify any settings where specifically inhibiting MPTP is acutely cytotoxic.

Physiologic role of the MPTP. Like cytochrome *c*, which is required for the essential function of electron transport, the ATP synthase of the MPTP is essential for ATP production. But why would such a dangerous conformation exist in the first place? Many reports have described transient, low-conductance opening of the pore—pore flicker—which might resemble a pressure release valve, allowing a small amount of swelling, release of Ca^{2+}, and a modest drop in mitochondrial membrane potential. These brief pore openings may serve to fine-tune mitochondrial energetics. In this way, pore flickering serves to distinguish between vigorous mitochondria and ones that are beginning to falter in their ability to restore membrane potential. These senescent mitochondria would be targets for mitophagy (discussed in a later section). Indeed, it was shown that in HL-1 cells subjected to starvation conditions, 70% of the mitochondrial mass is removed by autophagy in a few hours, but this is abrogated in cyclophilin D null cells [24]. Failure to "groom" mitochondria via the combined efforts of pore flicker and selective mitophagy would lead eventually to the accumulation of dysfunctional and damaged mitochondria. A second role for the MPTP was revealed by studies of platelet function in cyclophilin D knockout mice. Jobe et al. found that platelet activation and mitochondrial membrane depolarization were altered in CypD KO mice exposed to strong platelet activators [25]. However, if this were a dominant role, cyclophilin D would likely be tissue-specific rather than globally expressed. Thus it seems likely that there is a physiologic role for MPTP in all tissues.

Long term consequences. Given the serious consequences of sustained opening of the MPTP, and the fact that the cyclophilin D knockout mice have a minimal phenotype, one might wonder why not inhibit or eliminate cyclophilin D entirely. However, the physiologic flicker described above may serve an important role. CsA treatment is associated with increased occurrence of malignancies [26], which has been attributed to the immunosuppression and loss of tumor immune surveillance. However, other instances of impaired autophagy and mitophagy are associated with an increase in cancers, including mutations or deficiency of Atg5, Parkin and PINK1 [27–29]. Evidence linking cyclophilin D deficiency to cancer has not been reported. However, the constellation of findings is consistent with a paradigm in which a failure to eliminate damaged mitochondria may result in cumulative DNA damage from mitochondrial ROS, or chronic inflammation by mtDNA activation of innate immunity; the combination of ROS, DNA damage, and chronic inflammation will drive tumorigenesis.

Pore flicker and mitochondrial grooming is also important for cells in which mitochondria must function optimally, such as the retina, where function is diminished by CsA treatment and where autophagy protects against light-induced retinal damage [30, 31]. Mitophagy is essential in neurons, where much has been written about the importance of Parkin and PINK1 in maintaining neuronal integrity (reviewed in [32]), although less is known about the related importance of pore flicker in neurons.

Anti-Apoptotic Defenses

Bcl-2 family. The Bcl-2 family is a complex network of pro-apoptotic and anti-apoptotic proteins. Bcl-2 family proteins are involved in the regulation of the intrinsic pathway of apoptosis which is important for tissue homeostasis, embryo development and maturation of blood cells [33]. Bcl-2 family proteins are divided into three subgroups: (1) anti-apoptotic Bcl-2-like proteins including Bcl-2, Bcl-x_L, Bcl-W, Mcl-1, and Bfl-1/A1, which share structural homology in the Bcl-2 homology (BH) 1, 2, 3, and 4 domains. (2) Pro-apoptotic Bcl-2-like proteins, or effectors, which share homology with BH domains 1, 2, and 3, including Bax, Bak, and Bok/mtd. (3) pro-apoptotic BH3-only proteins which share homology with BH3 domain of Bcl-2 including Bim, Bad, Bid, Bik, and Noxa. BH3-only proteins can be further divided into two groups: 'sensitizer', such as Noxa, Bfm or Bik, which interact with the anti-apoptotic proteins and inhibit their function, and 'direct activators', such as Bid and Bim, which are able to directly bind and activate pro-apoptotic proteins Bax and Bak [34, 35].

Anti-apoptotic Bcl-2 proteins can regulate apoptosis by maintaining mitochondrial integrity and preventing mitochondrial outer membrane permeabilization (MOMP) [36]. During apoptosis, Bax becomes activated and translocates from cytosol to outer mitochondrial membrane. There, they oligomerize with mitochondria resident Bak proteins and allow the release of cytochrome *c* to the cytosol [37, 38]. Conversely, under physiological conditions Bax is retrotranslocated to cytosol by anti-apoptotic Bcl-2 through an unknown mechanism. This retrotranslocation ensures that Bax doesn't permanently attach to outer mitochondrial membrane and therefore it doesn't autoactivate [39].

Two major models have been suggested to regulate Bcl-2 family members and apoptosis. The derepression model suggests that anti-apoptotic Bcl-x_L continuously binds to Bax and/or Bak to prevent their oligomerization and subsequent MOMP [40]. Direct activation model states that Bax/Bak are activated by BH3 proteins, and anti-apoptotic Bcl-x_L sequesters and inhibits these activators (such as tBid and Bim) [41, 42]. Later, the "embedded together" model was proposed emphasizing on the interaction with the membrane as the critical step and stating that both models could be correct [43, 44]. Recently, a more comprehensive model incorporating two different regulation modes was proposed by Llambi et al. They created a chimeric tBid protein in which the BH3 domain was substituted with the BH3 domain of Bax or Bak, (tBID$^{BAX-BH3}$ and tBID$^{BAK-BH3}$ respectively) to see whether the activation of

BH3 only proteins was due to inhibition of anti-apoptotic Bcl-2 proteins or activation of Bax/Bak [45]. Four different scenarios were investigated using these constructs in which the anti-apoptotic protein can bind (1) both direct activator and effector, (2) the direct activator but not the effector, (3) the effector but not the direct activator, or (4) none. Among these, only scenario 2 and 3 were shown to be involved which were called Mode 1 and Mode 2 respectively. They showed that in healthy cells, chimeric BH3 proteins containing the Bax or Bak domains were inactive and inhibited by anti-apoptotic Bcl-2 proteins, referred to as Mode 1. When apoptosis was induced, though, Bak/Bak were activated but sequestered and inhibited by anti-apoptotic Bcl-2 proteins. Thus anti-apoptotic Bcl-2 proteins must either sequester the activating BH3-only proteins or the activated effectors Bax and Bak. Mode 1 inhibition proved to be less efficient since inhibition of MOMP required more anti-apoptotic protein. Interestingly, Mode 2 inhibition was associated with mitochondrial fragmentation [45].

Anti-apoptotic Bcl-2 proteins are overexpressed in a variety of cancers such as acute B-cell leukemia and follicular lymphoma. In fact, the first member of this family was discovered by detecting a translocation between chromosomes 14 and 18, t(14;18), which results in the fusion of the immunoglobin heavy chain and Bcl-2 loci in acute B-cell leukemia and follicular lymphoma cells, and subsequent overexpression of Bcl-2. This overexpression enhances the survival of these cells by inhibiting apoptosis [46, 47]. Different mechanisms have been suggested for the overexpression of anti-apoptotic Bcl-2 proteins such as chromosomal translocation, transcriptional activation of NF-κB, promoter hypomethylation, or post-translational mechanisms [48, 49]. Overexpression of anti-apoptotic Bcl-2 proteins facilitates tumorigenesis and tumor progression and results in a poor response to therapy. In fact, Mcl-1 or Bcl-x_L amplification have been identified in lung, breast and bone tumors [50, 51]. Conversely, downregulation of pro-apoptotic Bcl-2 and BH3 only proteins facilitates and accelerates the formation of tumors. Loss of genomic region containing Puma and Bok gene copies has been observed in a variety of cancers [52]. Bax loss of function has been related to colon cancer and combined loss of Bax and Bak has been observed in some AML patients [53, 54]. Due to their major role in tumor formation and progression, Bcl-2 proteins have been extensively studied as potential targets for cancer therapy (reviewed in [55]). BH3 mimetics are a new class of molecules that have been developed as a novel strategy for cancer therapy. They inhibit anti-apoptotic Bcl-2 proteins by binding to the BH3-groove of protein and disrupting its interaction with BH3 proteins, therefore activating apoptosis. ABT-737 and ABT-263 are two examples of these mimetics binding to Bcl-2, Bcl-xL, and Bcl-w, which are currently being tested in clinical trials [55].

IAPs and ARC. As mentioned before, inhibition of apoptosis results in increased survival of cancer cells and tumor progression. Inhibitors of apoptosis (IAPs) are another class of molecules involved in this process. All IAPs have one to three BIR domains (zinc-binding baculovirus IAP repeat), which is essential for their function. Some of them also have a carboxy-terminal RING domain with ubiquitin ligase activity and a caspase-associated recruitment domain (CARD) [56, 57]. To date, eight IAPs have been identified in human including NAIP, cIAP1, cIAP2, XIAP,

Survivin, Apollon/Bruce, ML-IAP and ILP-2. cIAP1, cIAP2, XIAP and ML-IAP are directly involved in apoptosis while the other members of IAP family control cell survival through regulation of other pathways such as cell cycle and inflammation [57].

XIAP controls the intrinsic pathway of apoptosis by directly binding to caspases 3, 7, and 9. This function is inhibited by SMAC/DIABLO which is released from mitochondria along with cytochrome *c* after induction of apoptosis [58, 59].

cIAP proteins use their ubiquitin ligase activity to modulate the extrinsic pathway of apoptosis. cIAP1/2 promote ubiquitination of a variety of molecules such as TRAF1, TRAF2, RIP, and NIK. Non-ubiquitinated RIP forms a cytosolic complex with FADD and caspase 8 which leads to induction of apoptosis [60].

IAPs regulate other pathways such as NF-κB, JNK, p38 MAPK, TGF-β, Myc, and PI3K/Akt and are involved in regulation of innate immunity and inflammation. For example, IAPs regulate both canonical and non-canonical NF-κB pathway. In the canonical pathway, cIAP1/2 bind TNF receptor through interaction with TRAF2. This binding in turn recruits more TRAF2 molecules, resulting in cIAP1/2 dimerization and activation and subsequent ubiquitination of RIP1, one of the main regulators of this pathway. In the non-canonical pathway TRAF2 and TRAF3 bring cIAPs and NIK (the key regulator of non-canonical pathway) together. cIAP1/2 promote ubiquitination of NIK which designates it for proteasomal degradation (reviewed in [61]).

As mentioned before, inhibition of apoptosis helps progression of tumors. In this regard, overexpression of IAPs has been observed in several cancers. For example, XIAP is overexpressed in some cases of breast cancer, melanoma and clear-cell renal carcinoma [62, 63]. Moreover, overexpression of survivin, cIAP1, cIAP2, and XIAP was associated with poor prognosis in multiple myeloma patients that undergo drug resistance [64]. Involvement of IAPs in various cancers makes them a target of choice for cancer therapy. Small molecule IAP agonist and antisense oligonucleotides have been used in this regard (reviewed in [61]).

ARC (apoptosis repressor with caspase recruitment (CARD) domain) is another class of apoptosis inhibitors which is mainly expressed in post-mitotic cells of heart, muscle, and brain [65, 66]. ARC exerts its function through the N-terminal caspase recruitment domain. It can dimerize via its CARD domain which results in loss of anti-apoptotic activity [66]. ARC inhibits the extrinsic pathway of apoptosis through interaction with the death effector domain (DED) of procaspase 8. It can also bind Fas and FADD and inhibit the formation of the Fas death-inducing signal complex (DISC). Moreover, when overexpressed, ARC can interact with procaspase 2 [67]. ARC is involved in the intrinsic pathway by interaction with BH3 proteins Bad and Puma through its CARD domain therefore acting as an anti-apoptotic Bcl-2 protein [67]. It also binds cytosolic Bax, inhibiting its spontaneous activation [68]. Mice lacking ARC develop normally and have normal heart function in resting conditions. When mice undergo stress such as ischemia or pressure overload, they develop cardiomyopathy faster along with an increase in cardiomyocyte apoptosis, affirming a role for ARC in resistance to apoptosis [69]. In contrast, high levels of ARC have been observed in various tumors suggesting its involvement in cancer development and progression [66, 70].

Connection to fission. Mitochondria are highly dynamic organelles, undergoing constant fusion and fission. Mitochondrial fragmentation or fission is one of the hallmarks of apoptosis. It requires the remodeling of inner and outer mitochondrial membranes. Several proteins are involved in fusion and fission including dynamin-related protein 1 (Drp1), Mitofusin (Mfn)1 and 2, and Opa1 [71].

Drp1 plays an important role in mitochondrial fragmentation. Under physiological conditions Drp1 moves between cytoplasm and outer mitochondrial membrane (OMM). When apoptosis is induced, Bax is activated and Drp1 is SUMOylated in a Bax/Bak dependent manner resulting in a stable association with OMM [72]. In fact, active Bax is found in distinct foci on OMM near Drp1 and Mfn2 in mitochondrial fission sites [73]. Evidence suggests that this phenomenon could also happen through a Drp1-independent mechanism since Drp1 inhibition only delays this process but does not completely inhibit it [74].

There are different hypotheses regarding the mechanism of apoptosis-induced mitochondrial fragmentation. One states that mitochondrial fragmentation is induced by release of proteins after MOMP and activating fission and/or inhibiting fusion. In fact, time-lapse studies have shown that mitochondrial fragmentation follows cytochrome *c* release [75]. A study by Arnoult et al. showed that Opa1 was released from mitochondria along with cytochrome *c*. Loss of Opa1 itself accelerated the release of cytochrome *c* [75]. It has been suggested that Opa1 is involved in maintaining mitochondrial cristae structure and its loss and subsequent disruption of cristae releases cytochrome *c* which is mostly sequestered in cristae. Opa1 depletion in turn blocks fusion, suggesting a mechanism for mitochondrial fragmentation during apoptosis [75].

Bax and Bak also play a role in regulation of mitochondrial dynamics. In physiological conditions, Bax and Bak regulate mitochondrial fusion possibly through interaction with mitofusin 2. When apoptosis is induced though, Bax activation and MOMP coincide with mitochondrial fragmentation [76].

ARC has been shown to inhibit mitochondrial fission. It directly binds Puma on mitochondria, blocking its function to promote Drp1 accumulation. In the study by Wang et al. it was shown that ARC inhibits mitochondrial fission induced by Doxorubicin. Translocation of Drp1 to mitochondria was enhanced upon Doxorubicin treatment which was attenuated by ARC expression. They further showed that Doxorubicin increased the level of Puma. Down-regulation of Puma decrease accumulation of Drp1 on mitochondria suggesting that Puma is required for this process. Further investigation showed that there is cross-talk between ARC and Puma which regulates mitochondrial fission through Drp1 [77].

Importance of mitochondrial fission in apoptosis is controversial. Some studies suggest that mitochondrial fission contributes to the release of cytochrome *c* to the cytosol. A study using a dominant-negative mutant of Drp1 which interferes with fission, showed inhibition of cytochrome *c* release and apoptosis [78]. Conversely, another study showed that although down-regulation of Drp1 or Fis1 blocks fission, it does not inhibit apoptosis. In this study, overexpression of Bax and Bak resulted in mitochondrial fragmentation and induction of apoptosis. Interestingly, cytochrome *c* release was blocked by co-expression of Bcl-x_L but not mitochondrial fission showing that fission is not necessarily required for apoptosis [76].

Humanin and MOTS-C. Retrograde signaling is relaying information from mitochondria to nucleus. It occurs through transient metabolites or mitochondrial-located nuclear protein such as Ca^{2+}, ROS (reactive oxygen species), NO (Nitric Oxide), and cytochrome *c* [79]. Recently, mitochondrial-derived peptides (MDP) have been discovered which can regulate cellular homeostasis as part of retrograde signaling [79].

Humanin (HN) was the first MDP discovered in the brain of a patient with sporadic Alzheimer's Disease (AD) [80]. It is encoded from an ORF within the 16 s rRNA region in mitochondria. If the translation happens in mitochondria, a 21 amino acid product will be obtained but if it gets exported from mitochondria and translated in the cytoplasm, the resulting product will have 24 amino acids. Both isoforms are functional and are found in a wide range of tissues, with heart expressing the highest levels, and also in body fluids such as plasma, cerebral, and seminal fluids [79, 81].

Humanin is involved in promoting cell survival, regulating insulin sensitivity, and preventing oxidative-stress induced damage [82]. Cytoprotection by Humanin was first demonstrated in F11 neuronal cells. Humanin protected F11 cells from cell death induced by exposure to APP, PS1, or PS2, which cause familial AD [80]. Humanin's role in conferring protection against other stressors such as cerebral and myocardial ischemia, oxygen and glucose deprivation, and hypoxia has been extensively studied [83]. Humanin promotes cell survival through interacting with intracellular molecules. It binds inactive Bax, blocking its conformational change and translocation from cytosol to mitochondria, thereby suppressing cytochrome *c* release [81]. Humanin also binds to Bid and its truncated form, tBid, and inhibits its interaction with Bak and Bax. Subsequently, it prevents Bax and Bak oligomerization and inhibits cytochrome *c* release [84]. Humanin also binds to extracellular receptors such as FPRL1 and CNTFR/WSX-1/gp130, thus activating the downstream ERK1/2 and STAT3 pathways [85, 86].

MOTS-C (Mitochondrial open reading frame of the twelve S rRNA type c) is a 16 amino acid peptide encoded from an open reading frame inside the 12S rRNA region on mtDNA. It is highly conserved among different species and is expressed in various tissues and can be detected in plasma as well. Unlike Humanin, MOTS-C bioactive form is only translated and produced in the cytoplasm because mitochondria-specific genetic code results in the formation of tandem start and stop codons [87]. Using Hela ρ0 and actinonin treatment, which deplete mtDNA and mtRNA respectively, it was shown that MOTS-C is strictly of mitochondrial origin [87].

MOTS-C treatment alters gene expression within 4 h which progresses to a distinct shift in gene expression profile by 72 h. This treatment mainly affects genes involved in cell metabolism and inflammation. In fact, global metabolomics profiling revealed that MOTS-C targets folate-methionine cycle and de novo purine synthesis [87]. Most recently, it has been shown that MOTS-C can prevent ovariectomy-induced osteoporosis through activation of AMPK and inhibition of osteoclastogenesis [88].

Accumulation of mtDNA damage occurs with age resulting in deterioration of mitochondrial function. As the level of Humanin and MOTS-C decrease with age, it could explain the metabolic dysfunction observed in older ages. In fact, treatment of aged mice with exogenous MOTS-C could improve glucose uptake [89]. Also, a specific polymorphism, m.1382A > C found in MOTS-C might be the reason for exceptional longevity in northeast Asian population [87]. All in all, these results point to a significant role for Humanin and MDP in delaying the onset of metabolic and age-related diseases.

Freezing Out the Host: Energy Insufficiency

Mitochondrial ATP production. While mitochondria are known to participate in a myriad of cellular functions such as regulating cell death through apoptosis, intracellular signaling, and iron/sulfur metabolite production, one of its most fundamental and important functions is the production of adenosine triphosphate (ATP). ATP is generated by the mitochondria through a series of catabolic reactions that harness the energy released from breaking down larger nutrients, such as carbohydrates, fats and proteins. Ultimately, theses macromolecules are oxidized into CO_2. NADH and $FADH_2$ are also generated during the Krebs cycle and their reducing power is used during oxidative phosphorylation to establish the protonmotive force needed to catalyze the generation of ATP. Electrons are transferred down the electron transport chain through a series of redox reactions that lead to increased H^+ in the mitochondrial intermembrane space forming a pH gradient and electrical potential across the mitochondrial inner membrane. ATP is generated as protons flow back into the matrix through the ATP synthase. In this chemiosmotic reaction, ADP is phosphorylated to form ATP.

Fuel inflexibility. An important feature animals have evolved is the versatility to maintain energy homeostasis despite variations in the fuel resources available. During periods of resource scarcity, it becomes critical to spare glucose for the brain which is limited in its ability to utilize fats. Tissues such as the skeletal muscle and heart are unique in their ability to quickly transition between fat-based fuels and carbohydrates, thus these tissues are key for the body's ability to spare glucose. In addition to systemic flexibility of fuel use, cellular partitioning of fuel also occurs, and the mitochondria play an essential role in this process.

This concept was first recognized by Sir Philip Randle in 1963 and subsequently referred to as the Randle glucose fatty-acid cycle [90, 91]. Substrates such as acetyl-CoA, NADH and ATP can allosterically inhibit the pyruvate dehydrogenase complex (PDC), limiting the entry of pyruvate into the Krebs Cycle. Conversely, these substrates activate pyruvate dehydrogenase kinases which further suppress the PDC through phosphorylation [92]. In this manner, utilization of fatty-acids leads to sparing of glucose by limiting the entry of pyruvate. Cellular citrate levels attenuate glycolysis and pyruvate utilization leading to an increase in G6P levels. This in turn inhibits hexokinase thus limiting glucose uptake into muscle tissues.

Seminal work by McGarry, Foster and colleagues elucidated the converse situation wherein glucose utilization leads to sparing of fat-derived resources (reviewed in [93, 94]). This situation happens upon feeding. Feeding leads to an increase in circulating glucose levels and the utilization of fat-derived fuels is diminished as malonyl-CoA levels rise. This metabolite is an allosteric inhibitor of CPT-1 and limits entry of long chain fatty acids into the mitochondria. The rise in glucose after feeding leads to increased pyruvate, which inhibits the activity of pyruvate dehydrogenase kinases, thereby allowing for increased flux of carbohydrates through the PDC. As flux of carbons through glucose increases into the mitochondria, so does efflux of citrate into the cytosol. This liberated citrate is a precursor for malonyl-CoA which prevents entry of fatty-acids into the mitochondria and thus their oxidation.

Mammals have evolved to survive episodic famine, in part by efficiently storing excess nutrients as fat, a pathway governed by so-called "thrifty" genes. This metabolic adaptation becomes maladaptive in western society's life of caloric excess. Continual nutrient overload leads to obesity and the other features of metabolic syndrome (MetS), including insulin resistance, dyslipidemia, and hypertension. Kelley and colleagues demonstrated that MetS impairs the ability of mitochondria to utilize different fuels [95]. A consequence of insulin resistance is that tissues lose the ability to respond appropriately to nutrient availability. Thus, even in the fed state these tissues (mainly skeletal muscle and heart), are unable to appropriately adjust fuel utilization to maintain efficient energy homeostasis for the entire organism. In this situation where there is an overabundance of all major fuels, the mitochondria are unable to shift appropriately to utilize fatty acids or glucose. This state of metabolic inflexibility (reviewed by Muoio [100]) has been observed in patients who are obese [96], are sedentary [99], have diabetes [97], or heart disease [98].

Spreading Rumors: Creating Inflammation

mtDNA and TLR9. The innate immune system is the first line of defense against external stimuli such as pathogens. This system can be activated through pattern-recognition receptors (PRRs). These receptors recognize pathogen-associated molecular patterns (PAMPs) and initiate the inflammatory response. They can also recognize the molecules released by host upon injury called danger-associated molecular patterns (DAMPs) [101].

Traumatic injury is the leading cause of death in younger adults. It results in the activation of neutrophils (PMN) and the systemic inflammatory response syndrome (SIRS). SIRS in the absence of infection is called sterile SIRS. It was long thought that bacterial gut release due to shock is the reason for SIRS but recent studies have shown that mechanical shock could release mtDNA as a DAMP in circulation and activate PMNs [101–103].

The first study showing that mtDNA activated immune response in vivo was performed in 2010. It was shown that mtDNA was elevated in the plasma of trauma patients [103]. Increased extracellular mtDNA has also been detected in the synovial fluid (SF) of inflamed joints of rheumatoid arthritis patients, in circulation in plasma of HIV patients and in the plasma obtained from animals and patients with non-alcoholic steatohepatitis (NASH) [103–106].

Toll-like receptor family (TLR) is a class of single, transmembrane PRRs [101]. TLR9 is a member of TLR family which is located in endolysosomes and senses bacterial DNA. Under physiological conditions, TLR9 is localized in the endoplasmic reticulum (ER). Upon stimulation, it is transferred to endosomes and binds to bacterial DNA which subsequently activates the downstream cascade of p38MAPK or NF-κB pathways [107]. mtDNA can in fact bind TLR9 receptor and trigger inflammatory response. It shares similarities with bacterial DNA since it contains unmethylated CpG islands and lacks histones, therefore it is believed to trigger inflammation in the same manner [103]. It has been shown that mtDNA isolated from hepatocytes of high-fat diet mice resulted in a greater activation of TLR9 compared to mtDNA obtained from chow diet fed mice [106].

Oxidized mtDNA and NLRP3. The Nod-like receptor (NLR) family is another subset of pattern recognition receptors. The Nod-like receptor protein 3 (NLRP3) is an important member of NLR family required for activation of caspase 1 inflammasome by different stimuli such as mitochondrial damage, and bacterial and viral RNA [108]. NLRP3 interacts with apoptosis-associate speck-like protein (ASC) to form oligomers which is able to activate pro-caspase1. Caspase 1 inflammasome in turn activates precursor interleukin-1β (IL-1β) and IL-18 to initiate immune responses. Under normal circumstances, NLRP3 is localized in endoplasmic reticulum. Upon activation, it translocates to the perinuclear region and colocalizes with ER and mitochondria [108].

Different mechanisms are suggested for activation of NLRP3 including reactive oxygen species (ROS) generation, lysosomal damage, and cytosolic K^+ efflux [108]. Mitochondria play a key role in activation of NLRP3 inflammasome since all the mechanisms suggested for this induction affect mitochondrial function and integrity [109]. A study by Kenichi et al. showed that Chlamydia pneumococcus (CP) infection in mice resulted in a decrease in mitochondrial membrane potential and oxygen consumption, suggesting that mitochondrial dysfunction might be linked to IL-1β secretion through activation of the NLRP3 inflammasome. It was shown that mtDNA is important for NLRP3 activation since cells lacking mitochondria (ρ0) were unable to secrete IL-1β in response to NLRP3 activation. Moreover, mtDNA was immunoprecipitated with NLRP3 inflammasome, showing an interaction between the two components [110].

Apoptosis is required for release of mtDNA since inhibiting apoptosis by overexpression of anti-apoptotic Bcl-2 reduced IL-1β secretion. Detailed analysis revealed that mtDNA bound to NLRP3 is in fact oxidized, since 8-OH-dG nucleotides were detected in association with NLRP3. To further confirm this phenomenon, mtDNA was analyzed using liquid chromatography-tandem mass spectrometry and a three-fold increase in the amount of 8-OH-dG was found. Treating cells with oxidized DNA increased IL-1β production, but not in the absence of NLRP3. All in all, this study proved that mitochondrial dysfunction induces apoptosis leading to production of oxidized mtDNA which in turn activates NLRP3 inflammasome [110].

It has also been shown that blocking autophagy results in accumulation of mitochondrially-generated ROS capable of activating the NLRP3 inflammasome [111]. LC3B-deficient macrophages had a higher level of active caspase 1 in response to LPS and ATP treatment. Also, higher levels of IL-1β were secreted into the medium. Similarly, treatment of Beclin1 heterozygous macrophages with LPS and ATP resulted in an increase in cleaved caspase 1 and IL-1β. Analysis of cells using transmission electron microscopy (TEM) revealed accumulation of swollen mitochondria in LC3B and Beclin1-deficient cells. This phenotype is characteristic of dysfunctional mitochondria capable of producing high levels of ROS [111].

The role of mitochondria in this process was further confirmed using ρ0 cells which showed an inhibition of caspase 1 activation and IL-1β production in response to LPS and ATP. On the other hand, treating cells with rotenone, which inhibits the mitochondrial respiratory chain at the level of Complex I and which generates large amounts of ROS, had the opposite effect to increase caspase 1 activation and IL-1β production. Co-treating the cells with mito-TEMPO, a scavenger of mitochondrial ROS, blocked the secretion of IL-1β, showing the involvement of mitochondrial ROS in this process [111]. These investigators further examined the role of cytosolic mtDNA. Treating macrophages with LPS and ATP resulted in an increase in mtDNA in the cytoplasm which was blocked by mito-TEMPO. Conversely, rotenone treatment together with LPS and ATP significantly increased cytosolic mtDNA in macrophages. This was further increased in LC3B- and Beclin1-deficient macrophages [111].

Involvement of mtDNA was further confirmed by treating the cells with DNaseI. These cells displayed a reduced secretion of IL-1β in response to LPS and ATP treatment. Inhibition of NLRP3 blocked the secretion of IL-1β and IL-18 in LC3B- and Beclin1-deficient macrophages treated with LPS and ATP or rotenone. Genetic deletion of NLRP3 had the same effect. This effect was due to inhibition of mtDNA release to cytosol in the absence of NLRP3. Same results were obtained in two murine models of sepsis [111]. These results show the involvement of mitochondrial DNA and ROS in inducing inflammation through NLRP3 pathway.

ROS and NF-κB. Nuclear factor κB (NF-κB), is a transcription factor with a major role in inflammation through inducing the expression of pro-IL-1β and NLRP3 inflammasome. NF-κB comprises of different transcription factors: p50 or NF-κB1, p52 or NF-κB2, RELA or p65, RELB and c-REL. These factor are sequestered in the cytoplasm by inhibitors of NF-κB (IKBs) (reviewed in [112]).

There are two different pathways mediating NF-κB activation: canonical and non-canonical [112]. Different stimuli such as ligands for antigen receptors, cytokine receptors, and pattern-recognition receptors activate the canonical pathway. IκB kinase (IKK) complex is a central molecule in this pathway which is composed of catalytic (IKKα and IKKβ) and regulatory (IKKγ) subunits. Upon activation, IκBα is phosphorylated by IKK, triggering its ubiquitination and proteasomal degradation finally leading to the nuclear translocation of NF-κB complexes such as RelA and c-Rel dimers [112]. In the non-canonical NF-κB pathway, the NF-κB2/RelB dimer is activated in response to selective receptor signals. In this pathway, a central step is stabilization of protein kinase NIK which together with IKKα induce processing of

NF-κB precursor protein, p100. This processing not only produces NF-κB2 but also results in nuclear translocation of NF-κB2/RelB dimer [113]. A very recent study has shown that NF-κB can prevent excessive inflammation by inducing delayed accumulation of the autophagy receptor p62/SQSTM1 [114].

Reactive oxygen species (ROS) such as hydrogen peroxide, superoxide anion, and hydroxyl radical, are byproducts of mitochondrial respiratory chain but can be produced by cytoplasmic enzymes as well. Under normal conditions, ROS are the main component of redox signaling important for maintaining tissue homeostasis by oxygen sensing, cell growth, migration and differentiation. Sustained production of ROS, however, can cause cell damage [115].

Production of ROS is essential for innate immunity and inflammation. During pathogen invasion, activated neutrophils emit a burst of ROS which are bactericidal [116]. ROS has been proposed to trigger the activation of the NLRP3 inflammasome. It has been shown that activators of NLRP3 such as uric acid crystals, silica, asbestos, and alum can trigger ROS production as well [117]. Different mechanisms have been suggested for ROS production by NLRP3 activators such as potassium efflux and frustrated phagocytosis [117, 118], however the exact mechanism is still unclear.

Recent evidence suggests that there is cross-talk between ROS and NF-κB pathways. ROS inhibits the phosphorylation of $I\kappa B\alpha$, therefore prevents activation of NF-κB [119]. In addition, S-glutathionylation of $IKK\beta$ by ROS inhibits its activity [120]. Finally, NIK is activated by ROS through inhibition of phosphatases and formation of NIK/TRAF6 complexes and the following recruitment to the IL-1 receptor [119]. On the other hand, the NF-κB pathway itself can regulate ROS levels by inducing the expression of antioxidant proteins such as SODs, GST, MT3, and FHC [115].

Bacteria-like molecules (hsp60, cardiolipin). As discussed before, mitochondria play a major role in activation of NLRP3 inflammasome. Different components of mitochondria are involved in this process. The role of mitochondrially generated ROS and mtDNA has been described in previous sections. Cardiolipin is another molecule which is able to activate the NLRP3 inflammasome. Cardiolipin is a phospholipid found in bacteria and the inner membrane of mitochondria. It has a very specific structure, highly acid with a head glycerol group and two phosphatidylglyceride backbone fragments instead of one. Cardiolipin interacts with components of oxidative phosphorylation (OXPHOS) complexes and stabilizes respiratory supercomplexes (see review [121]). Cardiolipin also plays a functional role in mitochondria. It is involved in apoptosis through interacting with cytochrome *c*. When peroxidized, cardiolipin releases cytochrome *c* to the cytosol where it binds to Apaf-1 which eventually results in the activation of caspase 9 [121].

Cardiolipin has been implicated in inflammation through direct interaction with NLRP3 inflammasome. A study by Iyre et al. has shown the involvement of cardiolipin in this process. Mitochondrial destabilization results in transfer of cardiolipin to the outer membrane. There, it acts as an endogenous PAMP, since it resembles a bacterial lipid. Cardiolipin binds to the leucine-rich repeat domain of NLRP3 and this interaction is sufficient for activation of the NLRP3 inflammasome. In fact, down-regulation of cardiolipin by using siRNA against cardiolipin synthase reduced caspase-1 activation and IL-1β production which occurs downstream of NLRP3 activation [8].

Heat-shock proteins (Hsp), as the name implies, were first discovered in response to high temperatures. They are highly conserved from prokaryotes to eukaryotes. Under physiological conditions, they act as chaperones involved in protein folding and transport [122]. Hsp60 is a mitochondrial molecular chaperone. It is involved in folding proteins into their functional structure and restoring the structure of misfolded or denatured proteins. It can also serve as a DAMP and activate the innate immune response if released in the extracellular environment [123].

In the central nervous system, inflammation is protective against pathogens and cell debris. On the other hand, reactive oxygen and nitrogen species generated as byproducts of the immune response can result in neuronal injury and cell death. In pathological conditions in the CNS, microglia (CNS-resident monocytes) are activated. Hsp60 released from damaged neurons binds to TLR4, a membrane-bound member of TLR family, on the surface of microglia. Myeloid differentiation factor 88 (MyD88) is an adaptor protein which is recruited to TLR4 upon its activation which eventually induces NF-κB and inflammation [124].

Rosenberger et al. showed an increase in Hsp60 expression in neurons and oligodendrocytes upon injury. It was also detected in the cerebrospinal fluid (CSF) in a mouse model of stroke. Intrathecal injection of Hsp60 in mouse resulted in neurodegeneration and demyelination. This effect requires both TLR4 and MyD88 since mice lacking these molecules didn't develop neural injury after Hsp60 injection [125]. Finally, a study by Kim et al. showed involvement of Hsp60 in apoptosis in cardiomyocytes. Hsp60 interacted with TLR4 receptor on the surface of cardiomyocytes, activating downstream NF-κB leading to apoptosis [126].

Unstable Allies: Fusion and Fission

Mitochondrial fission is governed by the small GTPase Drp1, which interacts with mitochondrial outer membrane receptors Fis1, Mff, and Mid49. Drp1 forms multimeric spirals that constrict the mitochondrion (reviewed in [17]). Fusion is mediated by the structurally related mitofusins 1 and 2 (Mfn1, Mfn2) which are anchored to the mitochondrial outer membrane by a transmembrane sequence, and by Opa1, which is responsible for inner membrane fusion. Mfn2 can also be found in ER membranes and is thought to also play a role in tethering ER to mitochondria to enhance calcium delivery to mitochondria [127]. Loss of ER tethering is thought to be the reason conditional knockout of Mfn2 protects against ischemia/reperfusion injury [20].

Mitochondria exist in a dynamic state, alternating between isolated organelles and extensive alliances. Like political alliances, fusion and fission are responses to environmental cues, energy availability, and cell programming. For instance, during cell division, mitochondria must be distributed between the daughter cells. While this may be symmetric, in the case of stem cells, mitochondria may be unequally apportioned between the cell that will undergo differentiation and the cell that will retain stemness, the latter retaining the "younger" mitochondria [128]. Mitochondrial fission itself is often asymmetric, resulting in two mitochondria that differ with respect to membrane potential; the low-potential mitochondrion is likely to be a target of mitophagy [129]. Loss of membrane potential is a trigger for Drp1 recruitment and generalized mitochondrial fission, but this is not the only stress that can trigger fission. Oligomycin preserves membrane potential (actually causes hyperpolarization) but results in fission and loss of mitochondrial mass (presumably by mitophagy) [130]. Oxidant stress also triggers fission [130]. It is plausible that cytosolic signals (kinases, cytosolic ATP, glucose metabolites) may regulate fission. Calcineurin regulates Drp1 by dephosphorylation, triggering its translocation to mitochondria; PKA phosphorylates Drp1 [131].

Even cardiomyocytes exhibit a certain degree of alternation between fusion and fission, although the extent of fusion is less dramatic than the complex branched networks of other cell types. Excluding pathologic settings of complete knockout of fusion or fission factors, the extent of fusion gives rise to mitochondria that may extend two sarcomere lengths, and fission may result in mitochondria about half a sarcomere length [132, 133]. This modest dynamic range may be an underestimate because transmission electron microscopy may cut through a mitochondrion that extends out of plane, much like the Loch Ness monster if only viewed at the plane of the ocean surface (Fig. 10.1). The limitations of electron microscopy are only exceeded by those of tissue homogenization and flow cytometric analysis, which is equivalent to putting poor Nessie through a wood chipper and scoring the size of the sawdust.

Given the rigid architecture of the cardiomyocyte, what can be accomplished by altering the extent of fusion and fission? Fission will alter the surface area to volume ratio, which may affect intra-mitochondrial volume and Ca^{2+} concentration, which in turn will affect OXPHOS and ATP production. Fusion will facilitate redistribution of components such as newly-synthesized proteins and mtDNA [134–136]. Fission is also essential for mitophagy, which recent studies in yeast have shown to be selective, in that matrix components are eliminated at different rates, implying some degree of segregation of specific components [18]. Dengjel and Abeliovich state "*These results are consistent with models in which phase separation within the mitochondrial matrix leads to unequal segregation of proteins during mitochondrial fission. Repeated fusion and fission cycles may thus lead to "distillation" of components that*

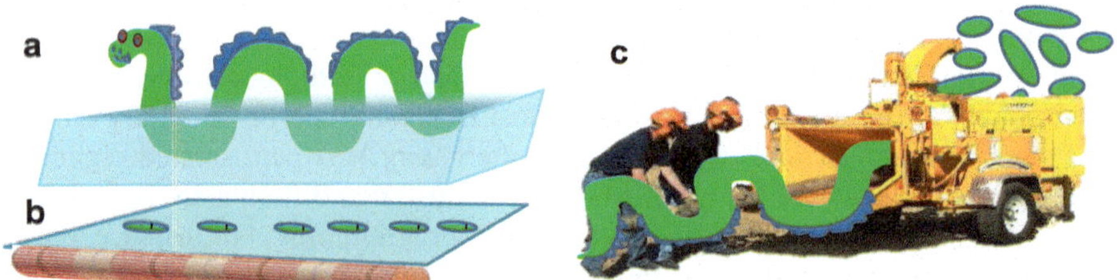

Fig. 10.1 Mitochondria in three dimensions, as a 2-D projection, and after homogenization

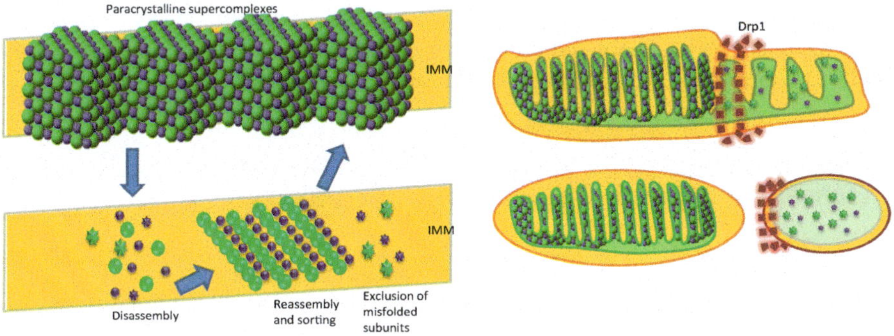

Fig. 10.2 Concept of OXPHOS complexes as a paracrystalline array, and mechanism of Drp-1 mediated asymmetric fission

are destined for degradation." [19] Similar studies remain to be conducted for inner membrane OXPHOS components, which are arranged in paracrystalline supercomplexes. If these supercomplexes are dynamic, undergoing cycles of assembly and disassembly, then there is an opportunity for nonfunctional subunits to be selectively excluded if misfolding precludes their incorporation during the next cycle of assembly. One question is how fission proteins might participate in the segregation process. One model we have proposed is that the supercomplexes confer rigidity to the mitochondrion, whereas disorganized membrane proteins may result in a more fluid membrane. The cinching action of Drp1 oligomers may be unable to constrict the supercomplex-rich segment of the mitochondrion but can easily pinch off the less rigid segment that would contain subunits and complexes that are too distorted to incorporate into supercomplexes. Repeated cycles of assembly/disassembly coupled with fission would serve to parsimoniously remove only the misfolded or damaged components (Fig. 10.2).

Fusion plays an equally important role in the cell. Regions of the mitochondria have been shown to be specialized for importing protein and additionally are enriched for mtDNA and mitofusin 2. The presence of Mfn2 suggests that these specialized mitochondria may 'share their wealth' by fusing with the rest of the network and delivering newly-imported proteins and newly-replicated mtDNA. Conditional deletion of Mfn1 and 2 in the heart results in mitochondrial dysfunction and heart failure [21]. As mitochondrial homeostasis depends upon both fission and fusion, it is not surprising that deletion of either element carries significant consequences for the heart.

The topic would not be complete without discussing Opa1, the factor responsible for inner membrane fusion and, after proteolytic processing, fission. Full-length Opa1 supports fusion and is important for maintenance of crista structure. Opa1 is processed by mitochondrial proteases Yme1L and Oma1, giving rise to several shorter forms of Opa1 that tend to promote fission, or at the very least, interfere with fusion. The short forms colocalize with Drp1 and ER-mitochondrial contact sites [22]. Loss of mitochondrial membrane potential is sufficient to activate Oma1, leading to processing of Opa1 [23]; elegantly, loss of membrane potential arrests the import and PARL-mediated processing of PINK1, resulting in accumulation of PINK1 and initiation of mitophagy (including recruitment of Drp1) [137, 138].

Defense: Isolate and Eliminate Troublemakers

Mitophagy. Mitophagy is the process cells use to clear damaged or unwanted mitochondria via autophagy. Loss of mitochondrial components required for oxidative phosphorylation to sustain the chemiosmotic gradient in mitochondria leads to membrane potential depolarization. Loss of mitochondrial membrane potential results in accumulation of PINK1 on the damaged mitochondria (discussed earlier). PINK1-mediated phosphorylation of ubiquitin and perhaps Parkin itself facilitates the activation and recruitment of the said E3 ubiquitin ligase to the mitochondria [139–143]. Parkin then ubiquitinates several outer mitochondrial membrane proteins causing recruitment of autophagy adapter proteins such as p62/SQSTM1, optineurin, and/or NDP52, all of which contain a ubiquitin binding domain (UBA) and an LC3 interacting region (LIR) [144–152]. The ubiquitin proteasome system plays an important role in the initial phase of mitophagy for degradation of some of the ubiquitinated outer membrane-associated proteins [152, 153]. Recruitment and binding of autophagy adapter proteins facilitates the formation of the autophagosome around the isolated mitochondrion allowing for autophagic engulfment and routing to the lysosome. It is important to note however that different models and/or stimulus may dictate which adapter proteins will be relevant to support mitophagy. Mitophagy has been shown to a critical factor in protecting the heart against ischemia/reperfusion damage [154, 155]. It is important to note that mitophagy can be triggered via mitochondria-specific unfolded protein response even without loss of membrane potential [156].

Mitochondrial Derived Vesicles. The importance of mitochondrial-derived vesicles (MDVs) as a mechanism of mitochondrial quality control is an emerging story. MDVs have been observed to target to peroxisomes [157], and late endosomes/lysosomes [158]. Little is currently understood about the role of MDV delivery to peroxisomes; however, the E3 ubiquitin ligase MULAN was identified as one such protein cargo [157, 159]. MDVs fated for transport to lysosomes contain an abundance of oxidized proteins and are destined for degradation [160]. Moreover, it was found that formation of these lysosome-destined MDVs requires PINK1 and Parkin [161]. McBride and colleagues offer the plausible suggestion that MDVs may act as a first line of defense to isolate and eliminate small portions of the mitochondrion which are characterized by protein import arrest and PINK1 accumulation. Elimination of these dysfunctional regions of mitochondria may serve to preserve the mitochondria and ultimately protect the whole cell [162].

Proteostasis. Proteostasis is a key element of mitochondrial quality control and delays the progression of ageing, development of cancer, and neurodegenerative diseases [163–166]. The buildup of misfolded proteins can impact OXPHOS function with deleterious consequences. Numerous mechanisms have evolved to ensure mitochondrial protein homeostasis. Proteostasis engages multiple systems sequentially and in proportion to the magnitude of the problem and its chronicity: molecular chaperones, sulfhydryl repair enzymes, proteases, protein sorting mechanisms, and mitophagy.

By assisting in the proper folding of new proteins, and refolding of misfolded ones, molecular chaperones contribute greatly to mitochondrial proteostasis. Cytosolic heat shock proteins Hsp70 and Hsp90 are important chaperones known to protect the hydrophobic regions of newly synthesized mitochondrial proteins as they are imported into mitochondria [167]. Hsp90 is a cytosolic chaperone that was recently shown to be important for facilitating mitochondrial protein import through the translocase of outer membrane (TOM) complex [168]. Mitochondrial Hsp70 (mtHsp70) and the complex associated with Hsp60 and Hsp10 were identified as important facilitators of folding of proteins entering the mitochondria [169, 170].

Mitochondrial proteases are estimated to degrade 6–12% of proteins per hour in yeast and deal with misfolded proteins or excess, unincorporated proteins to prevent their aggregation [171]. Mitochondrial proteases regulate protein half-life; it is noteworthy that the half-life of mitochondrial proteins vary widely, indicating that a variety of different mechanisms govern their turnover. Protein quality control is monitored by several ATP-dependent peptidases such as iAAA, mAAA (AAA—ATPasese associated with diverse cellular activities), Lon protease homologue (LONP), and Clp protease proteolytic subunit (CLPP). ATP-independent peptidases include Atp23 homolog and inner membrane associated HTRA2/OMI. PITRM1 is a key metalloprotesase involved in monitoring and degrading proteins in the mitochondrial matrix [175]. These proteases detect and degrade non-assembled or misfolded mitochondrial proteins [176]. Processing peptidases are essential partners in protein import into the mitochondria, responsible for removing the mitochondrial import and sorting signals of new proteins as they enter [172]. Key proteases in this category include the mitochondrial processing peptidase (β-MPP a.k.a. PMPCB) and the inner membrane proteases IMMP1L and IMMP2L [173, 174].

An excess of either mitochondrial-encoded or nuclear-encoded OXPHOS components that fail to be assembled into complexes, termed mitonuclear imbalance, activates the mitochondrial unfolded protein response (mtUPR) [177]. Activation of the mtUPR leads to the transcription of chaperones and proteases that act to reverse mitochondrial proteotoxic stress [178]. More severe proteotoxic stress leads to sequestration of material in regions of the mitochondria followed by their elimination as MDVs; when large portions of mitochondria contain defective OXPHOS components (e.g., unable to maintain membrane potential), they become targets of mitophagy (discussed above). Excellent reviews on general proteostasis, mitochondrial proteases, and the mitochondrial unfolded protein response and have been published by the groups of Hartl [179], Lopez-Otin [176] and Auwerx [178] respectively.

Conclusions

Mitochondria play a central role in the life (and death) of a cell, regulating metabolism, inflammation, and survival. The dialog between mitochondria and the rest of the cell proceeds via small molecules and ions, lipids, peptides that are mitochondria-encoded, and proteins that are normally sequestered within the mitochondria. Even mitochondrial DNA can serve as a signaling molecule, activating inflammation. Release of cytochrome *c* can trigger cell death, making this a perilous friendship, indeed. The cell has evolved multiple mechanisms to de-escalate damage including antiapoptotic Bcl-2 proteins, ARC, and IAPs. While the benefits of a high-efficiency system for ATP production cannot be denied, the alliance carries considerable risk, and homeostatic happiness is poised on a knife's edge. The real miracle is that it doesn't go awry more frequently.

Best of Frenemies

The best of frenemies, mitochondria cohabit the cell.
Friends today, but tomorrow things could go askance.
When the balance of Bax/bak and Bcl-xL.
Are by Bid, Bad and Noxa, thrown out of balance.

Within the cell, mitochondria squirm.
Between fission and fusion, and occasionally,
Instead of choosing life to affirm,
Opt instead to release cytochrome *c*.

Though mitochondria are great.
At synthesizing copious ATP,
Burning sugar and fats at a prodigious rate;
But they sometimes annoy NLRP3.

When mtDNA helps activate caspase 1.
With cardiolipin and inflammasome assembly.
For secretion of interleukin-1.
It's an inflammation calamity.

But love them we must.
Despite occasional strife.
If we are to be more animated than dust.
And enjoy eukaryotic life.

Roberta A. Gottlieb.
July 23, 2016.

References

1. Rubinstein AD, Eisenstein M, Ber Y, Bialik S, Kimchi A (2011) The autophagy protein atg12 associates with antiapoptotic bcl-2 family members to promote mitochondrial apoptosis. Mol Cell 44:698–709
2. Maiuri MC, Le Toumelin G, Criollo A, Rain JC, Gautier F, Juin P, Tasdemir E, Pierron G, Troulinaki K, Tavernarakis N, Hickman JA, Geneste O, Kroemer G (2007) Functional and physical interaction between bcl-x(l) and a bh3-like domain in beclin-1. EMBO J 26:2527–2539
3. Strappazzon F, Di Rita A, Cianfanelli V, D'Orazio M, Nazio F, Fimia GM, Cecconi F (2016) Prosurvival ambra1 turns into a proapoptotic bh3-like protein during mitochondrial apoptosis. Autophagy 12:963–75
4. Gardai SJ, Hildeman DA, Frankel SK, Whitlock BB, Frasch SC, Borregaard N, Marrack P, Bratton DL, Henson PM (2004) Phosphorylation of bax ser184 by akt regulates its activity and apoptosis in neutrophils. J Biol Chem 279:21085–21095
5. Scorrano L, Ashiya M, Buttle K, Weiler S, Oakes SA, Mannella CA, Korsmeyer SJ (2002) A distinct pathway remodels mitochondrial cristae and mobilizes cytochrome c during apoptosis. Dev Cell 2:55–67
6. Yuan H, Williams SD, Adachi S, Oltersdorf T, Gottlieb RA (2003) Cytochrome c dissociation and release from mitochondria by truncated bid and ceramide. Mitochondrion 2:237–244
7. Saelens X, Festjens N, Vande Walle L, van Gurp M, van Loo G, Vandenabeele P (2004) Toxic proteins released from mitochondria in cell death. Oncogene 23:2861–2874
8. Iyer SS, He Q, Janczy JR, Elliott EI, Zhong Z, Olivier AK, Sadler JJ, Knepper-Adrian V, Han R, Qiao L, Eisenbarth SC, Nauseef WM, Cassel SL, Sutterwala FS (2013) Mitochondrial cardiolipin is required for nlrp3 inflammasome activation. Immunity 39:311–323
9. Heiserman JP, Chen L, Kim BS, Kim SC, Tran AL, Siebenborn N, Knowlton AA (2015) Tlr4 mutation and hsp60-induced cell death in adult mouse cardiac myocytes. Cell Stress Chaperones 20:527–535
10. Hong EE, Okitsu CY, Smith AD, Hsieh CL (2013) Regionally specific and genome-wide analyses conclusively demonstrate the absence of cpg methylation in human mitochondrial DNA. Mol Cell Biol 33:2683–2690
11. Malouitre S, Dube H, Selwood D, Crompton M (2010) Mitochondrial targeting of cyclosporin a enables selective inhibition of cyclophilin-d and enhanced cytoprotection after glucose and oxygen deprivation. Biochem J 425:137–148
12. Vieira HL, Haouzi D, El Hamel C, Jacotot E, Belzacq AS, Brenner C, Kroemer G (2000) Permeabilization of the mitochondrial inner membrane during apoptosis: impact of the adenine nucleotide translocator. Cell Death Differ 7:1146–1154
13. Kokoszka JE, Waymire KG, Levy SE, Sligh JE, Cai J, Jones DP, MacGregor GR, Wallace DC (2004) The adp/atp translocator is not essential for the mitochondrial permeability transition pore. Nature 427:461–465
14. Baines CP, Kaiser RA, Sheiko T, Craigen WJ, Molkentin JD (2007) Voltage-dependent anion channels are dispensable for mitochondrial-dependent cell death. Nat Cell Biol 9:550–555
15. Baines CP, Kaiser RA, Purcell NH, Blair NS, Osinska H, Hambleton MA, Brunskill EW, Sayen MR, Gottlieb RA, Dorn GW, Robbins J, Molkentin JD (2005) Loss of cyclophilin d reveals a critical role for mitochondrial permeability transition in cell death. Nature 434:658–662
16. Giorgio V, von Stockum S, Antoniel M, Fabbro A, Fogolari F, Forte M, Glick GD, Petronilli V, Zoratti M, Szabo I, Lippe G, Bernardi P (2013) Dimers of mitochondrial atp synthase form the permeability transition pore. Proc Natl Acad Sci USA 110:5887–5892
17. Lemasters JJ, Nieminen A-L, Qian T, Trost LC, Elmore SP, Nishimura Y, Crowe RA, Cascio WE, Bradham CA, Brenner DA, Herman B (1998) The mitochondrial permeability transition in cell death: a common mechanism in necrosis, apoptosis and autophagy. Biochim Biophys Acta (BBA)—Bioenerg 1366:177–196

18. Gharanei M, Hussain A, Janneh O, Maddock HL (2013) Doxorubicin induced myocardial injury is exacerbated following ischaemic stress via opening of the mitochondrial permeability transition pore. Toxicol Appl Pharmacol 268:149–156
19. Friberg H, Ferrand-Drake M, Bengtsson F, Halestrap AP, Wieloch T (1998) Cyclosporin a, but not fk 506, protects mitochondria and neurons against hypoglycemic damage and implicates the mitochondrial permeability transition in cell death. J Neurosci 18:5151–5159
20. Hausenloy DJ, Maddock HL, Baxter GF, Yellon DM (2002) Inhibiting mitochondrial permeability transition pore opening: a new paradigm for myocardial preconditioning? Cardiovasc Res 55:534–543
21. Waldmeier PC, Feldtrauer JJ, Qian T, Lemasters JJ (2002) Inhibition of the mitochondrial permeability transition by the nonimmunosuppressive cyclosporin derivative nim811. Mol Pharmacol 62:22–29
22. Healy E, Dempsey M, Lally C, Ryan MP (1998) Apoptosis and necrosis: mechanisms of cell death induced by cyclosporine a in a renal proximal tubular cell line. Kidney Int 54:1955–1966
23. Camilleri B, Bridson JM, Halawa A (2016) Calcineurin inhibitor-sparing strategies in renal transplantation: where are we? a comprehensive review of the current evidence. Exp Clin Transplant 14:471–483
24. Carreira RS, Lee Y, Ghochani M, Gustafsson AB, Gottlieb RA (2010) Cyclophilin d is required for mitochondrial removal by autophagy in cardiac cells. Autophagy 6:462–472
25. Jobe SM, Wilson KM, Leo L, Raimondi A, Molkentin JD, Lentz SR, Di Paola J (2008) Critical role for the mitochondrial permeability transition pore and cyclophilin d in platelet activation and thrombosis. Blood 111:1257–1265
26. Durnian JM, Stewart RMK, Tatham R, Batterbury M, Kaye SB (2007) Cyclosporin-a associated malignancy. Clin Ophthalmol (Auckland, N.Z.) 1:421–430
27. O'Flanagan CH, Morais VA, Wurst W, De Strooper B, O'Neill C (2015) The parkinson's gene pink1 regulates cell cycle progression and promotes cancer-associated phenotypes. Oncogene 34:1363–1374
28. Lee J, Ham S, Lee MH, Kim SJ, Park JH, Lee SE, Chang JY, Joung KH, Kim TY, Kim JM, Sul HJ, Kweon GR, Jo YS, Kim KS, Shong YK, Gasparre G, Chung JK, Porcelli AM, Shong M (2015) Dysregulation of parkin-mediated mitophagy in thyroid hurthle cell tumors. Carcinogenesis 36:1407–1418
29. Xue F, Hu L, Ge R, Yang L, Liu K, Li Y, Sun Y, Wang K (2016) Autophagy-deficiency in hepatic progenitor cells leads to the defects of stemness and enhances susceptibility to neoplastic transformation. Cancer Lett 371:38–47
30. de Almeida FP, Saliba JB, Ribeiro JA, Siqueira RC, Fialho SL, Silva-Cunha A, Jorge R, Messias A (2015) In vivo release and retinal toxicity of cyclosporine-loaded intravitreal device. Doc Ophthalmol Adv Ophthalmol 131:207–214
31. Chen Y, Sawada O, Kohno H, Le YZ, Subauste C, Maeda T, Maeda A (2013) Autophagy protects the retina from light-induced degeneration. J Biol Chem 288:7506–7518
32. Han JY, Kim JS, Son JH (2014) Mitochondrial homeostasis molecules: regulation by a trio of recessive parkinson's disease genes. Exp Neurobiol 23:345–351
33. Garcia-Saez AJ (2012) The secrets of the bcl-2 family. Cell Death Differ 19:1733–1740
34. Brahmbhatt H, Oppermann S, Osterlund EJ, Leber B, Andrews DW (2015) Molecular pathways: leveraging the bcl-2 interactome to kill cancer cells–mitochondrial outer membrane permeabilization and beyond. Clin Cancer Res: Off J Am Assoc Cancer Res 21:2671–2676
35. Hata AN, Engelman JA, Faber AC (2015) The bcl2 family: key mediators of the apoptotic response to targeted anticancer therapeutics. Cancer Discov 5:475–487
36. Billen LP, Kokoski CL, Lovell JF, Leber B, Andrews DW (2008) Bcl-xl inhibits membrane permeabilization by competing with bax. PLoS Biol 6:e147
37. Annis MG, Soucie EL, Dlugosz PJ, Cruz-Aguado JA, Penn LZ, Leber B, Andrews DW (2005) Bax forms multispanning monomers that oligomerize to permeabilize membranes during apoptosis. EMBO J 24:2096–2103
38. Westphal D, Kluck RM, Dewson G (2014) Building blocks of the apoptotic pore: how bax and bak are activated and oligomerize during apoptosis. Cell Death Differ 21:196–205

39. Edlich F, Banerjee S, Suzuki M, Cleland MM, Arnoult D, Wang C, Neutzner A, Tjandra N, Youle RJ (2011) Bcl-x(l) retrotranslocates bax from the mitochondria into the cytosol. Cell 145:104–116
40. Willis SN, Fletcher JI, Kaufmann T, van Delft MF, Chen L, Czabotar PE, Ierino H, Lee EF, Fairlie WD, Bouillet P, Strasser A, Kluck RM, Adams JM, Huang DC (2007) Apoptosis initiated when bh3 ligands engage multiple bcl-2 homologs, not bax or bak. Science (New York, N.Y.) 315:856–859
41. Kuwana T, Bouchier-Hayes L, Chipuk JE, Bonzon C, Sullivan BA, Green DR, Newmeyer DD (2005) Bh3 domains of bh3-only proteins differentially regulate bax-mediated mitochondrial membrane permeabilization both directly and indirectly. Mol Cell 17:525–535
42. Kuwana T, Mackey MR, Perkins G, Ellisman MH, Latterich M, Schneiter R, Green DR, Newmeyer DD (2002) Bid, bax, and lipids cooperate to form supramolecular openings in the outer mitochondrial membrane. Cell 111:331–342
43. Leber B, Lin J, Andrews DW (2007) Embedded together: the life and death consequences of interaction of the bcl-2 family with membranes. Apoptosis: Int J Progr Cell Death 12:897–911
44. Leber B, Lin J, Andrews DW (2010) Still embedded together binding to membranes regulates bcl-2 protein interactions. Oncogene 29:5221–5230
45. Llambi F, Moldoveanu T, Tait SW, Bouchier-Hayes L, Temirov J, McCormick LL, Dillon CP, Green DR (2011) A unified model of mammalian bcl-2 protein family interactions at the mitochondria. Mol Cell 44:517–531
46. Bakhshi A, Jensen JP, Goldman P, Wright JJ, McBride OW, Epstein AL, Korsmeyer SJ (1985) Cloning the chromosomal breakpoint of t(14;18) human lymphomas: clustering around jh on chromosome 14 and near a transcriptional unit on 18. Cell 41:899–906
47. Tsujimoto Y, Finger LR, Yunis J, Nowell PC, Croce CM (1984) Cloning of the chromosome breakpoint of neoplastic b cells with the t(14;18) chromosome translocation. Science 226:1097–9
48. Catz SD, Johnson JL (2001) Transcriptional regulation of bcl-2 by nuclear factor kappa b and its significance in prostate cancer. Oncogene 20:7342–7351
49. Hanada M, Delia D, Aiello A, Stadtmauer E, Reed JC (1993) Bcl-2 gene hypomethylation and high-level expression in b-cell chronic lymphocytic leukemia. Blood 82:1820–1828
50. Jager R, Herzer U, Schenkel J, Weiher H (1997) Overexpression of bcl-2 inhibits alveolar cell apoptosis during involution and accelerates c-myc-induced tumorigenesis of the mammary gland in transgenic mice. Oncogene 15:1787–1795
51. Song L, Coppola D, Livingston S, Cress D, Haura EB (2005) Mcl-1 regulates survival and sensitivity to diverse apoptotic stimuli in human non-small cell lung cancer cells. Cancer Biol Ther 4:267–276
52. Beroukhim R, Mermel CH, Porter D, Wei G, Raychaudhuri S, Donovan J, Barretina J, Boehm JS, Dobson J, Urashima M, Mc Henry KT, Pinchback RM, Ligon AH, Cho YJ, Haery L, Greulich H, Reich M, Winckler W, Lawrence MS, Weir BA, Tanaka KE, Chiang DY, Bass AJ, Loo A, Hoffman C, Prensner J, Liefeld T, Gao Q, Yecies D, Signoretti S, Maher E, Kaye FJ, Sasaki H, Tepper JE, Fletcher JA, Tabernero J, Baselga J, Tsao MS, Demichelis F, Rubin MA, Janne PA, Daly MJ, Nucera C, Levine RL, Ebert BL, Gabriel S, Rustgi AK, Antonescu CR, Ladanyi M, Letai A, Garraway LA, Loda M, Beer DG, True LD, Okamoto A, Pomeroy SL, Singer S, Golub TR, Lander ES, Getz G, Sellers WR, Meyerson M (2010) The landscape of somatic copy-number alteration across human cancers. Nature 463:899–905
53. Glaser SP, Lee EF, Trounson E, Bouillet P, Wei A, Fairlie WD, Izon DJ, Zuber J, Rappaport AR, Herold MJ, Alexander WS, Lowe SW, Robb L, Strasser A (2012) Anti-apoptotic mcl-1 is essential for the development and sustained growth of acute myeloid leukemia. Genes Dev 26:120–125
54. Rampino N, Yamamoto H, Ionov Y, Li Y, Sawai H, Reed JC, Perucho M (1997) Somatic frameshift mutations in the bax gene in colon cancers of the microsatellite mutator phenotype. Science 275:967–969
55. Delbridge AR, Strasser A (2015) The bcl-2 protein family, bh3-mimetics and cancer therapy. Cell Death Differ 22:1071–1080

56. Deveraux QL, Reed JC (1999) IAP family proteins–suppressors of apoptosis. Genes Dev 13:239–252
57. Silke J, Meier P (2013) Inhibitor of apoptosis (IAP) proteins-modulators of cell death and inflammation. Cold Spring Harbor Perspect Biol 5:a008730
58. Du C, Fang M, Li Y, Li L, Wang X (2000) Smac, a mitochondrial protein that promotes cytochrome c-dependent caspase activation by eliminating iap inhibition. Cell 102:33–42
59. Vaux DL, Silke J (2003) Mammalian mitochondrial IAP binding proteins. Biochem Biophys Res Commun 304:499–504
60. Bertrand MJ, Milutinovic S, Dickson KM, Ho WC, Boudreault A, Durkin J, Gillard JW, Jaquith JB, Morris SJ, Barker PA (2008) Ciap1 and ciap2 facilitate cancer cell survival by functioning as e3 ligases that promote rip1 ubiquitination. Mol Cell 30:689–700
61. de Almagro MC, Vucic D (2012) The inhibitor of apoptosis (IAP) proteins are critical regulators of signaling pathways and targets for anti-cancer therapy. Exp Oncol 34:200–211
62. Jaffer S, Orta L, Sunkara S, Sabo E, Burstein DE (2007) Immunohistochemical detection of antiapoptotic protein x-linked inhibitor of apoptosis in mammary carcinoma. Hum Pathol 38:864–870
63. Kluger HM, McCarthy MM, Alvero AB, Sznol M, Ariyan S, Camp RL, Rimm DL, Mor G (2007) The x-linked inhibitor of apoptosis protein (xiap) is up-regulated in metastatic melanoma, and xiap cleavage by phenoxodiol is associated with carboplatin sensitization. J Transl Med 5:6
64. Nakagawa Y, Abe S, Kurata M, Hasegawa M, Yamamoto K, Inoue M, Takemura T, Suzuki K, Kitagawa M (2006) Iap family protein expression correlates with poor outcome of multiple myeloma patients in association with chemotherapy-induced overexpression of multidrug resistance genes. Am J Hematol 81:824–831
65. Koseki T, Inohara N, Chen S, Nunez G (1998) Arc, an inhibitor of apoptosis expressed in skeletal muscle and heart that interacts selectively with caspases. Proc Natl Acad Sci U S A 95:5156–5160
66. Ludwig-Galezowska AH, Flanagan L, Rehm M (2011) Apoptosis repressor with caspase recruitment domain, a multifunctional modulator of cell death. J Cell Mol Med 15:1044–1053
67. Nam YJ, Mani K, Ashton AW, Peng CF, Krishnamurthy B, Hayakawa Y, Lee P, Korsmeyer SJ, Kitsis RN (2004) Inhibition of both the extrinsic and intrinsic death pathways through nonhomotypic death-fold interactions. Mol Cell 15:901–912
68. Gustafsson AB, Tsai JG, Logue SE, Crow MT, Gottlieb RA (2004) Apoptosis repressor with caspase recruitment domain protects against cell death by interfering with bax activation. J Biol Chem 279:21233–21238
69. Donath S, Li P, Willenbockel C, Al-Saadi N, Gross V, Willnow T, Bader M, Martin U, Bauersachs J, Wollert KC, Dietz R, von Harsdorf R (2006) Apoptosis repressor with caspase recruitment domain is required for cardioprotection in response to biomechanical and ischemic stress. Circulation 113:1203–1212
70. Mercier I, Vuolo M, Jasmin JF, Medina CM, Williams M, Mariadason JM, Qian H, Xue X, Pestell RG, Lisanti MP, Kitsis RN (2008) Arc (apoptosis repressor with caspase recruitment domain) is a novel marker of human colon cancer. Cell Cycle 7:1640–7
71. Westermann B (2010) Mitochondrial fusion and fission in cell life and death. Nat Rev Mol Cell Biol 11:872–884
72. Wasiak S, Zunino R, McBride HM (2007) Bax/bak promote sumoylation of drp1 and its stable association with mitochondria during apoptotic cell death. J Cell Biol 177:439–450
73. Karbowski M, Lee YJ, Gaume B, Jeong SY, Frank S, Nechushtan A, Santel A, Fuller M, Smith CL, Youle RJ (2002) Spatial and temporal association of bax with mitochondrial fission sites, drp1, and mfn2 during apoptosis. J Cell Biol 159:931–938
74. Ishihara N, Nomura M, Jofuku A, Kato H, Suzuki SO, Masuda K, Otera H, Nakanishi Y, Nonaka I, Goto Y, Taguchi N, Morinaga H, Maeda M, Takayanagi R, Yokota S, Mihara K (2009) Mitochondrial fission factor drp1 is essential for embryonic development and synapse formation in mice. Nat Cell Biol 11:958–966

75. Arnoult D, Grodet A, Lee YJ, Estaquier J, Blackstone C (2005) Release of opa1 during apoptosis participates in the rapid and complete release of cytochrome c and subsequent mitochondrial fragmentation. J Biol Chem 280:35742–35750
76. Sheridan C, Delivani P, Cullen SP, Martin SJ (2008) Bax- or bak-induced mitochondrial fission can be uncoupled from cytochrome c release. Mol Cell 31:570–585
77. Wang JX, Li Q, Li PF (2009) Apoptosis repressor with caspase recruitment domain contributes to chemotherapy resistance by abolishing mitochondrial fission mediated by dynamin-related protein-1. Can Res 69:492–500
78. Arnoult D, Rismanchi N, Grodet A, Roberts RG, Seeburg DP, Estaquier J, Sheng M, Blackstone C (2005) Bax/bak-dependent release of ddp/timm8a promotes drp1-mediated mitochondrial fission and mitoptosis during programmed cell death. Curr Biol 15:2112–2118
79. Lee C, Yen K, Cohen P (2013) Humanin: a harbinger of mitochondrial-derived peptides? Trends Endocrinol Metab 24:222–228
80. Hashimoto Y, Niikura T, Tajima H, Yasukawa T, Sudo H, Ito Y, Kita Y, Kawasumi M, Kouyama K, Doyu M, Sobue G, Koide T, Tsuji S, Lang J, Kurokawa K, Nishimoto I (2001) A rescue factor abolishing neuronal cell death by a wide spectrum of familial alzheimer's disease genes and abeta. Proc Natl Acad Sci USA 98:6336–6341
81. Guo B, Zhai D, Cabezas E, Welsh K, Nouraini S, Satterthwait AC, Reed JC (2003) Humanin peptide suppresses apoptosis by interfering with bax activation. Nature 423:456–461
82. Voigt A, Jelinek HF (2016) Humanin: a mitochondrial signaling peptide as a biomarker for impaired fasting glucose-related oxidative stress. Physiol Rep 4:e12796
83. Muzumdar RH, Huffman DM, Calvert JW, Jha S, Weinberg Y, Cui L, Nemkal A, Atzmon G, Klein L, Gundewar S, Ji SY, Lavu M, Predmore BL, Lefer DJ (2010) Acute humanin therapy attenuates myocardial ischemia and reperfusion injury in mice. Arterioscler Thromb Vasc Biol 30:1940–1948
84. Zhai D, Luciano F, Zhu X, Guo B, Satterthwait AC, Reed JC (2005) Humanin binds and nullifies bid activity by blocking its activation of bax and bak. J Biol Chem 280:15815–15824
85. Hashimoto Y, Kurita M, Aiso S, Nishimoto I, Matsuoka M (2009) Humanin inhibits neuronal cell death by interacting with a cytokine receptor complex or complexes involving cntf receptor alpha/wsx-1/gp130. Mol Biol Cell 20:2864–2873
86. Ying G, Iribarren P, Zhou Y, Gong W, Zhang N, Yu ZX, Le Y, Cui Y, Wang JM (2004) Humanin, a newly identified neuroprotective factor, uses the g protein-coupled formylpeptide receptor-like-1 as a functional receptor. J Immunol 172:7078–7085
87. Fuku N, Pareja-Galeano H, Zempo H, Alis R, Arai Y, Lucia A, Hirose N (2015) The mitochondrial-derived peptide mots-c: a player in exceptional longevity? Aging Cell 14:921–923
88. Ming W, Lu G, Xin S, Huanyu L, Yinghao J, Xiaoying L, Chengming X, Banjun R, Li W, Zifan L (2016) Mitochondria related peptide mots-c suppresses ovariectomy-induced bone loss via ampk activation. Biochem Biophys Res Commun 476:412–419
89. Lee C, Zeng J, Drew BG, Sallam T, Martin-Montalvo A, Wan J, Kim SJ, Mehta H, Hevener AL, de Cabo R, Cohen P (2015) The mitochondrial-derived peptide mots-c promotes metabolic homeostasis and reduces obesity and insulin resistance. Cell Metab 21:443–454
90. Randle PJ, Garland PB, Hales CN, Newsholme EA (1963) The glucose fatty-acid cycle. Its role in insulin sensitivity and the metabolic disturbances of diabetes mellitus. Lancet 7891:785–9
91. Randle PJ (1998) Regulatory interactions between lipids and carbohydrates: the glucose fatty acid cycle after 35 years. Diabetes Metab Rev 14:263–283
92. Sugden MC, Holness MJ (2006) Mechanisms underlying regulation of the expression and activities of the mammalian pyruvate dehydrogenase kinases. Arch Physiol Biochem 112:139–149
93. McGarry JD (2002) Banting lecture 2001: dysregulation of fatty acid metabolism in the etiology of type 2 diabetes. Diabetes 51:7–18
94. Foster DW (2012) Malonyl-coa: the regulator of fatty acid synthesis and oxidation. J Clin Investig 122:1958–1959

95. Kelley DE, Goodpaster B, Wing RR, Simoneau JA (1999) Skeletal muscle fatty acid metabolism in association with insulin resistance, obesity, and weight loss. Am J Physiol 277:E1130–1141
96. Prior SJ, Ryan AS, Stevenson TG, Goldberg AP (2014) Metabolic inflexibility during submaximal aerobic exercise is associated with glucose intolerance in obese older adults. Obesity (Silver Spring) 22:451–457
97. Ukropcova B, Sereda O, de Jonge L, Bogacka I, Nguyen T, Xie H, Bray GA, Smith SR (2007) Family history of diabetes links impaired substrate switching and reduced mitochondrial content in skeletal muscle. Diabetes 56:720–727
98. Turer AT, Malloy CR, Newgard CB, Podgoreanu MV (2010) Energetics and metabolism in the failing heart: important but poorly understood. Curr Opin Clin Nutr Metab Care 13:458–465
99. Bergouignan A, Antoun E, Momken I, Schoeller DA, Gauquelin-Koch G, Simon C, Blanc S (1985) Effect of contrasted levels of habitual physical activity on metabolic flexibility. J Appl Physiol 2013(114):371–379
100. Muoio DM (2014) Metabolic inflexibility: when mitochondrial indecision leads to metabolic gridlock. Cell 159:1253–1262
101. Goulopoulou S, McCarthy CG, Webb RC (2016) Toll-like receptors in the vascular system: sensing the dangers within. Pharmacol Rev 68:142–167
102. Pugin J (2012) How tissue injury alarms the immune system and causes a systemic inflammatory response syndrome. Ann Intensive Care 2:27
103. Zhang Q, Raoof M, Chen Y, Sumi Y, Sursal T, Junger W, Brohi K, Itagaki K, Hauser CJ (2010) Circulating mitochondrial damps cause inflammatory responses to injury. Nature 464:104–107
104. Collins LV, Hajizadeh S, Holme E, Jonsson IM, Tarkowski A (2004) Endogenously oxidized mitochondrial DNA induces in vivo and in vitro inflammatory responses. J Leukoc Biol 75:995–1000
105. Cossarizza A, Pinti M, Nasi M, Gibellini L, Manzini S, Roat E, De Biasi S, Bertoncelli L, Montagna JP, Bisi L, Manzini L, Trenti T, Borghi V, Mussini C (2011) Increased plasma levels of extracellular mitochondrial DNA during HIV infection: a new role for mitochondrial damage-associated molecular patterns during inflammation. Mitochondrion 11:750–755
106. Garcia-Martinez I, Santoro N, Chen Y, Hoque R, Ouyang X, Caprio S, Shlomchik MJ, Coffman RL, Candia A, Mehal WZ (2016) Hepatocyte mitochondrial DNA drives nonalcoholic steatohepatitis by activation of tlr9. J Clin Investig 126:859–864
107. Latz E, Visintin A, Espevik T, Golenbock DT (2004) Mechanisms of tlr9 activation. J Endotoxin Res 10:406–412
108. Abderrazak A, Syrovets T, Couchie D, El Hadri K, Friguet B, Simmet T, Rouis M (2015) Nlrp3 inflammasome: from a danger signal sensor to a regulatory node of oxidative stress and inflammatory diseases. Redox Biol 4:296–307
109. Gurung P, Lukens JR, Kanneganti TD (2015) Mitochondria: diversity in the regulation of the nlrp3 inflammasome. Trends Mol Med 21:193–201
110. Shimada K, Crother TR, Karlin J, Dagvadorj J, Chiba N, Chen S, Ramanujan VK, Wolf AJ, Vergnes L, Ojcius DM, Rentsendorj A, Vargas M, Guerrero C, Wang Y, Fitzgerald KA, Underhill DM, Town T, Arditi M (2012) Oxidized mitochondrial DNA activates the nlrp3 inflammasome during apoptosis. Immunity 36:401–414
111. Nakahira K, Haspel JA, Rathinam VA, Lee SJ, Dolinay T, Lam HC, Englert JA, Rabinovitch M, Cernadas M, Kim HP, Fitzgerald KA, Ryter SW, Choi AM (2011) Autophagy proteins regulate innate immune responses by inhibiting the release of mitochondrial DNA mediated by the nalp3 inflammasome. Nat Immunol 12:222–230
112. Hayden MS, Ghosh S (2008) Shared principles in nf-kappab signaling. Cell 132:344–362
113. Sun SC (2012) The noncanonical nf-kappab pathway. Immunol Rev 246:125–140
114. Zhong Z, Umemura A, Sanchez-Lopez E, Liang S, Shalapour S, Wong J, He F, Boassa D, Perkins G, Ali SR, McGeough MD, Ellisman MH, Seki E, Gustafsson AB, Hoffman HM, Diaz-Meco MT, Moscat J, Karin M (2016) Nf-kappab restricts inflammasome activation via elimination of damaged mitochondria. Cell 164:896–910

115. Morgan MJ, Liu ZG (2011) Crosstalk of reactive oxygen species and nf-kappab signaling. Cell Res 21:103–115
116. Bogdan C, Rollinghoff M, Diefenbach A (2000) Reactive oxygen and reactive nitrogen intermediates in innate and specific immunity. Curr Opin Immunol 12:64–76
117. O'Neill LA (2008) Immunology. How frustration leads to inflammation. Science 320:619–620
118. Bergstrand H (1990) The generation of reactive oxygen-derived species by phagocytes. Agents Actions Suppl 30:199–211
119. Li Q, Engelhardt JF (2006) Interleukin-1beta induction of nfkappab is partially regulated by h2o2-mediated activation of nfkappab-inducing kinase. J Biol Chem 281:1495–1505
120. Finn NA, Kemp ML (2012) Pro-oxidant and antioxidant effects of n-acetylcysteine regulate doxorubicin-induced nf-kappa b activity in leukemic cells. Mol BioSyst 8:650–662
121. Paradies G, Paradies V, De Benedictis V, Ruggiero FM, Petrosillo G (2014) Functional role of cardiolipin in mitochondrial bioenergetics. Biochem Biophys Acta 1837:408–417
122. Kaul G, Thippeswamy H (2011) Role of heat shock proteins in diseases and their therapeutic potential. Indian Journal Microbiol 51:124–131
123. Grundtman C, Kreutmayer SB, Almanzar G, Wick MC, Wick G (2011) Heat shock protein 60 and immune inflammatory responses in atherosclerosis. Arterioscler Thromb Vasc Biol 31:960–968
124. Rosenberger K, Dembny P, Derkow K, Engel O, Kruger C, Wolf SA, Kettenmann H, Schott E, Meisel A, Lehnardt S (2015) Intrathecal heat shock protein 60 mediates neurodegeneration and demyelination in the cns through a tlr4- and myd88-dependent pathway. Mol Neurodegener 10:5
125. Rosenberger K, Dembny P, Derkow K, Engel O, Krüger C, Wolf S, Kettenmann H, Schott E, Meisel A, Lehnardt S (2015) Intrathecal heat shock protein 60 mediates neurodegeneration and demyelination in the cns through a tlr4- and myd88-dependent pathway. Mol Neurodegener 10:5
126. Kim SC, Stice JP, Chen L, Jung JS, Gupta S, Wang Y, Baumgarten G, Trial J, Knowlton AA (2009) Extracellular heat shock protein 60, cardiac myocytes, and apoptosis. Circ Res 105:1186–1195
127. de Brito OM, Scorrano L (2008) Mitofusin 2 tethers endoplasmic reticulum to mitochondria. Nature 456:605–610
128. Katajisto P, Dohla J, Chaffer CL, Pentinmikko N, Marjanovic N, Iqbal S, Zoncu R, Chen W, Weinberg RA, Sabatini DM (2015) Stem cells. Asymmetric apportioning of aged mitochondria between daughter cells is required for stemness. Science 348:340–343
129. Twig G, Elorza A, Molina AJ, Mohamed H, Wikstrom JD, Walzer G, Stiles L, Haigh SE, Katz S, Las G, Alroy J, Wu M, Py BF, Yuan J, Deeney JT, Corkey BE, Shirihai OS (2008) Fission and selective fusion govern mitochondrial segregation and elimination by autophagy. EMBO J 27:433–446
130. Leonard AP, Cameron RB, Speiser JL, Wolf BJ, Peterson YK, Schnellmann RG, Beeson CC, Rohrer B (2015) Quantitative analysis of mitochondrial morphology and membrane potential in living cells using high-content imaging, machine learning, and morphological binning. Biochem Biophys Acta 1853:348–360
131. Cereghetti GM, Stangherlin A, Martins de Brito O, Chang CR, Blackstone C, Bernardi P, Scorrano L (2008) Dephosphorylation by calcineurin regulates translocation of drp1 to mitochondria. Proc Natl Acad Sci U S Am 105:15803–8
132. Laguens RP, Gómez-Dumm CL (1967) Fine structure of myocardial mitochondria in rats after exercise for one-half to two hours. Circ Res 21:271–280
133. Dorn GW (2013) Mitochondrial dynamics in heart disease. Biochem Biophys Acta 1833:233–241
134. Hermann GJ, Thatcher JW, Mills JP, Hales KG, Fuller MT, Nunnari J, Shaw JM (1998) Mitochondrial fusion in yeast requires the transmembrane gtpase fzo1p. J Cell Biol 143:359–373
135. Rapaport D, Brunner M, Neupert W, Westermann B (1998) Fzo1p is a mitochondrial outer membrane protein essential for the biogenesis of functional mitochondria in saccharomyces cerevisiae. J Biol Chem 273:20150–20155

136. Stotland A, Gottlieb RA (2016) Alpha-mhc mitotimer mouse: in vivo mitochondrial turnover model reveals remarkable mitochondrial heterogeneity in the heart. J Mol Cell Cardiol 90:53–58
137. Jin SM, Lazarou M, Wang C, Kane LA, Narendra DP, Youle RJ (2010) Mitochondrial membrane potential regulates pink1 import and proteolytic destabilization by parl. J Cell Biol 191:933–942
138. Greene AW, Grenier K, Aguileta MA, Muise S, Farazifard R, Haque ME, McBride HM, Park DS, Fon EA (2012) Mitochondrial processing peptidase regulates pink1 processing, import and parkin recruitment. EMBO Rep 13:378–385
139. Shiba-Fukushima K, Imai Y, Yoshida S, Ishihama Y, Kanao T, Sato S, Hattori N (2012) Pink1-mediated phosphorylation of the parkin ubiquitin-like domain primes mitochondrial translocation of parkin and regulates mitophagy. Sci Rep 2:1002
140. Lazarou M, Sliter DA, Kane LA, Sarraf SA, Wang C, Burman JL, Sideris DP, Fogel AI, Youle RJ (2015) The ubiquitin kinase pink1 recruits autophagy receptors to induce mitophagy. Nature 524:309–314
141. Kazlauskaite A, Kondapalli C, Gourlay R, Campbell DG, Ritorto MS, Hofmann K, Alessi DR, Knebel A, Trost M, Muqit MM (2014) Parkin is activated by pink1-dependent phosphorylation of ubiquitin at ser65. Biochem J 460:127–139
142. Kane LA, Lazarou M, Fogel AI, Li Y, Yamano K, Sarraf SA, Banerjee S, Youle RJ (2014) Pink1 phosphorylates ubiquitin to activate parkin e3 ubiquitin ligase activity. J Cell Biol 205:143–153
143. Iguchi M, Kujuro Y, Okatsu K, Koyano F, Kosako H, Kimura M, Suzuki N, Uchiyama S, Tanaka K, Matsuda N (2013) Parkin-catalyzed ubiquitin-ester transfer is triggered by pink1-dependent phosphorylation. J Biol Chem 288:22019–22032
144. Wong YC, Holzbaur EL (2014) Optineurin is an autophagy receptor for damaged mitochondria in parkin-mediated mitophagy that is disrupted by an als-linked mutation. Proc Natl Acad Sci U S A 111:E4439–4448
145. Tanaka A, Cleland MM, Xu S, Narendra DP, Suen DF, Karbowski M, Youle RJ (2010) Proteasome and p97 mediate mitophagy and degradation of mitofusins induced by parkin. J Cell Biol 191:1367–1380
146. Narendra D, Tanaka A, Suen DF, Youle RJ (2008) Parkin is recruited selectively to impaired mitochondria and promotes their autophagy. J Cell Biol 183:795–803
147. Narendra D, Kane LA, Hauser DN, Fearnley IM, Youle RJ (2010) P62/sqstm1 is required for parkin-induced mitochondrial clustering but not mitophagy; vdac1 is dispensable for both. Autophagy 6:1090–1106
148. Matsuda N, Sato S, Shiba K, Okatsu K, Saisho K, Gautier CA, Sou YS, Saiki S, Kawajiri S, Sato F, Kimura M, Komatsu M, Hattori N, Tanaka K (2010) Pink1 stabilized by mitochondrial depolarization recruits parkin to damaged mitochondria and activates latent parkin for mitophagy. J Cell Biol 189:211–221
149. Geisler S, Holmstrom KM, Skujat D, Fiesel FC, Rothfuss OC, Kahle PJ, Springer W (2010) Pink1/parkin-mediated mitophagy is dependent on vdac1 and p62/sqstm1. Nat Cell Biol 12:119–131
150. Gegg ME, Cooper JM, Chau KY, Rojo M, Schapira AH, Taanman JW (2010) Mitofusin 1 and mitofusin 2 are ubiquitinated in a pink1/parkin-dependent manner upon induction of mitophagy. Hum Mol Genet 19:4861–4870
151. Chen Y, Dorn GW 2nd (2013) Pink1-phosphorylated mitofusin 2 is a parkin receptor for culling damaged mitochondria. Science 340:471–475
152. Chan NC, Salazar AM, Pham AH, Sweredoski MJ, Kolawa NJ, Graham RL, Hess S, Chan DC (2011) Broad activation of the ubiquitin-proteasome system by parkin is critical for mitophagy. Hum Mol Genet 20:1726–1737
153. Yoshii SR, Kishi C, Ishihara N, Mizushima N (2011) Parkin mediates proteasome-dependent protein degradation and rupture of the outer mitochondrial membrane. J Biol Chem 286:19630–19640

154. Huang C, Andres AM, Ratliff EP, Hernandez G, Lee P, Gottlieb RA (2011) Preconditioning involves selective mitophagy mediated by parkin and p62/sqstm1. PLoS One 6:e20975
155. Andres AM, Hernandez G, Lee P, Huang C, Ratliff EP, Sin J, Thornton CA, Damasco MV, Gottlieb RA (2014) Mitophagy is required for acute cardioprotection by simvastatin. Antioxid Redox Signal 21:1960–1973
156. Jin SM, Youle RJ (2013) The accumulation of misfolded proteins in the mitochondrial matrix is sensed by pink1 to induce park2/parkin-mediated mitophagy of polarized mitochondria. Autophagy 9:1750–1757
157. Neuspiel M, Schauss AC, Braschi E, Zunino R, Rippstein P, Rachubinski RA, Andrade-Navarro MA, McBride HM (2008) Cargo-selected transport from the mitochondria to peroxisomes is mediated by vesicular carriers. Curr Biol 18:102–108
158. Soubannier V, McLelland GL, Zunino R, Braschi E, Rippstein P, Fon EA, McBride HM (2012) A vesicular transport pathway shuttles cargo from mitochondria to lysosomes. Curr Biol 22:135–141
159. Braschi E, Zunino R, McBride HM (2009) Mapl is a new mitochondrial sumo e3 ligase that regulates mitochondrial fission. EMBO Rep 10:748–754
160. Soubannier V, Rippstein P, Kaufman BA, Shoubridge EA, McBride HM (2012) Reconstitution of mitochondria derived vesicle formation demonstrates selective enrichment of oxidized cargo. PloS one 7:e52830
161. McLelland GL, Soubannier V, Chen CX, McBride HM, Fon EA (2014) Parkin and pink1 function in a vesicular trafficking pathway regulating mitochondrial quality control. EMBO J 33:282–295
162. Sugiura A, McLelland GL, Fon EA, McBride HM (2014) A new pathway for mitochondrial quality control: mitochondrial-derived vesicles. EMBO J 33:2142–2156
163. Rugarli EI, Langer T (2012) Mitochondrial quality control: a matter of life and death for neurons. EMBO J 31:1336–1349
164. Osellame LD, Duchen MR (2014) Quality control gone wrong: mitochondria, lysosomal storage disorders and neurodegeneration. Br J Pharmacol 171:1958–1972
165. Lopez-Otin C, Blasco MA, Partridge L, Serrano M, Kroemer G (2013) The hallmarks of aging. Cell 153:1194–1217
166. Bulteau AL, Bayot A (2011) Mitochondrial proteases and cancer. Biochem Biophys Acta 1807:595–601
167. Young JC, Agashe VR, Siegers K, Hartl FU (2004) Pathways of chaperone-mediated protein folding in the cytosol. Nat Rev Mol Cell Biol 5:781–791
168. Zanphorlin LM, Lima TB, Wong MJ, Balbuena TS, Minetti CA, Remeta DP, Young JC, Barbosa LR, Gozzo FC, Ramos CH (2016) Heat shock protein 90 kda hsp90 has a second functional interaction site with the mitochondrial import receptor tom70. J Biol Chem 291:18620–31
169. Horst M, Oppliger W, Feifel B, Schatz G, Glick BS (1996) The mitochondrial protein import motor: dissociation of mitochondrial hsp70 from its membrane anchor requires atp binding rather than atp hydrolysis. Protein Sci 5:759–767
170. Bottinger L, Oeljeklaus S, Guiard B, Rospert S, Warscheid B, Becker T (2015) Mitochondrial heat shock protein (hsp) 70 and hsp10 cooperate in the formation of hsp60 complexes. J Biol Chem 290:11611–11622
171. Augustin S, Nolden M, Muller S, Hardt O, Arnold I, Langer T (2005) Characterization of peptides released from mitochondria: evidence for constant proteolysis and peptide efflux. J Biol Chem 280:2691–2699
172. Anand R, Langer T, Baker MJ (2013) Proteolytic control of mitochondrial function and morphogenesis. Biochem Biophys Acta 1833:195–204
173. Nunnari J, Fox TD, Walter P (1993) A mitochondrial protease with two catalytic subunits of nonoverlapping specificities. Science 262:1997–2004
174. Dvorakova-Hola K, Matuskova A, Kubala M, Otyepka M, Kucera T, Vecer J, Herman P, Parkhomenko N, Kutejova E, Janata J (2010) Glycine-rich loop of mitochondrial processing peptidase alpha-subunit is responsible for substrate recognition by a mechanism analogous to mitochondrial receptor tom20. J Mol Biol 396:1197–1210

175. Stahl A, Moberg P, Ytterberg J, Panfilov O, Brockenhuus Von Lowenhielm H, Nilsson F, Glaser E (2002) Isolation and identification of a novel mitochondrial metalloprotease (prep) that degrades targeting presequences in plants. J Biol Chem 277:41931–9
176. Quiros PM, Langer T, Lopez-Otin C (2015) New roles for mitochondrial proteases in health, ageing and disease. Nat Rev Mol Cell Biol 16:345–359
177. Houtkooper RH, Mouchiroud L, Ryu D, Moullan N, Katsyuba E, Knott G, Williams RW, Auwerx J (2013) Mitonuclear protein imbalance as a conserved longevity mechanism. Nature 497:451–457
178. Jovaisaite V, Mouchiroud L, Auwerx J (2014) The mitochondrial unfolded protein response, a conserved stress response pathway with implications in health and disease. J Exp Biol 217:137–143
179. Balchin D, Hayer-Hartl M, Hartl FU (2016) In vivo aspects of protein folding and quality control. Science 353:aac4354

Chapter 11
Mitochondrial Dysfunction and Mitophagy: Physiological Implications in Cardiovascular Health

Åsa B. Gustafsson

Abstract Heart failure is the typical end stage for cardiovascular diseases and presents a major public health burden. Heart failure occurs as a result of an excess loss of cardiac myocytes and increased fibrosis, which leads to a reduced capacity to sustain contractile function. Cardiac myocytes are highly enriched in mitochondria which are responsible for generating energy via oxidative phosphorylation. Although mitochondria are critical for myocyte function, they can become harmful when damaged or dysfunctional. Therefore, the quality of mitochondria must be carefully and continuously monitored to ensure cellular homeostasis. Mitochondrial autophagy or mitophagy refers to the selective engulfment of mitochondria by autophagosomes. This selective elimination of mitochondria in response to various bioenergetic or environmental cues is responsible for regulating both mitochondrial quality and quantity and is critical for maintaining a healthy population of mitochondria. Defects in this process affect cardiac homeostasis and contribute to cardiac aging and development of various myocardial pathologies. This chapter provides insights into the molecular mechanisms involved in regulating mitophagy in cells and examines the functional importance of mitophagy in the myocardium.

Keywords Autophagy · Mitophagy · Mitochondria · Heart failure · Cardiac myocytes · PINK1 · Parkin · BNIP3 · FUNDC1 · Deubiquitinases

Introduction

Cardiovascular diseases are a global epidemic and are the leading cause of morbidity and mortality in the developed world [5]. Heart failure is the most common end stage of cardiovascular diseases and results when there is an excess loss of cardiac myocytes coupled with increased fibrosis. To support their high energy demands, cardiac myocytes are highly enriched in mitochondria which are responsible for

Å. B. Gustafsson (✉)
Skaggs School of Pharmacy and Pharmaceutical Sciences, University of California San Diego, 9500 Gilman Drive #0758, La Jolla, California 92093-0758, USA
e-mail: asag@ucsd.edu

© Springer Nature Switzerland AG 2022

L. A. Kirshenbaum (ed.), *Biochemistry of Apoptosis and Autophagy*, Advances in Biochemistry in Health and Disease 18,
https://doi.org/10.1007/978-3-030-78799-8_11

producing ATP via oxidative phosphorylation [43]. However, damaged or dysfunctional mitochondria can be a major source of reactive oxygen species (ROS). They also release pro-death factors and mitochondrial DNA (mtDNA) that activate cell death pathways and induce activation of the inflammasome, respectively [3, 102, 129]. Mitochondrial dysfunction also disrupts functions of other organelles such a lysosomes, which are critical in degradation and recycling of cellular material [20]. Therefore, the quality of mitochondria must be carefully and continuously monitored to ensure cellular homeostasis. Indeed, defects in mitochondrial quality control pathways have been reported to contribute to pathologies such as Parkinson's disease, diabetic cardiomyopathy, heart failure and cardiac aging [19, 28, 29, 34, 38, 46, 61, 63, 110, 111].

Mitochondrial autophagy or mitophagy refers to the selective engulfment of mitochondria by autophagosomes. This specific elimination of damaged or excessive mitochondria is responsible for regulating both mitochondrial quality and quantity and is critical for maintaining a healthy population of mitochondria. Mitophagy is responsible for different aspects of mitochondrial clearance such as: (1) baseline turnover of old or excess mitochondria; (2) elimination of mitochondria during metabolic transition or differentiation; and (3) removal of damaged mitochondria in response to cellular stress. The mitophagy process requires crosstalk between mitochondria and the autophagy machinery. First, the dysfunctional mitochondrion is segregated from the network of healthy mitochondria and labeled for degradation. Second, mitophagy also requires the concurrent formation of autophagosomes which are responsible for engulfing the mitochondria that have been marked for degradation. Autophagy is initiated by the formation of a phagophore membrane (Fig. 11.1). The membrane is elongated and then closes around the labeled cargo. The autophagosome then delivers the cargo to a lysosome for degradation [28].

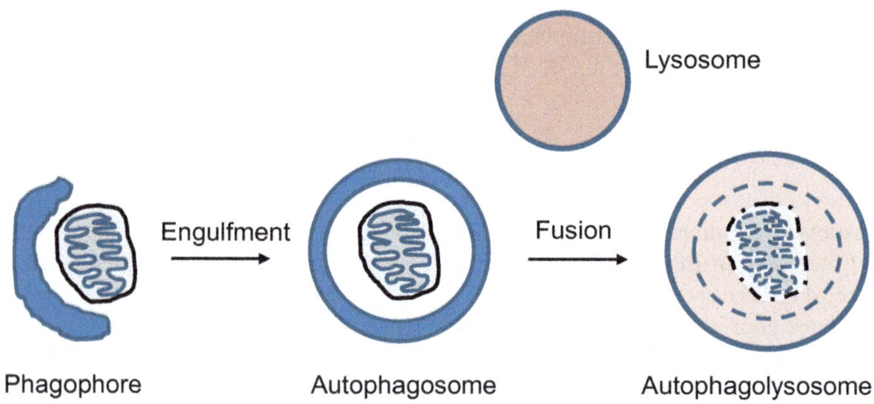

Fig. 11.1 Scheme of mitochondrial autophagy process. A phagophore is formed near the cargo to be degraded. The membrane elongates and then encloses around the cargo. The double membraned autophagosome then fuses with a lysosome leading to formation of the autophagolysosome and subsequent degradation of cargo

Mitophagy plays an important role in clearing mitochondria in cardiac myocytes in response to various bioenergetic or environmental cues. Defects in this mitochondrial quality control pathway affects cardiac homeostasis and contribute to development of cardiac dysfunction [38, 44, 63, 104, 111]. This chapter describes our current knowledge of the molecular mechanisms involved in regulating mitophagy in cells. It also examines the role of mitophagy in cardiovascular homeostasis and how defects contribute to development of heart failure.

PINK1/Parkin-Mediated Mitophagy

PINK1/Parkin-dependent mitophagy is one of the best-characterized pathways for the degradation of depolarized mitochondria (Fig. 11.2a). PINK1 is a serine/threonine kinase that contains an N-terminal mitochondrial targeting sequence (MTS) and a C-terminal kinase domain [83], whereas Parkin is a cytosolic E3 ubiquitin ligase that is recruited to depolarized mitochondria by PINK1 [76, 83]. Under normal conditions, PINK1 is imported into mitochondria in a $\Delta\Psi$m-dependent manner via the translocase of outer membrane (TOM) and translocase of inner membrane (TIM) complexes where it is subjected to proteolytic cleavage [68, 83]. This constitutive turnover of PINK1 ensures that the protein levels are maintained low under normal conditions.

Depolarization of the mitochondria abrogates the import of PINK1 and leads to its accumulation on the outer mitochondrial surface and activation of mitophagy [76, 83]. Studies have found that saturating the import mechanisms with excess PINK1, knock down of mitochondrial proteases responsible for PINK1 cleavage or deleting the MTS also leads to PINK1 accumulation on the outer mitochondrial membrane and activation of mitophagy [39, 76, 87]. Once PINK1 is stabilized on the outer mitochondrial membrane, it phosphorylates pre-existing ubiquitin proteins in the OMM at the conserved Serine 65 which then functions to recruit Parkin to the mitochondria [50, 53, 59]. PINK1 also phosphorylates Mitofusin 2 (Mfn2) which acts as a mitochondrial receptor for Parkin [13]. Interestingly, once Parkin has translocated to the mitochondria, it is also phosphorylated by PINK1 on Serine 65 in its ubiquitin-like domain which leads to its full activation [55, 99, 100]. After its phosphorylation and binding to phospho-ubiquitin, Parkin proceeds to attach polyubiquitin chains to numerous proteins in the outer mitochondrial membrane that in turn are phosphorylated by PINK1 in a feed-forward mechanism [90]. K63, K48, K11 and K6-mediated ubiquitination of various substrates by Parkin lead to the coordinated proteasomal and autophagic degradation of proteins and mitochondria [8, 58, 89, 90, 126]. The UPS-mediated degradation of certain proteins functions to facilitate mitophagy. For instance, mitochondrial fusion proteins Mfn1 and Mfn2 are Parkin substrates and proteasome-dependent degradation of Mfn1 and Mfn2 facilitates mitophagy by preventing re-fusion of depolarized mitochondrial fragments to the healthy mitochondrial network [109]. Degradation of MIRO1, a protein involved in connecting mitochondria to the microtubules, serves to disconnect the damaged

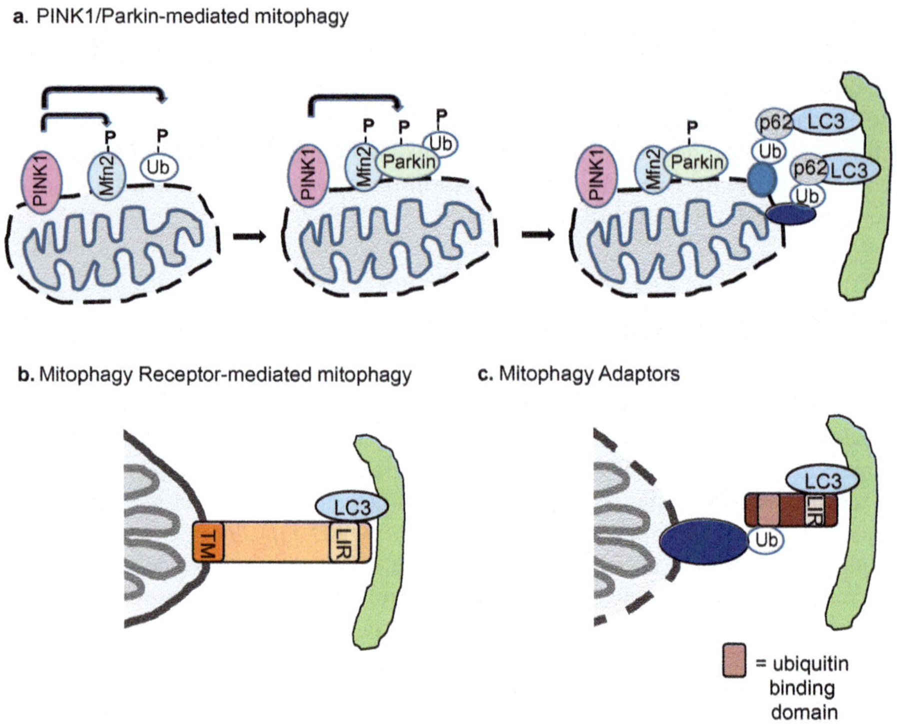

Fig. 11.2 Mitophagy pathways. **a** PINK1/Parkin-mediated mitophagy. PINK1-mediated phosphorylation of ubiquitin and Mfn2 leads to recruitment of Parkin. Phosphorylated Parkin ubiquitinates proteins in the outer mitochondrial membrane to label the mitochondrion for degradation. **b** Mitochondrial autophagy receptors in the outer mitochondrial membrane directly bind to LC3 on the autophagosome via their LC3 interacting region (LIR). TM = transmembrane domain. **c** Autophagy adaptors containing ubiquitin binding domains and LIR motifs tether the ubiquitinated mitochondrion to the autophagosome

mitochondrion from the network and prevents it from being transported in the cell [120].

The PINK1/Parkin pathway was initially identified to play a role Parkinson's Disease (PD) where loss-of-function mutations in *PINK1* or *PARKIN* contribute to development of early onset PD [56, 74, 114]. Subsequent studies have demonstrated that the PINK1/Parkin pathway is critical for efficient mitochondrial quality control in other systems as well, including the heart. For instance, PINK1-deficient mice are more susceptible to I/R injury compared to wild-type (WT) mice [101], whereas cardiac specific PINK1 transgenic mice have reduced myocardial I/R injury [119]. Similarly, the presence of Parkin is also critical for efficient mitophagy in response to acute mitochondrial stress in the myocardium. Mice that are deficient in Parkin are unable to efficiently clear their dysfunctional mitochondria in myocytes after a myocardial infarction which leads to rapid development of heart failure [63]. Moreover, Parkin-deficient mice accumulate abnormal mitochondria with age [61] and cardiac specific overexpression of Parkin ameliorates the functional decline in cardiac function that occurs with aging [44]. Parkin also plays a role in ischemic pre-conditioning (IPC) and Parkin-deficiency abolishes the cardioprotective effects of IPC [45]. Parkin-mediated mitophagy has also been reported to play an important role in the removal of fetal mitochondria in the neonatal heart. After birth, the heart undergoes a metabolic transition from utilizing carbohydrates to fatty acids for energy production. This change in metabolic substrate preference has been thought to involve transcriptional reprogramming of mitochondria [118]. However, a recent study discovered that the reprogramming of mitochondria involves removal of fetal mitochondria via Parkin-mediated mitophagy and that specific abrogation of Parkin in myocytes prevents the normal perinatal transition from fetal to adult mitochondria [38].

Furthermore, diabetic patients are at higher risk of developing cardiac defects and mitochondrial dysfunction is one of the characteristics of the diabetic cardiomyopathy. It was recently reported that high fat diet is associated with activation of Parkin-mediated mitophagy in hearts [111]. Parkin-deficiency leads to accumulation of dysfunctional mitochondria and exacerbated diabetic cardiomyopathy in mice [111], confirming the importance of mitophagy in maintaining mitochondrial health in the heart during high fat intake. Overall, these findings suggest that Parkin-mediated mitophagy plays important roles in adapting to metabolic changes and to acute mitochondrial stress.

Autophagy Adaptors

Ubiquitin serves as a label for degradation. Thus, targeting ubiquitinated mitochondria to autophagosomes is a critical step in mitophagy and involves the use of adaptor proteins such as p62/SQSTM1, NBR1, optineurin (OPTN), and NDP52 [65, 69, 91, 122]. Autophagy adaptors contain an LC3-interacting region and ubiquitin binding motifs to connect ubiquitinated cargo to autophagosomes (Fig. 11.2b). How autophagy adaptors promote the autophagosomal engulfment of select substrates is still poorly understood but recent studies have uncovered a role for the TANK-binding kinase 1 (TBK1) in regulating activity of the autophagy adaptors. Mitochondrial damage leads to activation of TBK1 and its phosphorylation of p62, OPTN, and NDP52 [42, 77, 78]. Inhibition of TBK1 abrogates the recruitment of these proteins to depolarized mitochondria [42]. Also, TBK1-mediated phosphorylation of OPTN enhances its binding to ubiquitin chains [42], while TBK1-mediated phosphorylation of p62 is required for the engulfment of mitochondria by autophagosomes. This suggests that TBK1-mediated phosphorylation is an important regulatory step in the mitophagy pathway. In addition, some of the autophagy adaptors have additional functions in mitophagy than just linking ubiquitinated mitochondria to LC3 on the autophagosome. They are also involved in recruiting components in autophagic machinery, such as ULK1, to the mitochondrion which ensures formation of the autophagosome adjacent to the damaged mitochondrion [69]. A recent study demonstrated that NDP52 and TBK1 cooperate to recruit ULK1 to ubiquitinated mitochondria [117].

Mitophagy Receptors

Mitophagy can also be directly initiated by proteins that are anchored in the outer mitochondrial membrane. Similar to the autophagy adaptors, these mitophagy receptors contain conserved LIR motifs and directly interact with ATG8/LC3 proteins on the autophagosomes eliminating the need for ubiquitination of proteins. Atg32 was the first protein to be identified as a mitophagy receptor in yeast [51, 86]. Subsequent studies have identified the presence of several mitophagy receptors in mammalian cells. The molecular mechanisms by which the mammalian receptors regulate mitophagy has been challenging to delineate since many of these proteins also have additional functions that are independent of their mitophagy functions. However, studies agree that post-translational modification plays a key role in the regulation of receptor-mediated mitophagy and that phosphorylation or dephosphorylation of these receptors function to modulate the interaction with LC3 [2, 10, 125, 132].

BNIP3 and NIX

The first mammalian mitophagy receptors identified were BCL2 and adenovirus E1B 19-kDa-interacting protein 3 (BNIP3) and BNIP3-like (BNIP3L, also known as NIX). They were originally characterized as atypical pro-apoptotic BH3-only proteins based on sequence homology with other BH3-only proteins [9, 116]. BNIP3 and NIX are both expressed in the heart where they induce apoptosis via the intrinsic or mitochondrial pathway [40, 95, 127]. Although these two proteins are closely functionally related, they induce cell death in response to different cellular stressor. For instance, BNIP3 is significantly upregulated in response to hypoxia and a major mediated of cell death in myocardial ischemic injury [24, 40, 95]. BNIP3 is also responsible for loss of myocytes in cardiotoxicity induced by the chemotherapy drug doxorubicin [21]. NIX is upregulated in response pathological hypertrophy and activation of Gαq and PKCα [32] which leads to loss of myocytes and development of heart failure [25, 127].

However, it was observed that BNIP3 and NIX are less efficient at inducing cell death compared to other BH3-only proteins. Subsequent studies identified that these two proteins are also inducers of mitophagy. BNIP3 was found to promote mitophagy in cardiac cells in response to I/R [40] and Nix was reported to be required for selective autophagic degradation of mitochondria during reticulocyte maturation [96, 98]. Both BNIP3 and NIX are anchored in the outer mitochondrial membrane via their C-terminal transmembrane domains and contain N-terminal LIRs that face the cytosol, thereby facilitating the clearance of mitochondria [41, 85, 97]. BNIP3 and Nix also form homodimers [62] and it was recently demonstrated that NIX homodimerization was required for recruitment of the autophagy machinery to the mitochondrion upon induction of mitophagy [75]. The exact function of BNIP3/NIX-mediated mitophagy is still being elucidated. In the myocardium, they appear to play a role in the normal turnover of mitochondria and BNIP3/NIX deficiency leads to development of age-dependent mitochondrial cardiomyopathy [26]. Although BNIP3 and NIX have dual functions as pro-death proteins and mitophagy receptors, it is currently unclear when and how they switch between these two functions.

There is clearly cross talk between the PINK1/Parkin and BNIP3/NIX mitophagy pathways. There are reports that NIX serves a critical role in Parkin-mediated mitophagy. Ding et al. initially reported that NIX is required for both the induction of autophagy and for recruiting Parkin to mitochondria [23]. In contrast, a recent study reported that NIX acts downstream in the PINK1/Parkin pathway to induce mitophagy [33]. This study found that NIX is a Parkin substrate and that ubiquitination of NIX by Parkin induces the recruitment of the autophagy adaptor NBR1, thereby targeting mitochondria for degradation [33]. A recent case study of an asymptomatic patient (i.e. no PD symptoms) with loss of function mutations in the Parkin gene reported that increased NIX protein levels compensated for the lack of Parkin-mediated mitophagy in cells [57]. Clearly, additional studies are needed to resolve the exact relationship between NIX and the PINK1/Parkin mitophagy pathways. Furthermore, although induction of BNIP3-mediated autophagy is reduced in Parkin-deficient myocytes [70], the relationship between BNIP3 and Parkin has not been extensively investigated. Interestingly, BNIP3-mediated mitophagy is unaffected in PINK1-deficient myocytes [60], suggesting that BNIP3 is not dependent on the presence of PINK1. On the other hand, BNIP3 can amplify PINK1/Parkin-mediated mitophagy by suppressing the proteolytic cleavage of PINK1 allowing it to accumulate on the OMM for Parkin recruitment [128].

FUNDC1

FUN14 domain-containing protein 1 (FUNDC1) is another protein in the outer mitochondrial membrane that promotes mitophagy by interacting with LC3 [72]. FUNDC1 functions as mitophagy receptor during hypoxia [72]. FUNDC1 levels are reduced in hearts of patients with heart failure [124]. Recent studies have identified an important role for FUNDC1-mediated mitophagy in protecting hearts against ischemia/reperfusion injury [130, 131] and in the differentiation of cardiac stem cells [66].

The regulation of FUNDC1 involves the coordination of multiple proteins. In particular, phosphorylation of FUNDC1 plays a key role in regulating its function and the interaction with LC3. ULK1 is a Ser/Threonine kinase and is required for mitophagy [64]. ULK1 translocates to damaged mitochondria where it phosphorylates FUNDC1 on Ser17 to activate mitophagy [125]. Phosphorylation at this site enhances the interaction between FUNDC1 and LC3. In contrast, phosphorylation of FUNDC1 at Ser 13 under normoxic conditions by casein kinase 2 (CK2), a constitutive Ser/Thr kinase, inhibits the interaction between FUNDC1 and LC3 [10]. The phosphoglycerate mutase family member 5 (PGAM5) is responsible for dephosphorylating FUNDC1 at Ser 13 [10] and under normoxic conditions, PGAM5 is sequestered and inhibited by Bcl2L1 [123]. During hypoxia, CK2 is inactivated and Bcl2L1 is degraded. This leads to release of PGAM5 and subsequent dephosphorylation of FUNDC1, thereby activating FUNDC1-mediated mitophagy [123]. Thus, coordinated ULK1-mediated phosphorylation of Ser17 and dephosphorylation of Ser13 by PGAM5 are required for activation of FUNDC1-mediated mitophagy. Interestingly, a recent study identified that PGAM5 is also a positive regulator of PINK1/Parkin-mediated mitophagy. Presenilin-associated rhomboid-like (PARL) is a mitochondrial inner membrane protease protein that is responsible for degrading PINK1 [83] and PGAM5 is responsible for stabilizing and protecting PINK1 against PARL cleavage after mitochondrial depolarization [73]. In the absence of PGAM5, PINK1/Parkin-mediated mitophagy is disrupted.

Other Mitophagy Receptors

Cardiolipin

Cardiolipin (CL) is a unique mitochondrial phospholipid in the inner membrane. CL is important for stabilizing the inner membrane and for optimal respiration [71]. Interestingly, CL has also been reported to function as a mitophagy receptor [15]. In response to mitochondrial stress, the intermembrane space protein hexameric nucleoside diphosphate kinase D (NDPK-D, NME4, or NM23-H4) facilitates the redistribution of CL to the outer mitochondrial surface where it then directly binds to LC3 [15, 48]. Of the eight different human ATG8 proteins, CL preferentially interacts with LC3B and their interaction involves both electrostatic forces and alterations in membrane properties by CL [1]. Interestingly, neurons have a much greater increase in externalized CL compared to HeLa cells after exposure to a mitochondrial uncoupler [15], suggesting that the mitophagy function of CL is limited to select cells. CL plays a fundamental role in the heart and genetic diseases with mutations in genes involved in CL biosynthesis, such as Barth syndrome, Sengers syndrome, and Dilated cardiomyopathy with ataxia, result in cardiomyopathy [30].

Ambra1

Activating Molecule in Beclin1-Regulated Autophagy 1 (AMBRA1) is also a regulator of mitophagy and can interact with Parkin to promote mitophagy [115]. It was reported that although Parkin translocation to depolarized mitochondria is independent of AMBRA1, the subsequent clearance of mitochondria requires the presence of AMBRA1. AMBRA1 is a component of the class III phosphatidylinositol 3-kinase (PI3K) complex that is essential for the initiation of phagophore formation [31].

Hence, the interaction between AMBRA1 and Parkin facilitates formation of the phagophore specifically adjacent to mitochondria that have been labeled for degradation, allowing for efficient removal of mitochondria. A more recent study reported that AMBRA1 can also function as a mitophagy receptor and directly interact with LC3 via its LC3-interacting region (LIR) motif to promote mitophagy. Interestingly, AMBRA1 induces mitophagy independently of PINK1/Parkin and can restore mitophagy in PINK1 and PARKIN-deficient cells, confirming the existence of a distinct AMBRA1-mediated mitophagy pathway [22, 107].

BCL-2-like protein 13

The Bcl2-like protein 13 (Bcl2-L-13, also known as Bcl-rambo) was identified as a putative mammalian homologue of the yeast mitophagy receptor Atg32 [51, 86] in a screening of the UniProt protein data base [81]. To meet the requirements of a mitophagy receptor, the authors used the following criteria in the screen: 1. localization to mitochondria, 2. presence of a LIR motif, 3. clusters of acidic amino acids, and 4. a single transmembrane domain. Based on these factors, the authors identified Bcl2-L-13 as a potential homologue of Atg32 and mitophagy receptor. Similar to BNIP3 and NIX, Bcl2-L-13 was originally identified to be a pro-apoptotic protein where overexpression induced caspase-mediated apoptosis [52]. The authors confirmed that Bcl2-L-13 promotes mitophagy by binding to LC3 via its LIR and that it induces degradation of mitochondria independently of Parkin. Bcl2-L-13, but not a LIR mutants, also partially restores mitophagy in Atg32-deficicient yeast cells [81]. Bcl-2-L-13-mediated mitophagy is also dependent on the serine/threonine kinase ULK1 [80].

Deubiquitinases as Critical Regulators of Mitophagy

The PINK1/Parkin pathway uses ubiquitin as the signal for autophagic degradation of mitochondria which is recognized by autophagy adaptor proteins. Hence, a balance between ubiquitination and deubiquitination dictates the level of mitophagy. Deubiquitinases (DUBs) are responsible for removing ubiquitin modifications from proteins [16]. Recent studies have described several DUBs, including USP8, USP15, USP30, USP33, USP35 and USP36, as important regulators of Parkin-mediated mitophagy [6, 17, 18, 27, 35, 84, 88, 121].

Ubiquitin specific protease 30 (USP30) is currently the only known DUB that is constitutively anchored in the outer mitochondrial membrane [82]. It counteracts mitophagy by removing ubiquitin from Parkin substrates such as MIRO1 and TOM20 (Fig. 11.3a) [6, 121]. USP30 overexpression in cells decreases the levels of ubiquitinated proteins in the outer mitochondrial membrane and hinders Parkin-mediated degradation of mitochondria. USP30 also delays the recruitment of Parkin to depolarized mitochondria likely due to a reduced pool of ubiquitin for PINK1 to phosphorylate. Because Parkin is recruited to mitochondria by phosphorylated ubiquitin, lower levels of ubiquitin in the presence of USP30 might lead to reduced recruitment

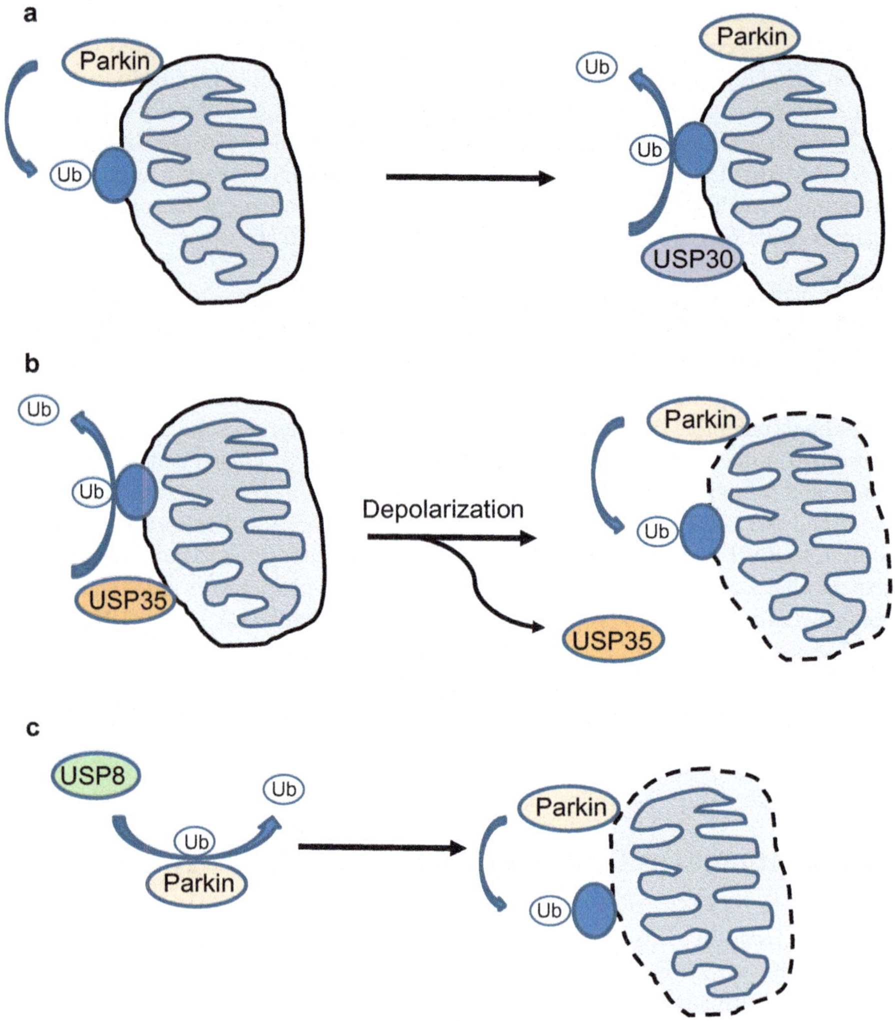

Fig. 11.3 Regulation of mitophagy by deubiquitylases. **a** USP30 opposes Parkin-mediated mitophagy by removing ubiquitin from Parkin substrates. **b** USP35 dissociates from the mitochondrion upon depolarization allowing Parkin-mediated ubiquitination to proceed. **c** USP8 removes ubiquitin from Parkin in the cytosol to promote its activation

of Parkin. USP30 is also a Parkin substrate and its ubiquitination leads to proteasomal degradation [6, 36], thereby removing the brake on mitophagy. Structural studies on USP30 also suggest that phosphorylated ubiquitin chains inhibit USP30 activity [36], suggesting that phosphorylation by PINK1 on depolarized mitochondria protects the ubiquitin from USP30-mediated hydrolysis and allows for recruitment of Parkin.

In addition, UPS35 is another DUB that is localized to respiring mitochondria but it quickly dissociates from depolarized mitochondria by an unknown mechanism (Fig. 11.3b). Interestingly, USP35 knockdown in cells also results in decreased levels of the mitochondrial fusion protein Mfn2, suggesting that the presence of USP35 at the mitochondria is important in maintaining Mfn2 levels [121]. It has been observed that mitochondria that undergo fusion during starvation are also protected from mitophagy [37, 94]. Thus, it is possible that the primary function of USP35 is to prevent unnecessary degradation of mitochondria, especially during energy limiting conditions such as starvation. Finally, USP15 also inhibits Parkin-mediated mitophagy [17]. Although USP15 is primarily localized to the cytosol, a portion is found at the mitochondria where it counteracts Parkin-mediated mitophagy. USP15 antagonizes Parkin-mediated accumulation of K48- and K63-linked poly-ubiquitin chains on mitochondria after mitochondrial depolarization [17]. Importantly, although *USP15* knockdown does not enhance mitophagy in the complete absence of Parkin, it restores mitophagy in fibroblasts from PD patient with *PARK2* mutations that lead to reduced Parkin activity. Many clinical *PARK2* mutations do not completely abolish the E3 ligase activity of Parkin [106], so this suggests that USP15 could be a therapeutic target for PD in patients carrying specific PARK2 mutations.

Moreover, DUBs can act as positive regulators of Parkin-mediated mitophagy. The activity of many E3 ubiquitin ligases, including Parkin, is regulated by auto-ubiquitination [7, 112]. In contrast to the DUBs described above, UPS8 does not act on known Parkin substrates at the mitochondria [27]. Instead, USP8 promotes activation of Parkin by removing its K6-linked ubiquitin chains (Fig. 11.3c). Interestingly, USP8-mediated de-ubiquitination of Parkin is required for recruitment of Parkin to mitochondria and induction of mitophagy [27]. Hence, these studies demonstrate that induction of PINK1/Parkin-mediated mitophagy requires the coordination of several DUBs. Future studies will provide insights into whether the DUBs are involved in regulating mitophagy in the myocardium.

Coordination Between Mitochondrial Dynamics and Mitophagy

Mitophagy is closely coordinated with mitochondrial dynamics for efficient clearance of dysfunctional mitochondria. Mitochondria undergo asymmetrical fission to segregate damaged mitochondria from the healthy network so that they can be degraded [113]. In contrast, mitochondrial fusion protects mitochondria from unnecessary mitophagy [37]. Mitochondrial dynamics are regulated by dynamin family GTPases. Drp1 is a cytosolic protein that is recruited to mitochondria where it coordinates with Mff and Fis1 to induce fission. Mfn1 and Mfn2 are required for fusion of the outer mitochondrial membrane whereas OPA1 is responsible for fusion of inner membrane.

The balance between mitochondrial fission and fusion dictates mitochondrial morphology and an imbalance in either process leads to abnormal mitochondrial function and cardiac dysfunction. For instance, disruption of fusion by cardiac specific deletion of both Mfn1 and Mfn2 in adult hearts leads to mitochondrial fragmentation, respiratory defects, and development of lethal dilated cardiomyopathy [14]. Similarly, deletion of Mfn1/2 in cardiac myocytes during midgestation also results in accumulation of abnormal mitochondria in the postnatal heart and development of a lethal cardiomyopathy before 16 days of age [92].

Disrupting fission by conditional deletion of *Drp1* in myocytes leads to mitochondrial elongation and abrogation of mitophagy. It also results in mitochondrial dysfunction, increased oxidative stress and cardiac dysfunction [46, 49]. Interestingly, another study found that conditional cardiomyocyte-specific *Drp1* deletion leads to the expected mitochondrial enlargement, and development of a lethal dilated cardiomyopathy [105]. However, this group did not observe a defect in mitophagy. Instead, they found that loss of Drp1 resulted in excessive Parkin-mitophagy [104, 105]. In addition, Mff is a mitochondrial membrane protein that acts a receptor for Drp1 and mice deficient in Mff develop dilated cardiomyopathy [11]. Similar to the findings by Song et al., these mice have reduced mitochondrial numbers and respiratory chain activity along with increased mitophagy. Remarkably, reducing mitochondrial fusion by deletion of *Mfn1* completely rescued the lethal cardiomyopathy observed in Mff-deficient mice.

Thus, the studies in mice clearly demonstrate the importance of functional fission and fusion in the heart. Interestingly, simultaneously disruption of fission and fusion in cardiac specific Mfn1/Mfn2/Drp1 triple knockout mice leads to a different cardiac phenotype with longer survival and development of a distinct pathological cardiac hypertrophy compared to Drp1 and Mfn1/2 deficient mice [103]. Disruption of both fission and fusion also leads to impaired mitophagy resulting in excess abundance of mitochondria. Overall, these studies clearly demonstrate that a proper balance between mitochondrial fusion and fission is critical for maintaining mitochondrial homeostasis and cardiac function.

There is evidence that some of the mitophagy proteins are directly involved in regulating mitochondrial dynamics. Studies have found that mitophagy regulators such as PINK1, Parkin, BNIP3 and FUNDC1 can directly regulate mitochondrial dynamics proteins and morphology. For instance, AKAP1 anchors PKA at the mitochondria where it phosphorylates Drp1 on Ser 637 to inhibit its pro-fission activity [54]. When PINK1 accumulates on the outer mitochondrial membrane, it disrupts PKA-mediated Drp1 inhibition by displacing PKA from AKAP1, thereby allowing for Drp1-mediated fission to proceed [93]. However, further studies are needed to determine how PINK1 disrupts the AKAP1–PKA axis upon mitochondrial damage. Moreover, Mfn2 has been reported to prevent PINK1/Parkin-mediated mitophagy by tethering mitochondria to the ER [4, 79]. For mitophagy to proceed, Mfn2 must be ubiquitinated by Parkin which leads to its proteasomal degradation and release of mitochondria from ER [79].

In addition, the mitophagy receptor FUNDC1 interacts with both Drp1 and OPA1 to regulate mitochondrial morphology [12]. Under baseline conditions, FUNDC1 exists in a complex with OPA1 which favors a fused mitochondrial morphology. In response to mitochondrial stress, the interaction between FUNDC1 and OPA1 is reduced whereas the interaction between FUNDC1 and Drp1 is increased favoring a fragmented mitochondrial morphology. In addition, it has been reported that the phosphatase PGAM5 which dephosphorylates FUNDC1 to promote its activation, also dephosphorylates and activates Drp1 [108]. Thus, this would lead to simultaneous activation of mitochondrial fission and mitophagy. BNIP3 has also been reported to interact with OPA1 to regulate mitochondrial morphology. However, whereas the interaction between FUNDC1 and OPA1 promotes mitochondrial fusion, the interaction between BNIP3 and OPA1 induces mitochondrial fission [67]. Overexpression of BNIP3 also leads to recruitment of Drp1 to mitochondria and induction of fission [70]. Inhibition of Drp1 leads to abrogation of BNIP3-mediated mitophagy, suggesting that this mitophagy receptors requires Drp1 for mitophagy [70].

Concluding Remarks

The presence of dysfunctional mitochondria is a prominent feature in various cardiovascular diseases which can contribute to loss of myocyte and development of heart failure. Therefore, there is a growing interest in understanding how mitophagy can be targeted to delay or inhibit mitochondrial deterioration. Proper modulation of mitophagy to eliminate dysfunctional mitochondria while maintaining efficient functional mitochondrial mass in response to stress could delay or prevent the development of heart failure in patients with cardiovascular diseases or in aging.

However, there are still many unanswered questions in this field and increased understanding of the distinct and complex mechanisms that exist for selective mitophagy is crucial before they can be targeted therapeutically. For instance, how the mitophagy receptors sense the mitochondrial stresses within mitochondria to activate mitophagy remains largely unknown. Also, it is unclear how they coordinate with each other and/or with the PINK1/Parkin pathway to ensure efficient mitophagy in response to mitochondrial damage. Another open question is whether there are additional E3 ubiquitin ligases that can also activate mitophagy similar to Parkin. Clearly, delineating the molecular mechanisms regulating mitophagy is of great importance and could potentially identify new therapeutic approaches or targets to prevent development of heart failure in patients.

Acknowledgements Å.B. Gustafsson is supported by grants from NIH (R01HL138560 and R01HL132300), and TRDRP (27IR-0013 and 28IP-0025).

References

1. Anton Z, Landajuela A, Hervas JH, Montes LR, Hernandez-Tiedra S, Velasco G, Goni FM, Alonso A (2016) Human Atg8-cardiolipin interactions in mitophagy: specific properties of LC3B, GABARAPL2 and GABARAP. Autophagy 12:2386–2403
2. Aoki Y, Kanki T, Hirota Y, Kurihara Y, Saigusa T, Uchiumi T, Kang D (2011) Phosphorylation of Serine 114 on Atg32 mediates mitophagy. Mol Biol Cell 22:3206–3217
3. Baines CP (2010) The cardiac mitochondrion: nexus of stress. Annu Rev Physiol 72:61–80
4. Basso V, Marchesan E, Peggion C, Chakraborty J, von Stockum S, Giacomello M, Ottolini D, Debattisti V, Caicci F, Tasca E, Pegoraro V, Angelini C, Antonini A, Bertoli A, Brini M, Ziviani E (2018) Regulation of ER-mitochondria contacts by Parkin via Mfn2. Pharmacol Res 138:43–56
5. Benjamin EJ, Muntner P, Alonso A, Bittencourt MS, Callaway CW, Carson AP, Chamberlain AM, Chang AR, Cheng S, Das SR, Delling FN, Djousse L, Elkind MSV, Ferguson JF, Fornage M, Jordan LC, Khan SS, Kissela BM, Knutson KL, Kwan TW, Lackland DT, Lewis TT, Lichtman JH, Longenecker CT, Loop MS, Lutsey PL, Martin SS, Matsushita K, Moran AE, Mussolino ME, O'Flaherty M, Pandey A, Perak AM, Rosamond WD, Roth GA, Sampson UKA, Satou GM, Schroeder EB, Shah SH, Spartano NL, Stokes A, Tirschwell DL, Tsao CW, Turakhia MP, VanWagner LB, Wilkins JT, Wong SS, Virani E (2019) American heart association council on, C. prevention statistics, and S. stroke statistics. 2019. heart disease and stroke statistics-2019 update: a report from the American heart association. Circulation 139:e56–e528
6. Bingol B, Tea JS, Phu L, Reichelt M, Bakalarski CE, Song Q, Foreman O, Kirkpatrick DS, Sheng M (2014) The mitochondrial deubiquitinase USP30 opposes parkin-mediated mitophagy. Nature 510:370–375
7. Byrd RA, Weissman AM (2013) Compact Parkin only: insights into the structure of an autoinhibited ubiquitin ligase. EMBO J 32:2087–2089
8. Chan NC, Salazar AM, Pham AH, Sweredoski MJ, Kolawa NJ, Graham RL, Hess S, Chan DC (2011) Broad activation of the ubiquitin-proteasome system by Parkin is critical for mitophagy. Hum Mol Genet 20:1726–1737
9. Chen G, Cizeau J, Vande VC, Park JH, Bozek G, Bolton J, Shi L, Dubik D, Greenberg A (1999) Nix and Nip3 form a subfamily of pro-apoptotic mitochondrial proteins. J Biol Chem 274:7–10
10. Chen G, Han Z, Feng D, Chen Y, Chen L, Wu H, Huang L, Zhou C, Cai X, Fu C, Duan L, Wang X, Liu L, Liu X, Shen Y, Zhu Y, Chen Q (2014) A regulatory signaling loop comprising the PGAM5 phosphatase and CK2 controls receptor-mediated mitophagy. Mol Cell 54:362–377
11. Chen H, Ren S, Clish C, Jain M, Mootha V, McCaffery JM, Chan DC (2015) Titration of mitochondrial fusion rescues Mff-deficient cardiomyopathy. J Cell Biol 211:795–805
12. Chen M, Chen Z, Wang Y, Tan Z, Zhu C, Li Y, Han Z, Chen L, Gao R, Liu L, Chen Q (2016) Mitophagy receptor FUNDC1 regulates mitochondrial dynamics and mitophagy. Autophagy 12:689–702
13. Chen Y, Dorn GW 2nd (2013) PINK1-phosphorylated mitofusin 2 is a Parkin receptor for culling damaged mitochondria. Science 340:471–475
14. Chen Y, Liu Y, Dorn GW 2nd (2011) Mitochondrial fusion is essential for organelle function and cardiac homeostasis. Circ Res 109:1327–1331
15. Chu CT, Ji J, Dagda RK, Jiang JF, Tyurina YY, Kapralov AA, Tyurin VA, Yanamala N, Shrivastava IH, Mohammadyani D, Qiang Wang KZ, Zhu J, Klein-Seetharaman J, Balasubramanian K, Amoscato AA, Borisenko G, Huang Z, Gusdon AM, Cheikhi A, Steer EK, Wang R, Baty C, Watkins S, Bahar I, Bayir H, Kagan VE (2013) Cardiolipin externalization to the outer mitochondrial membrane acts as an elimination signal for mitophagy in neuronal cells. Nat Cell Biol 15:1197–1205
16. Clague MJ, Barsukov I, Coulson JM, Liu H, Rigden DJ, Urbe S (2013) Deubiquitylases from genes to organism. Physiol Rev 93:1289–1315

17. Cornelissen T, Haddad D, Wauters F, Van Humbeeck C, Mandemakers W, Koentjoro B, Sue C, Gevaert K, De Strooper B, Verstreken P, Vandenberghe W (2014) The deubiquitinase USP15 antagonizes Parkin-mediated mitochondrial ubiquitination and mitophagy. Hum Mol Genet 23:5227–5242
18. Cunningham CN, Baughman JM, Phu L, Tea JS, Yu C, Coons M, Kirkpatrick DS, Bingol B, Corn JE (2015) USP30 and parkin homeostatically regulate atypical ubiquitin chains on mitochondria. Nat Cell Biol 17:160–169
19. Deas E, Wood NW, Plun-Favreau H (2011) Mitophagy and Parkinson's disease: the PINK1-parkin link. Biochim Biophys Acta 1813:623–633
20. Demers-Lamarche J, Guillebaud G, Tlili M, Todkar K, Belanger N, Grondin M, Nguyen AP, Michel J, Germain M (2016) Loss of mitochondrial function impairs lysosomes. J Biol Chem 291:10263–10276
21. Dhingra R, Guberman M, Rabinovich-Nikitin I, Gerstein J, Margulets V, Gang H, Madden N, Thliveris J, Kirshenbaum LA (2020) Impaired NF-kappaB signalling underlies cyclophilin D-mediated mitochondrial permeability transition pore opening in doxorubicin cardiomyopathy. Cardiovasc Res 116:1161–1174
22. Di Rita A, Peschiaroli A, Strobbe D, Hu Z, Gruber J, Nygaard M, Lambrughi M, Melino G, Papaleo E, Dengjel J, El Alaoui S, Campanella M, Dotsch V, Rogov VV, Strappazzon F, Cecconi F (2018) HUWE1 E3 ligase promotes PINK1/PARKIN-independent mitophagy by regulating AMBRA1 activation via IKKalpha. Nat Commun 9:3755
23. Ding WX, Ni HM, Li M, Liao Y, Chen X, Stolz DB, Dorn GW 2nd, Yin XM (2010) Nix is critical to two distinct phases of mitophagy, reactive oxygen species-mediated autophagy induction and Parkin-ubiquitin-p62-mediated mitochondrial priming. J Biol Chem 285:27879–27890
24. Diwan A, Krenz M, Syed FM, Wansapura J, Ren X, Koesters AG, Li H, Kirshenbaum LA, Hahn HS, Robbins J, Jones WK, Dorn GW (2007) Inhibition of ischemic cardiomyocyte apoptosis through targeted ablation of Bnip3 restrains postinfarction remodeling in mice. J Clin Invest 117:2825–2833
25. Diwan A, Wansapura J, Syed FM, Matkovich SJ, Lorenz JN, Dorn GW 2nd (2008) Nix-mediated apoptosis links myocardial fibrosis, cardiac remodeling, and hypertrophy decompensation. Circulation 117:396–404
26. Dorn GW (2010) Mitochondrial pruning by Nix and BNip3: an essential function for cardiac-expressed death factors, (2nd edn) J Cardiovasc Transl Res 3:374–383
27. Durcan TM, Tang MY, Perusse JR, Dashti EA, Aguileta MA, McLelland GL, Gros P, Shaler TA, Faubert D, Coulombe B, Fon EA (2014) USP8 regulates mitophagy by removing K6-linked ubiquitin conjugates from parkin. EMBO J 33:2473–2491
28. Dutta D, Calvani R, Bernabei R, Leeuwenburgh C, Marzetti E (2012) Contribution of impaired mitochondrial autophagy to cardiac aging: mechanisms and therapeutic opportunities. Circ Res 110:1125–1138
29. Eisenberg T, Abdellatif M, Schroeder S, Primessnig U, Stekovic S, Pendl T, Harger A, Schipke J, Zimmermann A, Schmidt A, Tong M, Ruckenstuhl C, Dammbrueck C, Gross AS, Herbst V, Magnes C, Trausinger G, Narath S, Meinitzer A, Hu Z, Kirsch A, Eller K, Carmona-Gutierrez D, Buttner S, Pietrocola F, Knittelfelder O, Schrepfer E, Rockenfeller P, Simonini C, Rahn A, Horsch M, Moreth K, Beckers J, Fuchs H, Gailus-Durner V, Neff F, Janik D, Rathkolb B, Rozman J, de Angelis MH, Moustafa T, Haemmerle G, Mayr M, Willeit P, von Frieling-Salewsky M, Pieske B, Scorrano L, Pieber T, Pechlaner R, Willeit J, Sigrist SJ, Linke WA, Muhlfeld C, Sadoshima J, Dengjel J, Kiechl S, Kroemer G, Sedej S, Madeo F (2016) Cardioprotection and lifespan extension by the natural polyamine spermidine. Nat Med 22:1428–1438
30. El-Hattab AW, Scaglia F (2016) Mitochondrial cardiomyopathies. Front Cardiovasc Med 3:25
31. Fimia GM, Stoykova A, Romagnoli A, Giunta L, Di Bartolomeo S, Nardacci R, Corazzari M, Fuoco C, Ucar A, Schwartz P, Gruss P, Piacentini M, Chowdhury K, Cecconi F (2007) Ambra1 regulates autophagy and development of the nervous system. Nature 447:1121–1125

32. Galvez AS, Brunskill EW, Marreez Y, Benner BJ, Regula KM, Kirschenbaum LA, Dorn GW 2nd (2006) Distinct pathways regulate proapoptotic nix and BNip3 in cardiac stress. J Biol Chem 281:1442–1448
33. Gao F, Chen D, Si J, Hu Q, Qin Z, Fang M, Wang G (2015) The mitochondrial protein BNIP3L is the substrate of PARK2 and mediates mitophagy in PINK1/PARK2 pathway. Hum Mol Genet 24:2528–2538
34. Geisler S, Holmstrom KM, Treis A, Skujat D, Weber SS, Fiesel FC, Kahle PJ, Springer W (2010) The PINK1/Parkin-mediated mitophagy is compromised by PD-associated mutations. Autophagy 6:871–878
35. Geisler S, Jager L, Golombek S, Nakanishi E, Hans F, Casadei N, Terradas AL, Linnemann C, Kahle PJ (2019) Ubiquitin-specific protease USP36 knockdown impairs Parkin-dependent mitophagy via downregulation of Beclin-1-associated autophagy-related ATG14L. Exp Cell Res 384:111641
36. Gersch M, Gladkova C, Schubert AF, Michel MA, Maslen S, Komander D (2017) Mechanism and regulation of the Lys6-selective deubiquitinase USP30. Nat Struct Mol Biol 24:920–930
37. Gomes LC, Di Benedetto G, Scorrano L (2011) During autophagy mitochondria elongate, are spared from degradation and sustain cell viability. Nat Cell Biol 13:589–598
38. Gong G, Song M, Csordas G, Kelly DP, Matkovich SJ, Dorn GW (2015) Parkin-mediated mitophagy directs perinatal cardiac metabolic maturation in mice, (2nd edn) Science 350:aad2459
39. Greene AW, Grenier K, Aguileta MA, Muise S, Farazifard R, Haque ME, McBride HM, Park DS, Fon EA (2012) Mitochondrial processing peptidase regulates PINK1 processing, import and Parkin recruitment. EMBO Rep 13:378–385
40. Hamacher-Brady A, Brady NR, Logue SE, Sayen MR, Jinno M, Kirshenbaum LA, Gottlieb RA, Gustafsson AB (2007) Response to myocardial ischemia/reperfusion injury involves Bnip3 and autophagy. Cell Death Differ 14:146–157
41. Hanna RA, Quinsay MN, Orogo AM, Giang K, Rikka S, Gustafsson AB (2012) Microtubule-associated protein 1 light chain 3 (LC3) interacts with Bnip3 protein to selectively remove endoplasmic reticulum and mitochondria via autophagy. J Biol Chem 287:19094–19104
42. Heo JM, Ordureau A, Paulo JA, Rinehart J, Harper JW (2015) The PINK1-PARKIN mitochondrial ubiquitylation pathway drives a program of OPTN/NDP52 recruitment and TBK1 activation to promote mitophagy. Mol Cell 60:7–20
43. Hom J, Sheu SS (2009) Morphological dynamics of mitochondria—A special emphasis on cardiac muscle cells. J Mol Cell Cardiol 46:811–820
44. Hoshino A, Mita Y, Okawa Y, Ariyoshi M, Iwai-Kanai E, Ueyama T, Ikeda K, Ogata T, Matoba S (2013) Cytosolic p53 inhibits Parkin-mediated mitophagy and promotes mitochondrial dysfunction in the mouse heart. Nat Commun 4:2308
45. Huang C, Andres AM, Ratliff EP, Hernandez G, Lee P, Gottlieb RA (2011) Preconditioning involves selective mitophagy mediated by Parkin and p62/SQSTM1. PLoS ONE 6:e20975
46. Ikeda Y, Shirakabe A, Maejima Y, Zhai P, Sciarretta S, Toli J, Nomura M, Mihara K, Egashira K, Ohishi M, Abdellatif M, Sadoshima J (2015) Endogenous Drp1 mediates mitochondrial autophagy and protects the heart against energy stress. Circ Res 116:264–278
47. Jin SM, Lazarou M, Wang C, Kane LA, Narendra DP, Youle RJ (2010) Mitochondrial membrane potential regulates PINK1 import and proteolytic destabilization by PARL. J Cell Biol 191:933–942
48. Kagan VE, Jiang J, Huang Z, Tyurina YY, Desbourdes C, Cottet-Rousselle C, Dar HH, Verma M, Tyurin VA, Kapralov AA, Cheikhi A, Mao G, Stolz D, St Croix CM, Watkins S, Shen Z, Li Y, Greenberg ML, Tokarska-Schlattner M, Boissan M, Lacombe ML, Epand RM, Chu CT, Mallampalli RK, Bayir H, Schlattner U (2016) NDPK-D (NM23-H4)-mediated externalization of cardiolipin enables elimination of depolarized mitochondria by mitophagy. Cell Death Differ
49. Kageyama Y, Hoshijima M, Seo K, Bedja D, Sysa-Shah P, Andrabi SA, Chen W, Hoke A, Dawson VL, Dawson TM, Gabrielson K, Kass DA, Iijima M, Sesaki H (2014) Parkin-independent mitophagy requires Drp1 and maintains the integrity of mammalian heart and brain. EMBO J 33:2798–2813

50. Kane LA, Lazarou M, Fogel AI, Li Y, Yamano K, Sarraf SA, Banerjee S, Youle RJ (2014) PINK1 phosphorylates ubiquitin to activate Parkin E3 ubiquitin ligase activity. J Cell Biol 205:143–153
51. Kanki T, Wang K, Cao Y, Baba M, Klionsky DJ (2009) Atg32 is a mitochondrial protein that confers selectivity during mitophagy. Dev Cell 17:98–109
52. Kataoka T, Holler N, Micheau O, Martinon F, Tinel A, Hofmann K, Tschopp J (2001) Bcl-rambo, a novel Bcl-2 homologue that induces apoptosis via its unique C-terminal extension. J Biol Chem 276:19548–19554
53. Kazlauskaite A, Kondapalli C, Gourlay R, Campbell DG, Ritorto MS, Hofmann K, Alessi DR, Knebel A, Trost M, Muqit MM (2014) Parkin is activated by PINK1-dependent phosphorylation of ubiquitin at Ser65. Biochem J 460:127–139
54. Kim H, Scimia MC, Wilkinson D, Trelles RD, Wood MR, Bowtell D, Dillin A, Mercola M, Ronai ZA (2011) Fine-tuning of Drp1/Fis1 availability by AKAP121/Siah2 regulates mitochondrial adaptation to hypoxia. Mol Cell 44:532–544
55. Kim Y, Park J, Kim S, Song S, Kwon SK, Lee SH, Kitada T, Kim JM, Chung J (2008) PINK1 controls mitochondrial localization of Parkin through direct phosphorylation. Biochem Biophys Res Commun 377:975–980
56. Kitada T, Asakawa S, Hattori N, Matsumine H, Yamamura Y, Minoshima S, Yokochi M, Mizuno Y, Shimizu N (1998) Mutations in the parkin gene cause autosomal recessive juvenile parkinsonism. Nature 392:605–608
57. Koentjoro B, Park JS, Sue CM (2017) Nix restores mitophagy and mitochondrial function to protect against PINK1/Parkin-related Parkinson's disease. Sci Rep 7:44373
58. Korolchuk VI, Menzies FM, Rubinsztein DC (2010) Mechanisms of cross-talk between the ubiquitin-proteasome and autophagy-lysosome systems. FEBS Lett 584:1393–1398
59. Koyano F, Okatsu K, Kosako H, Tamura Y, Go E, Kimura M, Kimura Y, Tsuchiya H, Yoshihara H, Hirokawa T, Endo T, Fon EA, Trempe JF, Saeki Y, Tanaka K, Matsuda N (2014) Ubiquitin is phosphorylated by PINK1 to activate parkin. Nature 510:162–166
60. Kubli DA, Cortez MQ, Moyzis AG, Najor RH, Lee Y, Gustafsson AB (2015) PINK1 is dispensable for mitochondrial recruitment of Parkin and activation of mitophagy in cardiac myocytes. PLoS One 10:e0130707
61. Kubli DA, Quinsay MN, Gustafsson AB (2013a) Parkin deficiency results in accumulation of abnormal mitochondria in aging myocytes. Commun Integr Biol 6:e24511
62. Kubli DA, Quinsay MN, Huang C, Lee Y, Gustafsson AB (2008) Bnip3 functions as a mitochondrial sensor of oxidative stress during myocardial ischemia and reperfusion. Am J Physiol Heart Circ Physiol 295:H2025–H2031
63. Kubli DA, Zhang X, Lee Y, Hanna RA, Quinsay MN, Nguyen CK, Jimenez R, Petrosyan S, Murphy AN, Gustafsson AB (2013) Parkin protein deficiency exacerbates cardiac injury and reduces survival following myocardial infarction. J Biol Chem 288:915–926
64. Kundu M, Lindsten T, Yang CY, Wu J, Zhao F, Zhang J, Selak MA, Ney PA, Thompson CB (2008) Ulk1 plays a critical role in the autophagic clearance of mitochondria and ribosomes during reticulocyte maturation. Blood 112:1493–1502
65. Lamark T, Kirkin V, Dikic I, Johansen T (2009) NBR1 and p62 as cargo receptors for selective autophagy of ubiquitinated targets. Cell Cycle 8:1986–1990
66. Lampert MA, Orogo AM, Najor RH, Hammerling BC, Leon LJ, Wang BJ, Kim T, Sussman MA, Gustafsson AB (2019) BNIP3L/NIX and FUNDC1-mediated mitophagy is required for mitochondrial network remodeling during cardiac progenitor cell differentiation. Autophagy 1–17
67. Landes T, Emorine LJ, Courilleau D, Rojo M, Belenguer P, Arnaune-Pelloquin L (2010) The BH3-only Bnip3 binds to the dynamin Opa1 to promote mitochondrial fragmentation and apoptosis by distinct mechanisms. EMBO Rep 11:459–465
68. Lazarou M, Jin SM, Kane LA, Youle RJ (2012) Role of PINK1 binding to the TOM complex and alternate intracellular membranes in recruitment and activation of the E3 ligase Parkin. Dev Cell 22:320–333

69. Lazarou M, Sliter DA, Kane LA, Sarraf SA, Wang C, Burman JL, Sideris DP, Fogel AI, Youle RJ (2015) The ubiquitin kinase PINK1 recruits autophagy receptors to induce mitophagy. Nature 524:309–314
70. Lee Y, Lee HY, Hanna RA, Gustafsson AB (2011) Mitochondrial autophagy by Bnip3 involves Drp1-mediated mitochondrial fission and recruitment of Parkin in cardiac myocytes. Am J Physiol Heart Circ Physiol 301:H1924-1931
71. Lesnefsky EJ, Chen Q, Hoppel CL (2016) Mitochondrial metabolism in aging heart. Circ Res 118:1593–1611
72. Liu L, Feng D, Chen G, Chen M, Zheng Q, Song P, Ma Q, Zhu C, Wang R, Qi W, Huang L, Xue P, Li B, Wang X, Jin H, Wang J, Yang F, Liu P, Zhu Y, Sui S, Chen Q (2012) Mitochondrial outer-membrane protein FUNDC1 mediates hypoxia-induced mitophagy in mammalian cells. Nat Cell Biol 14:177–185
73. Lu W, Karuppagounder SS, Springer DA, Allen MD, Zheng L, Chao B, Zhang Y, Dawson VL, Dawson TM, Lenardo M (2014) Genetic deficiency of the mitochondrial protein PGAM5 causes a Parkinson's-like movement disorder. Nat Commun 5:4930
74. Lucking CB, Durr A, Bonifati V, Vaughan J, De Michele G, Gasser T, Harhangi BS, Meco G, Denefle P, Wood NW, Agid Y, Brice A (2000) Association between early-onset Parkinson's disease and mutations in the parkin gene. N Engl J Med 342:1560–1567
75. Marinkovic M, Sprung M, Novak I (2020) Dimerization of mitophagy receptor BNIP3L/NIX is essential for recruitment of autophagic machinery. Autophagy1–12
76. Matsuda N, Sato S, Shiba K, Okatsu K, Saisho K, Gautier CA, Sou YS, Saiki S, Kawajiri S, Sato F, Kimura M, Komatsu M, Hattori N, Tanaka K (2010) PINK1 stabilized by mitochondrial depolarization recruits Parkin to damaged mitochondria and activates latent Parkin for mitophagy. J Cell Biol 189:211–221
77. Matsumoto G, Shimogori T, Hattori N, Nukina N (2015) TBK1 controls autophagosomal engulfment of polyubiquitinated mitochondria through p62/SQSTM1 phosphorylation. Hum Mol Genet 24:4429–4442
78. Matsumoto G, Wada K, Okuno M, Kurosawa M, Nukina N (2011) Serine 403 phosphorylation of p62/SQSTM1 regulates selective autophagic clearance of ubiquitinated proteins. Mol Cell 44:279–289
79. McLelland GL, Goiran T, Yi W, Dorval G, Chen CX, Lauinger ND, Krahn AI, Valimehr S, Rakovic A, Rouiller I, Durcan TM, Trempe JF, Fon EA (2018) Mfn2 ubiquitination by PINK1/parkin gates the p97-dependent release of ER from mitochondria to drive mitophagy. Elife 7
80. Murakawa T, Okamoto K, Omiya S, Taneike M, Yamaguchi O, Otsu K (2019) A mammalian mitophagy receptor, Bcl2-L-13, recruits the ULK1 complex to induce mitophagy. Cell Rep 26:338–345, e336
81. Murakawa T, Yamaguchi O, Hashimoto A, Hikoso S, Takeda T, Oka T, Yasui H, Ueda H, Akazawa Y, Nakayama H, Taneike M, Misaka T, Omiya S, Shah AM, Yamamoto A, Nishida K, Ohsumi Y, Okamoto K, Sakata Y, Otsu K (2015) Bcl-2-like protein 13 is a mammalian Atg32 homologue that mediates mitophagy and mitochondrial fragmentation. Nat Commun 6:7527
82. Nakamura N, Hirose S (2008) Regulation of mitochondrial morphology by USP30, a deubiquitinating enzyme present in the mitochondrial outer membrane. Mol Biol Cell 19:1903–1911
83. Narendra DP, Jin SM, Tanaka A, Suen DF, Gautier CA, Shen J, Cookson MR, Youle RJ (2010) PINK1 is selectively stabilized on impaired mitochondria to activate Parkin. PLoS Biol 8:e1000298
84. Niu K, Fang H, Chen Z, Zhu Y, Tan Q, Wei D, Li Y, Balajee AS, Zhao Y (2020) USP33 deubiquitinates PRKN/parkin and antagonizes its role in mitophagy. Autophagy 16:724–734
85. Novak I, Kirkin V, McEwan DG, Zhang J, Wild P, Rozenknop A, Rogov V, Lohr F, Popovic D, Occhipinti A, Reichert AS, Terzic J, Dotsch V, Ney PA, Dikic I (2010) Nix is a selective autophagy receptor for mitochondrial clearance. EMBO Rep 11:45–51

86. Okamoto K, Kondo-Okamoto N, Ohsumi Y (2009) Mitochondria-anchored receptor Atg32 mediates degradation of mitochondria via selective autophagy. Dev Cell 17:87–97
87. Okatsu K, Kimura M, Oka T, Tanaka K, Matsuda N (2015) Unconventional PINK1 localization to the outer membrane of depolarized mitochondria drives Parkin recruitment. J Cell Sci 128:964–978
88. Ordureau A, Paulo JA, Zhang J, An H, Swatek KN, Cannon JR, Wan Q, Komander D, Harper JW (2020) Global landscape and dynamics of Parkin and USP30-dependent ubiquitylomes in ineurons during mitophagic signaling. Mol Cell 77:1124–1142, e1110
89. Ordureau A, Paulo JA, Zhang W, Ahfeldt T, Zhang J, Cohn EF, Hou Z, Heo JM, Rubin LL, Sidhu SS, Gygi SP, Harper JW (2018) Dynamics of PARKIN-dependent mitochondrial ubiquitylation in induced neurons and model systems revealed by digital snapshot proteomics. Mol Cell 70:211–227, e218
90. Ordureau A, Sarraf SA, Duda DM, Heo JM, Jedrychowski MP, Sviderskiy VO, Olszewski JL, Koerber JT, Xie T, Beausoleil SA, Wells JA, Gygi SP, Schulman BA, Harper JW (2014) Quantitative proteomics reveal a feedforward mechanism for mitochondrial PARKIN translocation and ubiquitin chain synthesis. Mol Cell 56:360–375
91. Pankiv S, Clausen TH, Lamark T, Brech A, Bruun JA, Outzen H, Overvatn A, Bjorkoy G, Johansen T (2007) p62/SQSTM1 binds directly to Atg8/LC3 to facilitate degradation of ubiquitinated protein aggregates by autophagy. J Biol Chem 282:24131–24145
92. Papanicolaou KN, Kikuchi R, Ngoh GA, Coughlan KA, Dominguez I, Stanley WC, Walsh K (2012) Mitofusins 1 and 2 are essential for postnatal metabolic remodeling in heart. Circ Res 111:1012–1026
93. Pryde KR, Smith HL, Chau KY, Schapira AH (2016) PINK1 disables the anti-fission machinery to segregate damaged mitochondria for mitophagy. J Cell Biol 213:163–171
94. Rambold AS, Kostelecky B, Elia N, Lippincott-Schwartz J (2011) Tubular network formation protects mitochondria from autophagosomal degradation during nutrient starvation. Proc Natl Acad Sci U S A 108:10190–10195
95. Regula KM, Ens K, Kirshenbaum LA (2002) Inducible expression of BNIP3 provokes mitochondrial defects and hypoxia-mediated cell death of ventricular myocytes. Circ Res 91:226–231
96. Sandoval H, Thiagarajan P, Dasgupta SK, Schumacher A, Prchal JT, Chen M, Wang J (2008) Essential role for Nix in autophagic maturation of erythroid cells. Nature 454:232–235
97. Schwarten M, Mohrluder J, Ma P, Stoldt M, Thielmann Y, Stangler T, Hersch N, Hoffmann B, Merkel R, Willbold D (2009) Nix directly binds to GABARAP: a possible crosstalk between apoptosis and autophagy. Autophagy 5:690–698
98. Schweers RL, Zhang J, Randall MS, Loyd MR, Li W, Dorsey FC, Kundu M, Opferman JT, Cleveland JL, Miller JL, Ney PA (2007) NIX is required for programmed mitochondrial clearance during reticulocyte maturation. Proc Natl Acad Sci U S A 104:19500–19505
99. Sha D, Chin LS, Li L (2010) Phosphorylation of parkin by Parkinson disease-linked kinase PINK1 activates parkin E3 ligase function and NF-kappaB signaling. Hum Mol Genet 19:352–363
100. Shiba-Fukushima K, Imai Y, Yoshida S, Ishihama Y, Kanao T, Sato S, Hattori N (2012) PINK1-mediated phosphorylation of the Parkin ubiquitin-like domain primes mitochondrial translocation of Parkin and regulates mitophagy. Sci Rep 2:1002
101. Siddall HK, Yellon DM, Ong SB, Mukherjee UA, Burke N, Hall AR, Angelova PR, Ludtmann MH, Deas E, Davidson SM, Mocanu MM, Hausenloy DJ (2013) Loss of PINK1 increases the heart's vulnerability to ischemia-reperfusion injury. PLoS One 8:e62400
102. Sliter DA, Martinez J, Hao L, Chen X, Sun N, Fischer TD, Burman JL, Li Y, Zhang Z, Narendra DP, Cai H, Borsche M, Klein C, Youle RJ (2018) Parkin and PINK1 mitigate STING-induced inflammation. Nature 561:258–262
103. Song M, Franco A, Fleischer JA, Zhang L, Dorn GW (2017) Abrogating mitochondrial dynamics in mouse hearts accelerates mitochondrial senescence. Cell Metab
104. Song M, Gong G, Burelle Y, Gustafsson AB, Kitsis RN, Matkovich SJ, Dorn GW (2015a) Interdependence of Parkin-mediated mitophagy and mitochondrial fission in adult mouse hearts. Circ Res

105. Song M, Mihara K, Chen Y, Scorrano L, Dorn GW (2015) Mitochondrial fission and fusion factors reciprocally orchestrate mitophagic culling in mouse hearts and cultured fibroblasts. Cell Metab 21:273–285
106. Sriram SR, Li X, Ko HS, Chung KK, Wong E, Lim KL, Dawson VL, Dawson TM (2005) Familial-associated mutations differentially disrupt the solubility, localization, binding and ubiquitination properties of parkin. Hum Mol Genet 14:2571–2586
107. Strappazzon F, Nazio F, Corrado M, Cianfanelli V, Romagnoli A, Fimia GM, Campello S, Nardacci R, Piacentini M, Campanella M, Cecconi F (2014) AMBRA1 is able to induce mitophagy via LC3 binding, regardless of PARKIN and p62/SQSTM1. Cell Death Differ
108. Sugo M, Kimura H, Arasaki K, Amemiya T, Hirota N, Dohmae N, Imai Y, Inoshita T, Shiba-Fukushima K, Hattori N, Cheng J, Fujimoto T, Wakana Y, Inoue H, Tagaya M (2018) Syntaxin 17 regulates the localization and function of PGAM5 in mitochondrial division and mitophagy. EMBO J 37
109. Tanaka A, Cleland MM, Xu S, Narendra DP, Suen DF, Karbowski M, Youle RJ (2010) Proteasome and p97 mediate mitophagy and degradation of mitofusins induced by Parkin. J Cell Biol 191:1367–1380
110. Thai PN, Seidlmayer LK, Miller C, Ferrero M, Dorn GW II, Schaefer S, Bers DM, Dedkova EN (2019) Mitochondrial quality control in aging and heart failure: influence of ketone bodies and mitofusin-stabilizing peptides. Front Physiol 10:382
111. Tong M, Saito T, Zhai P, Oka SI, Mizushima W, Nakamura M, Ikeda S, Shirakabe A, Sadoshima J (2019) Mitophagy is essential for maintaining cardiac function during high fat diet-induced diabetic cardiomyopathy. Circ Res 124:1360–1371
112. Trempe JF, Sauve V, Grenier K, Seirafi M, Tang MY, Menade M, Al-Abdul-Wahid S, Krett J, Wong K, Kozlov G, Nagar B, Fon EA, Gehring K (2013) Structure of parkin reveals mechanisms for ubiquitin ligase activation. Science 340:1451–1455
113. Twig G, Elorza A, Molina AJ, Mohamed H, Wikstrom JD, Walzer G, Stiles L, Haigh SE, Katz S, Las G, Alroy J, Wu M, Py BF, Yuan J, Deeney JT, Corkey BE, Shirihai OS (2008) Fission and selective fusion govern mitochondrial segregation and elimination by autophagy. Embo J 27:433–446
114. Valente EM, Abou-Sleiman PM, Caputo V, Muqit MM, Harvey K, Gispert S, Ali Z, Del Turco D, Bentivoglio AR, Healy DG, Albanese A, Nussbaum R, Gonzalez-Maldonado R, Deller T, Salvi S, Cortelli P, Gilks WP, Latchman DS, Harvey RJ, Dallapiccola B, Auburger G, Wood NW (2004) Hereditary early-onset Parkinson's disease caused by mutations in PINK1. Science 304:1158–1160
115. Van Humbeeck C, Cornelissen T, Hofkens H, Mandemakers W, Gevaert K, De Strooper B, Vandenberghe W (2011) Parkin interacts with Ambra1 to induce mitophagy. J Neurosci 31:10249–10261
116. Vande VC, Cizeau J, Dubik D, Alimonti J, Brown T, Israels S, Hakem R, Greenberg AH (2000) BNIP3 and genetic control of necrosis-like cell death through the mitochondrial permeability transition pore. Mol Cell Biol 20:5454–5468
117. Vargas JNS, Wang C, Bunker E, Hao L, Maric D, Schiavo G, Randow F, Youle RJ (2019) Spatiotemporal control of ULK1 activation by NDP52 and TBK1 during selective autophagy. Mol Cell 74:347–362, e346
118. Vega RB, Horton JL, Kelly DP (2015) Maintaining ancient organelles: mitochondrial biogenesis and maturation. Circ Res 116:1820–1834
119. Wang K, Zhou LY, Wang JX, Wang Y, Sun T, Zhao B, Yang YJ, An T, Long B, Li N, Liu CY, Gong Y, Gao JN, Dong YH, Zhang J, Li PF (2015) E2F1-dependent miR-421 regulates mitochondrial fragmentation and myocardial infarction by targeting Pink1. Nat Commun 6:7619
120. Wang X, Winter D, Ashrafi G, Schlehe J, Wong YL, Selkoe D, Rice S, Steen J, LaVoie MJ, Schwarz TL (2011) PINK1 and Parkin target Miro for phosphorylation and degradation to arrest mitochondrial motility. Cell 147:893–906
121. Wang Y, Serricchio M, Jauregui M, Shanbhag R, Stoltz T, Di Paolo CT, Kim PK, McQuibban GA (2015) Deubiquitinating enzymes regulate PARK2-mediated mitophagy. Autophagy 11:595–606

122. Wong YC, Holzbaur EL (2014) Optineurin is an autophagy receptor for damaged mitochondria in parkin-mediated mitophagy that is disrupted by an ALS-linked mutation. Proc Natl Acad Sci U S A 111:E4439-4448
123. Wu H, Xue D, Chen G, Han Z, Huang L, Zhu C, Wang X, Jin H, Wang J, Zhu Y, Liu L, Chen Q (2014) The BCL2L1 and PGAM5 axis defines hypoxia-induced receptor-mediated mitophagy. Autophagy 10:1712–1725
124. Wu S, Lu Q, Wang Q, Ding Y, Ma Z, Mao X, Huang K, Xie Z, Zou MH (2017) Binding of FUN14 domain containing 1 with inositol 1,4,5-trisphosphate receptor in mitochondria-associated endoplasmic reticulum membranes maintains mitochondrial dynamics and function in hearts in vivo. Circulation 136:2248–2266
125. Wu W, Tian W, Hu Z, Chen G, Huang L, Li W, Zhang X, Xue P, Zhou C, Liu L, Zhu Y, Zhang X, Li L, Zhang L, Sui S, Zhao B, Feng D (2014) ULK1 translocates to mitochondria and phosphorylates FUNDC1 to regulate mitophagy. EMBO Rep 15:566–575
126. Yoshii SR, Kishi C, Ishihara N, Mizushima N (2011) Parkin mediates proteasome-dependent protein degradation and rupture of the outer mitochondrial membrane. J Biol Chem 286:19630–19640
127. Yussman MG, Toyokawa T, Odley A, Lynch RA, Wu G, Colbert MC, Aronow BJ, Lorenz JN, Dorn GW (2002) Mitochondrial death protein Nix is induced in cardiac hypertrophy and triggers apoptotic cardiomyopathy. Nat Med 8:725–730
128. Zhang T, Xue L, Li L, Tang C, Wan Z, Wang R, Tan J, Tan Y, Han H, Tian R, Billiar TR, Tao WA, Zhang Z (2016) BNIP3 protein suppresses PINK1 Kinase proteolytic cleavage to promote mitophagy. J Biol Chem 291:21616–21629
129. Zhong Z, Umemura A, Sanchez-Lopez E, Liang S, Shalapour S, Wong J, He F, Boassa D, Perkins G, Ali SR, McGeough MD, Ellisman MH, Seki E, Gustafsson AB, Hoffman HM, Diaz-Meco MT, Moscat J, Karin M (2016) NF-kappaB restricts inflammasome activation via elimination of damaged mitochondria. Cell 164:896–910
130. Zhou H, Zhu P, Guo J, Hu N, Wang S, Li D, Hu S, Ren J, Cao F, Chen Y (2017) Ripk3 induces mitochondrial apoptosis via inhibition of FUNDC1 mitophagy in cardiac IR injury. Redox Biol 13:498–507
131. Zhou H, Zhu P, Wang J, Zhu H, Ren J, Chen Y (2018) Pathogenesis of cardiac ischemia reperfusion injury is associated with CK2alpha-disturbed mitochondrial homeostasis via suppression of FUNDC1-related mitophagy. Cell Death Differ 25:1080–1093
132. Zhu Y, Massen S, Terenzio M, Lang V, Chen-Lindner S, Eils R, Novak I, Dikic I, Hamacher-Brady A, Brady NR (2013) Modulation of serines 17 and 24 in the LC3-interacting region of Bnip3 determines pro-survival mitophagy versus apoptosis. J Biol Chem 288:1099–1113

Chapter 12
Proteotoxicity and Autophagy in Neurodegenerative and Cardiovascular Diseases

Kevin M. Alexander, Isabel Morgado, and Ronglih Liao

Abstract Aberrant protein folding and subsequent aggregation play a central role in a broad range of diseases that can affect nearly every tissue and organ. In particular, protein aggregation has been implicated in several neurodegenerative and cardiovascular diseases, such as Alzheimer's disease and cardiac amyloidosis, that are associated with significant morbidity and mortality. Increasing evidence highlights that misfolded protein oligomers exert significant proteotoxicity and result in cell death, serving as a main driver of disease pathogenesis. Autophagy is a natural, precisely regulated process responsible for the sequestration and clearance of misfolded proteins and damaged organelles to counteract the effects of proteotoxicity. Through a variety of mechanisms, autophagic pathways are often impaired in protein aggregation diseases. Moreover, augmenting autophagy has been demonstrated to ameliorate the end organ damage caused by protein aggregates. Therefore, understanding the interactions between protein aggregation and autophagy are crucial to treat proteotoxic neurodegenerative and cardiovascular diseases.

Keywords Proteotoxicity · Autophagy · Amyloidosis · Alzheimer's disease · Parkinson's disease · Transthyretin · Immunoglobulin light chain · β amyloid · α-synuclein

Introduction

Protein aggregation is a major event underlying many severe and fatal disorders. For reasons not fully understood, a protein can fail to adopt its functional conformation and instead misfold and self-associate into aggregates that are toxic to cells. This type of toxicity caused by proteins is referred to as proteotoxicity [1, 2].

Kevin M. Alexander and Isabel Morgado contributed equally to the manuscript.

K. M. Alexander · I. Morgado · R. Liao (✉)
Stanford Cardiovascular Institute, Stanford University School of Medicine, Stanford, CA, USA
e-mail: rliao@stanford.edu

© Springer Nature Switzerland AG 2022

L. A. Kirshenbaum (ed.), *Biochemistry of Apoptosis and Autophagy*, Advances in Biochemistry in Health and Disease 18,
https://doi.org/10.1007/978-3-030-78799-8_12

Often, protein aggregation can produce amyloid-like structures, which have been implicated in an increasing number of neurologic disorders including Alzheimer's disease, familial amyloid polyneuropathy and cardiomyopathies, such as wild-type transthyretin amyloidosis (ATTRwt) or light chain amyloidosis (AL) [3–7].

Factors that contribute to amyloid formation include genetic mutations [8, 9] and conditions that disturb cellular function, such as aging or stress [10]. However, there is no established causative agent or event that clearly triggers amyloid formation, and it is likely that several factors are required for amyloid formation to occur. Once triggered, the amyloid aggregation pathway culminates in the formation of long rigid amyloid fibrils that accumulate in vivo as plaques, but consistent data suggest that the main proteotoxic agents are not the final plaques but rather the intermediate amyloid oligomers [11–17]. Their exact mode of action is unclear and focus of intensive research but, it is likely, to be linked with the disturbance of protein homeostasis in the cell, which is dependent on specific events, such as protein synthesis, processing, transport, assembly and clearance [18].

For example, aggregation and concomitant proteotoxicity can arise when amino acid mutations render a protein susceptible to aggregation [3, 5, 19] or, for reasons extrinsic to the protein, a chaperone fails to facilitate proper protein folding during synthesis [20, 21]. Alternatively, if the cellular quality control mechanisms that normally remove abnormally folded proteins become compromised or less efficient due to aging, stress, or environmental conditions (e.g., changes in pH, temperature, osmotic pressure, or ionic strength), aggregation and/or accumulation of damaged proteins can occur [22, 23].

In tissues or organs where the cells are mostly post-mitotic and cannot easily divide to reduce aggregate deposition, the risk of proteotoxicity significantly increases. Accordingly, these tissues have a very limited ability to counteract proteotoxic-driven cell death. This is the case for cardiomyocytes and neurons and explains why the heart and the central and peripheral nervous systems, which have a low regeneration capacity, are particularly affected by protein aggregation and amyloid diseases [24]. In neurodegenerative diseases, such as Alzheimer's disease, proteotoxicity occurs via aggregation of the peptide Aβ, which ultimately forms cerebral plaques [25, 26]. Prion disease secondary to prion protein [27] or Parkinson's disease due to α-synuclein [28] are additional examples of neurodegenerative amyloid diseases. In the case of amyloid-related cardiomyopathies, important causes arise from systemic amyloidoses, such as AL [5, 29] or ATTR amyloidosis [16, 30, 31], in which either a wild type or mutant form of the protein can form aggregates. The presence of these aggregates leads to severe heart conditions, such as restrictive cardiomyopathy and subsequent heart failure.

The ultimate outcome of proteotoxicity is cell death, due to severe disruption of protein homeostasis. To prevent and/or control the formation of proteotoxic aggregates, cells rely on quality control mechanisms [22]. Such mechanisms involve molecular chaperones [1, 21], small specialized proteins that assist and assure correct nascent protein folding, and also two major clearance/degradation systems, the autophagy and the ubiquitin–proteasome system [7, 22, 24], which remove damaged and aggregated proteins. Autophagy focuses on the degradation of pre-formed aggregates and organelles by transporting them to lysosomes within autophagic vacuoles. On the other hand, the ubiquitin–proteasome system involves an organized cascade of enzymes that use ubiquitin molecules to label proteins to target them for degradation by the proteasome. Disruption of these specific mechanisms in cardiomyocytes and neurons and their contribution to cell death in cardiomyopathies and neuropathies will be further detailed in this chapter. We will also explore the different amyloid species involved in protein aggregation, in particular Aβ, ATTR, and AL amyloidoses and discuss their specific contributions to proteotoxicity and cell death.

Protein Aggregation and Proteotoxicity

Protein Amyloid Structures: Proteotoxic Oligomers

The misfolding of proteins into potentially toxic aggregates is a gradual, nucleation-dependent process [32] initiated by the monomeric protein or peptide, culminating in the formation of mature amyloid fibrils. Such fibrils are characterized by a rigid beta sheet-rich structure organized in a cross-beta pattern [33–35] that can be identified by X-ray and also have the ability to bind molecules like thioflavin-T and Congo red, commonly used as amyloid markers.

The structure of amyloid fibrils, in particular the ones derived from the peptide Aβ, which is involved in Alzheimer's disease, have been well described using nuclear magnetic resonance, cryo-electron microscopy, Fourier transform infared spectroscopy, circular dichroism, and X-ray crystallography [35–40].

In vivo, amyloid fibrils co-deposit with additional components (e.g., heparin sulfates, lipids) to form plaques in organs and tissues. The detection of such plaques is the common diagnostic strategy for amyloid diseases, but curiously there is often a poor correlation between the presence and extent of amyloid deposition and disease severity [13, 41]. However, a large body of evidence indicates that intermediate species generated during amyloid formation are the actual proteotoxic species and correlate with pathogenesis and cell death [13, 42, 43]. During protein self-assembly, a range of structurally different species are formed, and they have been broadly classified as oligomers and protofibrils (i.e., elongated oligomers) [12, 13, 17, 25, 44–46] (Fig. 12.1). Significant effort has been directed toward identifying, characterizing, and understanding the mode of action of these toxic agents. However, this has proven to be a difficult task. Due to their transient nature and dynamic structure, amyloid intermediates cannot be easily isolated.

Important progress has been achieved using the above-mentioned biophysical techniques to study oligomers produced in vitro [25, 38, 44–50]. Another important tool is conformation-specific antibodies generated against specific amyloid conformations, rather than simply an amino acid sequence. In some cases, these antibodies have been found to inhibit fibril formation, impair toxicity, and isolate specific intermediate conformations (e.g., protofibrils), allowing for their structural characterization [11, 51–55].

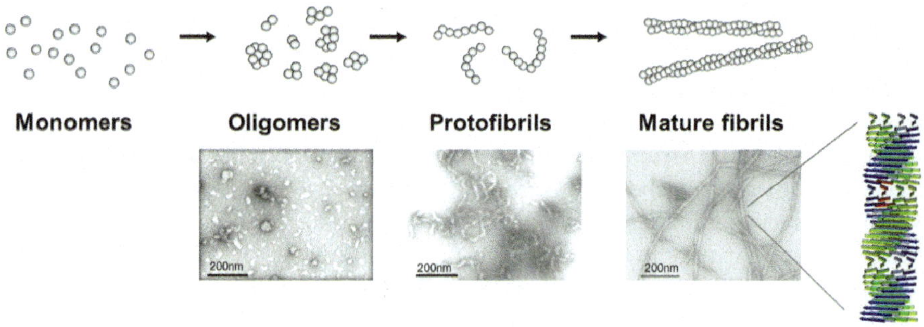

Fig. 12.1 Schematic representation of the protein fibril formation process. Protein monomers self-assembly into oligomer that elongate into curvilinear protofibrils which eventually become mature fibrils. Transmission electron microscopy images illustrate the appearance of each species (lower panel). In the mature state fibrils are characterized by a rigid β-sheet-rich structure organized in a cross-β pattern where stacked β-sheets organize in a helical fashion (as represented in colors on the right)

Oligomer Structure and Toxicity

The intrinsic metastability of oligomers and the different methodologies employed for their preparation in vitro have led to a large plethora of polymorphic species being described in the literature. These oligomers have different sizes, morphologies, secondary structures, stability, and degrees of toxicity. For example, oligomers have been described as Aβ-derived diffusible ligands (ADDLs) [25] due to their solubility or annular protofibrils [45] due to an annular shape. The latter are proposed to be derived from pre-fibrillar oligomers [56] that appear as spheres of 3–10 nm in diameter and develop into ring-shaped, pore-like structures [57]. Annual protofibrils have been observed for different amyloid forming proteins, such as Aβ [45], α-synuclein [58], and islet amyloid polypeptide precursor [59], and have been suggested to be a factor in toxic membrane pore formation [60, 61].

The general consensus from reports on oligomer structures is that they begin as small pre-fibrils structures (early aggregates), often composed of dimers or up to hexamers [62], that are mostly unstructured [63], molten [64], and have a high degree of hydrophobic surface exposure [65]. Regardless of the native structure of the precursor protein (even those containing α-helices [66]), these oligomers usually undergo a progressive transition into a stable β-sheet structure. Along with this structural rearrangement, oligomers also increase in size and stability [67] and become more compact [68]. The β-sheet structure becomes more organized, and hydrophobic clusters become progressively less exposed and more prone to form long, rigid and stable structures similar to fibrils. These structural transitions may correlate well with a decrease in toxicity as the oligomers become fibrils. Oligomers are now well established as being the major pathogenic agents in different amyloid diseases by interfering with cell function as supported both by cellular assays and mouse models [11–15, 69, 70]. Although exact mechanisms through which oligomers exert proteotoxicity and cause cell death remains controversial, it is believed that their observed structural plasticity and surface hydrophobicity allows them to easily interact with cellular membranes. However, as they develop into fibrils, their structure becomes more rigid and hydrophobic and likely less prone to interact with cell membranes and exert toxicity. This idea is consistent with the observation that fibrils seem to be less toxic (or non-toxic) compared with oligomers [41]. The toxicity of oligomeric structures will be further detailed below for Aβ, TTR, and immunoglobulin light chain.

Oligomers in Alzheimer's Disease, TTR Amyloidosis, and AL Amyloidosis

In Alzheimer's disease, amyloid fibrils derived from the peptide Aβ interfere with neurons [25, 26], while in AL amyloidosis, light chain-derived amyloids can interfere with cardiomyocytes leading to severe cardiomyopathy [5, 29]. Furthermore, TTR-derived amyloid can affect both the nervous system (i.e., familial amyloid polyneuropathy) and cardiomyocytes (i.e., ATTRwt) [30, 31]. While Aβ is an intrinsically disordered peptide, TTR and immunoglobulin light chains are globular proteins with complex, well-defined three-dimensional folding; thus, they possess differing mechanisms of aggregation [71].

Aβ Oligomers

Aβ is a small 39–43 amino acid peptide derived from the endoproteolysis of the transmembrane protein amyloid precursor protein (APP). Monomeric Aβ has poor solubility, is mostly unstructured, and has a very low content of α-helices in aqueous solution [72, 73]. Its propensity to aggregate can be influenced not only by its amino acid sequence and post-translational modifications but also the solution characteristics (e.g., peptide concentration, temperature, type of buffer, and the presence of lipids). These factors can easily prompt a conformational transition into a β-sheet structure [74–78], causing and Aβ to self-associate into transient intermediate species including dimers, oligomers and protofibrils. This culminates in the formation of mature amyloid fibrils, such as those found in AD plaques [75, 79]. These fibrils are composed of intermolecular β-sheets running perpendicular to the fibril axis to form a highly stable "cross-β" conformation [33, 34].

Aβ oligomers are known to interfere with synapses and disrupt neuronal function [15, 80, 81], probably because they can easily transfer from the cerebral interstitial fluid into synaptic membranes, disrupting their structure [70]. In fact, it is believed that cell membranes can serve as a scaffold for Aβ amyloid aggregation. This process compromises the integrity and permeability of the membrane [82–84] and can lead to pore formation [45, 57, 85, 86]. These perturbations can affect intra- and extracellular homeostasis (e.g., Ca^{2+} imbalance) and neuronal signal transduction [57, 87] that lead to cell death. These mechanisms will be discussed further in Sect. 3. Interestingly, other studies show that adsorption of mature fibrils into membranes has no significant impact on their integrity [88, 89] which might explain the low toxicity of fibrils.

TTR and Light Chain Oligomers

TTR and immunoglobulin light chains are natively folded proteins, and studies have suggested that a partial loss of the globular structure is an essential, rate-limiting step to prompt aggregation and subsequent amyloid formation [90, 91].

TTR is a homotetramer rich in β-sheets, and its major function is to transport vitamin A and thyroid hormone in the circulation [92]. Kinetic destabilization of TTR tetramers triggers subsequent misfolding, self-association, and monomer aggregation. Often, point mutations causing a single amino acid substitution can lead to increased TTR tetramer and/or monomer instability. More than 125 different amino acid mutations [93] have been associated with TTR amyloidosis. However, amyloidosis can also occur with wild-type TTR (i.e., ATTRwt cardiomyopathy). The mechanism for ATTRwt aggregation remains unclear.

Similar to Aβ, TTR oligomers are also believed to be the proteotoxic species rather than mature fibrils. In early studies, TTR monomers and very small oligomeric aggregates (less than 100 kDa) were found to be highly cytotoxic to neuronal cells, while mature fibrils and high molecular weight aggregates were innocuous [16]. These results have been reproduced by other groups [4, 94, 95]. For example, Sorgjerd et al. [17] suggested that small cytotoxic TTR oligomers comprised of only 20–30 monomers and wild-type TTR became toxic upon storage at 4 °C due to conformational changes in the tetramer.

Again, previous evidence suggests that the structure and conformation of the amyloid species dictates its proteotoxicity. Biophysical methods have been employed to uncover such structures. A recent in vitro observation using atomic force microscopy suggests that, upon acid-induced tetramer destabilization, the early, supposedly toxic TTR aggregates appear as annular oligomers that can assemble into a "double-stack of octameric rings" and further transition into spheroid oligomers and extended protofibrils [96]. Such annular morphology has been previously associated with cytotoxic pore formation in cellular membranes [41, 60, 97]. Interestingly, protofibrils can also disassemble into small oligomers [27], and studies [98] have reported a dynamic equilibrium between TTR protofibrils and unfolded monomers which seem to remain present throughout the aggregation process and are in constant exchange with the more mature amyloid states. In this way, fibrils can be a source of toxicity by storing and leaking oligomers, triggering new structural transitions into toxic species.

In AL amyloidosis, blood marrow cells produce abnormal antibody light chains that are intrinsically unstable and aggregation prone [91, 99]. Like TTR, light chain monomers are also composed of an extensive β-sheet framework that potentially contributes to its amyloid propensity. Structural studies [5, 19, 100, 101] indicate that light chains can form dimers with different conformations depending on somatic mutations present in the dimer interface [5]. The importance of the contact region between the dimers for amyloid formation is further supported by three-dimensional models produced for amyloid-forming light chain proteins [101, 102], indicating that light chains can form dimers with different conformations depending on somatic mutations. Overexpression of certain mutated forms of light chains can promote the formation of partially folded states that initiate the amyloidogenic process [5]. Besides the mutations described for most amyloidogenic proteins, environmental factors, such as concentration, pH, temperature, and ionic strength that promote structural destabilization, can easily contribute to amyloid formation [103].

Although structural information about light chain oligomers is scarce, early studies using atomic force microscopy have proposed a model for amyloid formation using a variable domain from patient-derived light chain fibrils [104]. Subsequent studies demonstrated that light chain fibrillogenesis is involved in the formation of smaller and thinner filaments (2.4 nm) and subsequently larger protofibrils (4.0 nm diameter) and mature fibrils (>6.0 nm diameter). Such structural transitions likely involve rearrangement of the β-strands in the monomer into β-sheets consistent with previous predictions for TTR [105].

Several studies have demonstrated that soluble light chain species isolated from patients with AL amyloidosis can be cytotoxic in vitro by disturbing cell homeostasis in renal and cardiac cells [106, 107], inducing cellular stress, impairing contractility in cardiomyocytes, and leading to apoptosis [107, 108]. Further studies with recombinant and structurally characterized counterparts of the toxic species revealed that the predominant cytotoxic species are monomers and dimers of the light chain variable domain [109].

Proteotoxicity and Autophagy in Neurologic Diseases

Alzheimer's Disease

Alzheimer's disease is the most common neurodegenerative disorder and is characterized by progressive cognitive dysfunction and dementia. The accumulation of Aβ plaques and neurofibrillary tangles, composed of hyperphosphorylated tau protein, are key components of pathogenesis [110]. These toxic protein species are known to cause oxidative and endoplasmic reticulum stress, resulting in mitochondrial damage and neuronal death [111, 112]. Moreover, Alzheimer's disease has been associated with impaired clearance of autophagosomes, containing these misfolded proteins and damaged mitochondria, due to defects in autophagy and lysosomal function [113].

Genetic studies of families with early-onset Alzheimer's disease provided some of the initial evidence for important factors in disease pathogenesis. Mutations in amyloid precursor protein (APP) and presenilin 1 and 2 have been associated with a highly penetrant, autosomal dominant form of Alzheimer's disease [114–116]. APP is an integral membrane protein found in neurons. It can be cleaved by β- and γ-secretases to form A β peptides of various lengths, including the amyloidogenic $A\beta_{42}$ [117]. APP mutations have been shown to increase Aβ production, leading to increased amyloid substrate [118]. Presenilin 1 and 2 are subunits of the γ-secretase complex, and mutations in these proteins can increase amyloidogenic $A\beta_{42}$ and cause early-onset Alzheimer's disease [117, 119].

In subsequent studies, both $A\beta_{42}$ and altered presenilin function have been demonstrated to impair autophagy and lysosomal function. In a Drosophila model, $A\beta_{42}$ caused to reduced autophagosome clearance in neurons, and leakage of these vesicles led to cytosolic acidification and cell damage [120]. In hippocampal neurons, presenilin 1 deficiency impairs autophagic vesicle fusion to lysosomes, thus blunting autophagic flux [121]. Furthermore, in fibroblasts derived from Alzheimer's patients, pathogenic presenilin 1 mutations have been shown to disrupt autophagy through reduce lysosome acidification and proteolysis [122].

More recently, genome-wide association and observational studies in Alzheimer's patients have identified additional proteins associated with the disease, some of which directly impact autophagy [123–125]. Phosphatidylinositol binding clathrin assembly protein (PICALM) is an assembly protein that facilitates endocytosis of clathrin-containing vesicles [126]. In Alzheimer's disease, PICALM has a demonstrated role in clearance of tau protein [127]. PICALM controls vesicle fusion through interactions with multiple soluble NSF attachment protein receptors. Consequently, impaired PICALM disrupts autophagy by affecting autophagosome formation and clearance. Another protein, beclin 1, appears to be involved in autophagy and Alzheimer's disease. Beclin 1 is an adaptor protein that interacts with numerous proteins, including phosphatidylinositol 3-kinase VPS34 and autophagy related 14, to promote autophagosome formation and subsequent autophagy [128]. Beclin 1 is reduced in the brains of patients with early-stage Alzheimer's disease [125]. Moreover, a loss of function beclin 1 mutation reduced autophagosome formation in vitro, and constitutively active beclin 1 in a mouse model increased autophagy, reduced cerebral Aβ aggregation, and halted cognitive decline [129, 130]. Taken together, these data suggest that beclin 1 has an important role in countering the proteotoxicity associated with Alzheimer's disease and may be a potential therapeutic target [131].

Parkinson's Disease

Parkinson's disease is the second most common neurodegenerative disease behind Alzheimer's disease. Parkinson's disease is predominantly a movement disorder and is characterized by deposition of α-synuclein aggregates, or Lewy bodies, in dopaminergic neurons leading to cytotoxicity. These aggregates have been thoroughly linked to oxidative stress and mitochondrial dysfunction. Consequently, autophagy and lysosomal function have been investigated for both their role in disease pathogenesis and as potential therapeutic targets.

Genetic studies have revealed multiple causal mutations that lead to heritable forms of Parkinson's disease [132]. Many of these mutations affect specific aspects of autophagy, including mitochondrial degradation (i.e., mitophagy), lysosomal function, and autophagosome generation. Several genetic mutations in proteins important in regulating mitophagy, including parkin, phosphatase and tensin homologue induced putative kinase 1 (PINK1), and F-box only protein 7 (FBXO7), have been associated with an autosomal recessive, early-onset form of Parkinson's disease [133–135]. Parkin is an E3 ubiquitin ligase that catalyzes the formation of polyubiquitin chains on substrates in damaged mitochondria [136]. This leads to binding of autophagy factors and lysosome-mediated degradation [137]. PINK1 is a kinase that translocates to the outer membrane in depolarized mitochondria [138]. Then, PINK1 phosphorylates both parkin and ubiquitin to enhance mitophagy [139]. FBX07 belongs to a family of proteins that recognize E3 ubiquitin ligase-protein complexes to facilitate ubiquitination [140]. FBX07 has been shown to directly bind and translocate parkin to damaged mitochondria [141]. FBXO7 is also involved in parkin-mediated ubiquitination of mitofusin, a mitochondrial fusion protein [141].

Lysosomal function, a key regulator of protein and mitochondrial clearance, has also been shown to play a role in the response to α-synuclein proteotoxicity. Indeed, mutations in several lysosomal proteins have been described to cause heritable forms of Parkinson's disease. Mutations in leucine-rich repeat kinase 2 (LRRK2), a protein with multiple domains and functions, have been associated with an autosomal dominant form of Parkinson's disease [142]. LRRK2 regulates endosome to lysosome trafficking at least in part by localizing trans-Golgi network-derived vesicles to lysosomes for degradation [143]. Moreover, LRRK2 mutants were associated with increased damage from oxidative stress in neuronal cells, and rodent models of LRRK2 mutations display abnormally reduce neurite length [144, 145]. ATP13A2 is a lysosomal P-type ATPase responsible for metal cation homeostasis [146]. In humans, mutations in ATP13A have been associated with early-onset Parkinson's disease, and fibroblasts derived from these patients demonstrate reduced autophagosome clearance and reduced lysosomal acidification [147, 148]. Glucocerebrosidase is a lysosomal enzyme involved in sphingolipid metabolism. Deficiency of this enzyme is known to cause the lysosomal storage disorder Gaucher disease [149]. In addition, glucocerebroside mutations have been associated with Parkinson's disease, in which accumulation of the glycolipid substrate leads to lysosomal dysfunction and impaired α-synuclein clearance [150, 151].

Proteotoxicity and Autophagy in Cardiovascular Diseases

AL Cardiac Amyloidosis

The pathophysiology of AL cardiac amyloidosis has long been considered to occur through gradual extracellular fibril infiltration of the myocardium, ultimately leading to passive myocardial restriction and dysfunction. However, both in vitro and observational human studies have demonstrated the lack of a strong association between the burden of fibril deposition and cardiac outcomes. Indeed, mouse hearts infused with light chains from cardiac amyloidosis developed diastolic dysfunction in the absence of significant myocardial infiltration [152]. Moreover, AL amyloidosis patients receiving chemotherapy who achieve a reduction in circulating free light chains have an improvement in cardiac function without any change in the amount of amyloid deposition, suggesting the importance of direct light chain proteotoxicity in disease progression (Fig. 12.2) [153–155].

Several groups have begun to elucidate the molecular mechanisms responsible for this active light chain toxicity. Early studies have shown an increase in reactive

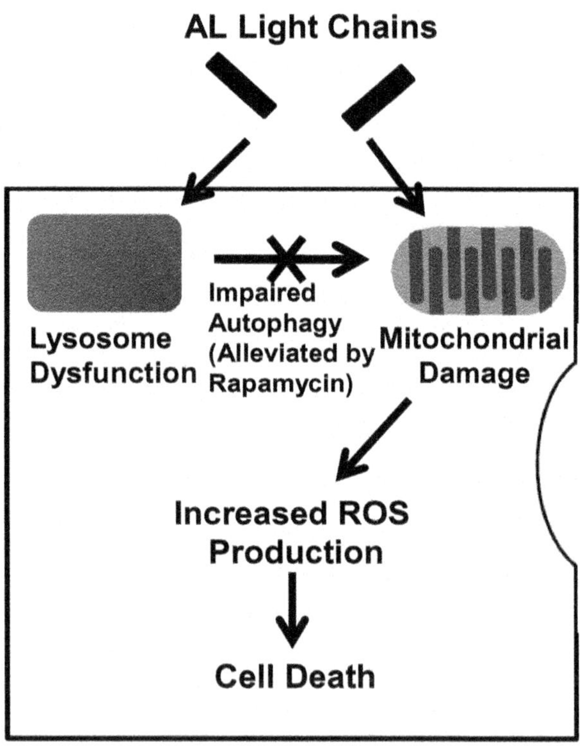

Fig. 12.2 Cellular mechanisms of light chain proteotoxicity. In AL amyloidosis, amyloidogenic light chains cause cellular toxicity through lysosomal dysfunction and mitochondrial damage. The latter leads to increased production of reactive oxygen species and subsequent oxidative stress and cell death. An inducer of autophagy, such as rapamycin, can attenuate light chain-mediated proteotoxicity

oxygen species and impaired contractility and relaxation in isolated rat cardiomyocytes treated with light chains from patients [106]. The importance of increased reactive oxygen species is supported by the attenuation of this myocardial dysfunction by treatment with an antioxidant (i.e., a superoxide dismutase/catalase mimetic). Another group showed atrial coronary arterioles and adipose arterioles exposed to amyloidogenic light chains developed endothelial dysfunction, increased reactive oxygen species production, and apoptosis [156, 157]. This observation may explain the coronary microvascular dysfunction seen clinically in patients with AL amyloidosis [158].

Subsequent experiments using rat cardiomyocytes and in vivo mouse models discovered the role of stress responsive p38 mitogen-activated protein kinase (MAPK) in triggering reactive oxygen species production and apoptosis [107, 159]. Through a non-canonical signaling pathway, p38 MAPK activates downstream target proteins, including stanniocalcin1. Stanniocalcin1 is important in calcium homeostasis, and increased expression leads to the loss mitochondrial membrane potential and apoptosis [160]. Stanniocalcin 1 is upregulated in myocardium from AL amyloidosis patients. Furthermore, gene silencing of stanniocalcin 1 prevents light chain-mediated cardiotoxicity in isolated cardiomyocytes, and stanniocalcin 1 overexpression mimicked the cardiac changes caused by light chains.

Light chains also affect mitochondrial function by altering metabolism [161]. This was demonstrated using purified human amyloidogenic and non-amyloidogenic light chains in a series of co-immunoprecipitation experiments with adult rat cardiomyocytes. Proteomic analysis was performed to identify the light chain-protein interactome. Several protein interactions that were unique to the amyloidogenic light chains were discovered. These proteins—peroxisomal acyl-coenzyme A oxidase 1, acyl-coenzyme A dehydrogenase family member 9, voltage-dependent anion channel, and optic atrophy 1-like protein—have crucial roles in cell survival and fatty acid β-oxidation pathways in mitochondria and peroxisomes. These data along the previously discussed experiments demonstrating stanniocalcin1-mediated mitochondrial toxicity suggest that impaired mitochondrial function and energy handling are key factors in light chain-induced cell death.

Another important feature of light chain-mediated proteotoxicity is impaired clearance of these damaged mitochondria via autophagy, or more specifically mitophagy. Lysosomes are responsible for degrading these defective mitochondria and appear also to be directly affected by light chain toxicity. Indeed, amyloidogenic light chains were shown to internalize into mouse cardiomyocytes and localize to lysosomal compartments [162]. Reactive oxygen species appear to be an important link between light chain toxicity and impaired autophagy. Multiple studies have demonstrated that reactive oxygen species regulate several components of crucial autophagy signaling pathways [163–165]. Therefore, the above-mentioned mechanisms of light chain-mediated reactive oxygen species productions have deleterious effects on normal autophagy.

The effects of light chains on autophagy and lysosome function were further delineated in a series of experiments [166]. Rat cardiomyocytes exposed to toxic light chains had impaired autophagic flux based on decreased expression of the autophagy marker LC3-II and increased expression on p62, a marker of autophagic clearance. Moreover, rapamycin, an mTOR signaling pathway antagonist and inducer of autophagy, ameliorated this light chain-mediated disturbance in autophagy in rat cardiomyocytes and an in vivo zebrafish model. Also, in the presence of toxic light chains, lysosomes had decreased acidity and expression of gene and transcription regulators of lysosome function, including lysosomal-specific ATPase, cathepsin D, and transcription factor EB.

Other Cardiomyopathies

The importance of autophagy has been described for several other protein aggregation-mediated cardiomyopathies. Misfolding of desmin, a type III intermediate filament protein, or its chaperone and heat shock protein family member α-β-crystallin, have been implicated in cardiotoxicity and subsequent dilated cardiomyopathy [167, 168]. In early experiments using a transgenic mutant α-β-crystallin mouse model of desmin-related cardiomyopathy, treatment with geranylgeranylacetone, a potent inducer of multiple heat shock proteins attenuated myocardial fibrosis and improved cardiac function. In addition, geranylgeranylacetone inhibited apoptosis in these transgenic mice [169]. Furthermore, in a proteotoxic mouse model of heart failure (desmin-related cardiomyopathy), upregulation of autophagy via overexpression of autophagy-related 7 ameliorated the cardiomyopathy phenotype [170].

These preclinical studies highlight the importance of mitochondrial dysfunction and impaired autophagy in AL amyloidosis and other proteotoxicity-mediated cardiomyopathies. Thus far, there has been a paucity of clinical trials targeting autophagy in cardiovascular disease. Currently, a clinical trial is investigating the use of the mTOR inhibitor rapamycin, an inducer of autophagy, to ameliorate cardiac allograft remodeling after heart transplantation (ClinicalTrials.gov Identifier NCT01889992). Further mechanistic studies and subsequent clinical trials are needed to identify additional therapeutic targets.

Conclusions

Protein aggregation is increasingly recognized as a fundamental component of pathogenesis in many diseases. A variety of precursor proteins have the ability to misfold and form oligomers that exert significant cytotoxicity. In particular, post-mitotic cells, such as neurons and cardiomyocytes are vulnerable to proteotoxic cell death. Consequently, proteotoxicity has been implicated in several neurodegenerative and cardiovascular diseases, including Alzheimer's disease, Parkinson's disease, and cardiac amyloidosis. Autophagy represents a natural defense for protecting cells against proteotoxic species, and impaired autophagy is often a key feature of protein aggregation diseases. Interventions that augment autophagy are a potential strategy to prevent and treat disease-related proteotoxicity.

References

1. Douglas PM, Summers DW, Cyr DM (2009) Molecular chaperones antagonize proteotoxicity by differentially modulating protein aggregation pathways. Prion 3(2):51–58
2. Ross CA, Poirier MA (2004) Protein aggregation and neurodegenerative disease. Nat Med 10(Suppl):S10–S17
3. Willis MS, Patterson C (2013) Proteotoxicity and cardiac dysfunction—Alzheimer's disease of the heart? N Engl J Med 368(5):455–464
4. Mendes Sousa M et al (2001) Deposition of transthyretin in early stages of familial amyloidotic polyneuropathy. Am J Pathol 159(6):1993–2000
5. Ramirez-Alvarado M (2012) Amyloid formation in light chain amyloidosis. Curr Top Med Chem 12(22):2523–2533
6. Pinney JH et al (2013) Senile systemic amyloidosis: clinical features at presentation and outcome. J Am Heart Assoc **2**(2)
7. Sandri M, Robbins J (2014) Proteotoxicity: an underappreciated pathology in cardiac disease. J Mol Cell Cardiol 71:3–10
8. Thomas PK (1975) Genetic factors in amyloidosis. J Med Genet 12(4):317–326
9. Spinney L (2014) Alzheimer's disease: the forgetting gene. Nature 510(7503):26–28
10. Douglas PM, Dillin A (2010) Protein homeostasis and aging in neurodegeneration. J Cell Biol 190(5):719–729
11. Kayed R et al (2003) Common structure of soluble amyloid oligomers implies common mechanism of pathogenesis. Science 300:486–489
12. Cleary JP et al (2005) Natural oligomers of the amyloid-B protein specifically disrupt cognitive function. Nat Neurosci 8:79–84
13. Haass C, Selkoe DJ (2007) Soluble protein oligomers in neurodegeneration: lessons from the Alzheimer's amyloid [β]-peptide. Nat Rev Mol Cell Bio 8(2):101–112
14. Demuro A et al (2005) Calcium dysregulation and membrane disruption as a ubiquitous neurotoxic mechanism of soluble amyloid oligomers. J Biol Chem 280(17):17294–17300
15. Zhao LN et al (2012) The toxicity of amyloid β oligomers. Int J Mol Sci 13(6):7303–7327
16. Reixach N et al (2004) Tissue damage in the amyloidoses: transthyretin monomers and nonnative oligomers are the major cytotoxic species in tissue culture. Proc Natl Acad Sci USA 101(9):2817–2822
17. Sörgjerd K et al (2008) Prefibrillar transthyretin oligomers and cold stored native tetrameric transthyretin are cytotoxic in cell culture. Biochem Biophys Res Commun 377(4):1072–1078
18. Powers ET, Balch WE (2013) Diversity in the origins of proteostasis networks [mdash] a driver for protein function in evolution. Nat Rev Mol Cell Biol 14(4):237–248
19. Peterson FC et al (2010) A single mutation promotes amyloidogenicity through a highly promiscuous dimer interface. Structure 18(5):563–570
20. Morimoto RI (2008) Proteotoxic stress and inducible chaperone networks in neurodegenerative disease and aging. Genes Dev 22(11):1427–1438
21. Hightower LE (1991) Heat shock, stress proteins, chaperones, and proteotoxicity. Cell 66(2):191–197
22. Kastle M, Grune T (2012) Interactions of the proteasomal system with chaperones: protein triage and protein quality control. Prog Mol Biol Transl Sci 109:113–160
23. Latonen L (2011) Nucleolar aggresomes as counterparts of cytoplasmic aggresomes in proteotoxic stress. Proteasome inhibitors induce nuclear ribonucleoprotein inclusions that accumulate several key factors of neurodegenerative diseases and cancer. Bioessays 33(5):386–395
24. He C, Klionsky DJ (2006) Autophagy and Neurodegeneration. ACS Chem Biol 1(4):211–213
25. Klein WL (2002) Abeta toxicity in Alzheimer's disease: globular oligomers (ADDLs) as new vaccine and drug targets. Neurochem Int 41(5):345–352
26. Finder VH, Glockshuber R (2007) Amyloid-beta aggregation. Neurodegener Dis 4(1):13–27
27. Safar JG (2012) Molecular pathogenesis of sporadic prion diseases in man. Prion 6(2):108–115

28. Ebrahimi-Fakhari D, Wahlster L, McLean PJ (2012) Protein degradation pathways in Parkinson's disease: curse or blessing. Acta Neuropathol 124(2):153–172
29. Gallo G et al (1996) Light chain cardiomyopathy. Structural analysis of the light chain tissue deposits. Am J Pathol 148(5):1397–1406
30. Ando Y et al (2013) Guideline of transthyretin-related hereditary amyloidosis for clinicians. Orphanet J Rare Dis 8(1):1–18
31. Benson M (2009) Genetics: clinical implications of TTR amyloidosis. In: Richardson SJ, Cody V (eds) Recent advances in transthyretin evolution, structure and biological functions, Springer, Berlin
32. Garvey M, Morgado I (2013) Peptide concentration alters intermediate species in amyloid β fibrillation kinetics. Biochem Biophys Res Commun 433(3):276–280
33. Sunde M et al (1997) Common core structure of amyloid fibrils by synchrotron X-ray diffraction. J Mol Biol 273(3):729–739
34. Lu J-X et al (2013) Molecular structure of β-amyloid fibrils in Alzheimer's disease brain tissue. Cell 154(6). https://doi.org/10.1016/j.cell.2013.08.035
35. Fandrich M (2007) On the structural definition of amyloid fibrils and other polypeptide aggregates. Cell Mol Life Sci 64(16):2066–2078
36. Schmidt A et al (2016) Cryo-EM reveals the steric zipper structure of a light chain-derived amyloid fibril. Proc Natl Acad Sci
37. Serpell LC (2000) Alzheimer's amyloid fibrils: structure and assembly. Biochimica et Biophysica Acta (BBA)—Mol Basis Dis 1502(1): 16–30
38. Fändrich M, Schmidt M, Grigorieff N (2011) Recent progress in understanding Alzheimer's β-amyloid structures. Trends Biochem Sci 36(6):338–345
39. Nelson R, Eisenberg D (2006) Structural models of amyloid-like fibrils. Adv Protein Chem 73:235–282
40. Toyama BH, Weissman JS (2011) Amyloid structure: conformational diversity and consequences. Annu Rev Biochem 80:557–585
41. Caughey B, Lansbury PT (2003) Protofibrils, pores, fibrils, and neurodegeneration: separating the responsible protein aggregates from the innocent bystanders. Annu Rev Neurosci 26:267–298
42. Lesne S et al (2006) A specific amyloid-beta protein assembly in the brain impairs memory. Nature 440(7082):352–357
43. Shankar GM et al (2008) Amyloid-beta protein dimers isolated directly from Alzheimer's brains impair synaptic plasticity and memory. Nat Med 14(8):837–842
44. Bemporad F, Chiti F (2012) Protein misfolded oligomers: experimental approaches, mechanism of formation, and structure-toxicity relationships. Chem Biol 19(3):315–327
45. Kayed R et al (2009) Annular protofibrils are a structurally and functionally distinct type of amyloid oligomer. J Biol Chem 284(7):4230–4237
46. tBreydo L, Uversky VN (2015) Structural, morphological, and functional diversity of amyloid oligomers. FEBS Lett 589(19 Pt A):2640–2648
47. Scheidt HA et al (2011) Solid-state NMR spectroscopic investigation of Ab protofibrils: implication of a b-sheet remodeling upon maturation into terminal amyloid fibrils. Angew Chem Int Ed Engl 50(12):2837–2840
48. Lindgren M, Sörgjerd K, Hammarström P (2005) Detection and characterization of aggregates, prefibrillar amyloidogenic oligomers, and protofibrils using fluorescence spectroscopy. Biophys J 88(6):4200–4212
49. Stefani M (2012) Structural features and cytotoxicity of amyloid oligomers: implications in Alzheimer's disease and other diseases with amyloid deposits. Prog Neurobiol 99(3):226–245
50. Pham JD et al (2013) Structures of oligomers of a peptide from beta-amyloid. J Am Chem Soc 135(33):12460–12467
51. Morgado I et al (2012) Molecular basis of beta-amyloid oligomer recognition with a conformational antibody fragment. Proc Natl Acad Sci U S A 109(31):12503–12508
52. Haupt C et al (2011) Amyloid fibril recognition with the conformational B10 antibody fragment depends on electrostatic interactions. J Mol Biol 405(2):341–348

53. Paduch M et al (2013) Generating conformation-specific synthetic antibodies to trap proteins in selected functional states. Methods 60(1):3–14
54. O'Nuallain B et al (2011) A monoclonal antibody against synthetic Abeta dimer assemblies neutralizes brain-derived synaptic plasticity-disrupting Abeta. J Neurochem 119(1):189–201
55. Kreutzer AG et al (2016) X-ray crystallographic structures of a trimer, dodecamer, and annular pore formed by an Aβ17–36 β-hairpin. J Am Chem Soc 138(13):4634–4642
56. Chiti F et al (2003) Rationalization of the effects of mutations on peptide and protein aggregation rates. Nature 424(6950):805–808
57. Lashuel HA et al (2002) Neurodegenerative disease: amyloid pores from pathogenic mutations. Nature 418(6895):291
58. Conway KA, Harper JD, Lansbury PT Jr (2000) Fibrils formed in vitro from alpha-synuclein and two mutant forms linked to Parkinson's disease are typical amyloid. Biochemistry 39(10):2552–2563
59. Janson J et al (1999) The mechanism of islet amyloid polypeptide toxicity is membrane disruption by intermediate-sized toxic amyloid particles. Diabetes 48(3):491–498
60. Lashuel HA, Lansbury PT Jr (2006) Are amyloid diseases caused by protein aggregates that mimic bacterial pore-forming toxins? Q Rev Biophys 39(2):167–201
61. Glabe CG, Kayed R (2006) Common structure and toxic function of amyloid oligomers implies a common mechanism of pathogenesis. Neurology 66(2 Suppl 1):S74–S78
62. Bitan G et al (2003) Amyloid beta-protein (Abeta) assembly: Abeta 40 and Abeta 42 oligomerize through distinct pathways. Proc Natl Acad Sci U S A 100(1):330–335
63. Carulla N et al (2009) Experimental characterization of disordered and ordered aggregates populated during the process of amyloid fibril formation. Proc Natl Acad Sci 106(19):7828–7833
64. Wei L et al (2011) The molecular basis of distinct aggregation pathways of islet amyloid polypeptide. J Biol Chem 286(8):6291–6300
65. Dusa A et al (2006) Characterization of oligomers during alpha-synuclein aggregation using intrinsic tryptophan fluorescence. Biochemistry 45(8):2752–2760
66. Plakoutsi G et al (2005) Evidence for a mechanism of amyloid formation involving molecular reorganisation within native-like precursor aggregates. J Mol Biol 351(4):910–922
67. Calamai M et al (2005) Reversal of protein aggregation provides evidence for multiple aggregated states. J Mol Biol 346(2):603–616
68. Kaylor J et al (2005) Characterization of oligomeric intermediates in alpha-synuclein fibrillation: FRET studies of Y125W/Y133F/Y136F alpha-synuclein. J Mol Biol 353(2):357–372
69. Alberdi E et al (2010) Amyloid β oligomers induce Ca^{2+} dysregulation and neuronal death through activation of ionotropic glutamate receptors. Cell Calcium 47(3):264–272
70. Hong S et al (2014) Soluble Aβ oligomers are rapidly sequestered from brain ISF in vivo and bind GM1 ganglioside on cellular membranes. Neuron 82(2):308–319
71. Eisele YS et al (2015) Targeting protein aggregation for the treatment of degenerative diseases. Nat Rev Drug Discov 14(11):759–780
72. Tomaselli S et al (2006) The α-to-β conformational transition of Alzheimer's Aβ-(1–42) peptide in aqueous media is reversible: a step by step conformational analysis suggests the location of β conformation seeding. ChemBioChem 7(2):257–267
73. Vivekanandan S et al (2011) A partially folded structure of amyloid-beta(1–40) in an aqueous environment. Biochem Biophys Res Commun 411(2):312–316
74. Fezoui Y, Teplow DB (2002) Kinetic studies of amyloid β-protein fibril assembly: differential effects of α-helix stabilization. J Biol Chem 277(40):36948–36954
75. Morgado I, Fändrich M (2011) Assembly of Alzheimer's Ab peptide into nanostructured amyloid fibrils. Curr Opin Colloid Interface Sci 16(6):508–514
76. Morgado I, Garvey M (2015) Lipids in amyloid-beta processing, aggregation, and toxicity. Adv Exp Med Biol 855:67–94
77. Benseny-Cases N, Cocera M, Cladera J (2007) Conversion of non-fibrillar β-sheet oligomers into amyloid fibrils in Alzheimer's disease amyloid peptide aggregation. Biochem Biophys Res Commun 361:916–921

78. Dahse K et al (2010) DHPC strongly affects the structure and oligomerization propensity of Alzheimer's Aβ(1–40) peptide. J Mol Biol 403(4):643–659
79. Benseny-Cases N, Cocera M, Cladera J (2007) Conversion of non-fibrillar beta-sheet oligomers into amyloid fibrils in Alzheimer's disease amyloid peptide aggregation. Biochem Biophys Res Commun 361:916–921
80. Walsh DM et al (2002) Naturally secreted oligomers of amyloid beta protein potently inhibit hippocampal long-term potentiation in vivo. Nature 416(6880):535–539
81. Walsh DM, Selkoe DJ (2004) Oligomers on the brain: the emerging role of soluble protein aggregates in neurodegeneration. Protein Pept Lett 11:213–228
82. McLaurin J, Chakrabartty A (1996) Membrane disruption by Alzheimer Î²-amyloid peptides mediated through specific binding to either phospholipids or gangliosides: Implications For Neurotoxicity. J Biol Chem 271(43):26482–26489
83. Murphy RM (2007) Kinetics of amyloid formation and membrane interaction with amyloidogenic proteins. Biochimica et Biophysica Acta (BBA)—Biomembranes 1768(8): 1923–1934
84. Relini A et al (2009) The two-fold aspect of the interplay of amyloidogenic proteins with lipid membranes. Chem Phys Lipid 158(1):1–9
85. Prangkio P et al (2012) Multivariate analyses of amyloid-β oligomer populations indicate a connection between pore formation and cytotoxicity. PLoS ONE 7(10): e47261
86. Kawahara M et al (2011) Membrane incorporation, channel formation, and disruption of calcium homeostasis by Alzheimer's β-Amyloid Protein. Int J Alzheimer's Dis 17
87. Soto C (2003) Unfolding the role of protein misfolding in neurodegenerative diseases. Nat Rev Neurosci 4(1):49–60
88. Yip CM, McLaurin J (2001) Amyloid-β peptide assembly: a critical step in fibrillogenesis and membrane disruption. Biophys J 80(3):1359–1371
89. Bokvist M et al (2004) Two Types of Alzheimer's β-Amyloid (1–40) peptide membrane interactions: aggregation preventing transmembrane anchoring versus accelerated surface fibril formation. J Mol Biol 335(4):1039–1049
90. Colon W, Kelly JW (1992) Partial denaturation of transthyretin is sufficient for amyloid fibril formation in vitro. Biochemistry 31(36):8654–8660
91. Hurle MR et al (1994) A role for destabilizing amino acid replacements in light-chain amyloidosis. Proc Natl Acad Sci U S A 91(12):5446–5450
92. Monaco HL, Rizzi M, Coda A (1995) Structure of a complex of two plasma proteins: transthyretin and retinol-binding protein. Science 268(5213):1039–1041
93. Saraiva MJ (1995) Transthyretin mutations in health and disease. Hum Mutat 5(3):191–196
94. Gonçalves NP, Teixeira-Coelho M, Saraiva MJ (2015) Protective role of anakinra against transthyretin-mediated axonal loss and cell death in a mouse model of familial amyloidotic polyneuropathy. J Neuropathol Exp Neurol 74(3):203–217
95. Andersson K et al (2002) Only amyloidogenic intermediates of transthyretin induce apoptosis. Biochem Biophys Res Commun 294(2):309–314
96. Pires RH et al (2012) Distinct annular oligomers captured along the assembly and disassembly pathways of transthyretin amyloid protofibrils. PLoS One 7(9): e44992
97. Demuro A, Smith M, Parker I (2011) Single-channel Ca($^{2+}$) imaging implicates Abeta1-42 amyloid pores in Alzheimer's disease pathology. J Cell Biol 195(3):515–524
98. Groenning M et al (2015) Considerably unfolded transthyretin monomers preceed and exchange with dynamically structured amyloid protofibrils. Sci Rep 5:11443
99. Kastritis E et al (2010) Bortezomib with or without dexamethasone in primary systemic (light chain) amyloidosis. J Clin Oncol 28(6):1031–1037
100. Schormann N et al (1995) Tertiary structure of an amyloid immunoglobulin light chain protein: a proposed model for amyloid fibril formation. Proc Natl Acad Sci U S A 92(21):9490–9494
101. Randles EG et al (2009) Structural alterations within native amyloidogenic immunoglobulin light chains. J Mol Biol 389(1):199–210
102. Schormann N et al (1995) Tertiary structure of an amyloid immunoglobulin light chain protein: a proposed model for amyloid fibril formation. Proc Natl Acad Sci USA 92(21):9490–9494

103. Chiti F et al (1999) Designing conditions for in vitro formation of amyloid protofilaments and fibrils. Proc Natl Acad Sci USA 96(7):3590–3594
104. Ionescu-Zanetti C et al (1999) Monitoring the assembly of IG light-chain amyloid fibrils by atomic force microscopy. Proc Natl Acad Sci 96(23):13175–13179
105. Blake C, Serpell L (1996) Synchrotron X-ray studies suggest that the core of the transthyretin amyloid fibril is a continuous beta-sheet helix. Structure 4(8):989–998
106. Brenner DA et al (2004) Human amyloidogenic light chains directly impair cardiomyocyte function through an increase in cellular oxidant stress. Circ Res 94(8):1008–1010
107. Shi J et al (2010) Amyloidogenic light chains induce cardiomyocyte contractile dysfunction and apoptosis via a non-canonical p38alpha MAPK pathway. Proc Natl Acad Sci U S A 107(9):4188–4193
108. Teng J et al (2004) Different types of glomerulopathic light chains interact with mesangial cells using a common receptor but exhibit different intracellular trafficking patterns. Lab Invest 84(4):440–451
109. Sikkink LA, Ramirez-Alvarado M (2010) Cytotoxicity of amyloidogenic immunoglobulin light chains in cell culture. Cell Death Dis 1(11): e98
110. Vinters HV (2015) Emerging concepts in Alzheimer's disease. Annu Rev Pathol 10:291–319
111. Chakrabarti S et al (2013) Oxidative stress and amyloid beta toxicity in Alzheimer's disease: intervention in a complex relationship by antioxidants. Curr Med Chem 20(37):4648–4664
112. Santos LE, Ferreira ST (2017) Crosstalk between endoplasmic reticulum stress and brain inflammation in Alzheimer's disease. Neuropharmacology
113. Zhang Y et al (2017) The role of ubiquitin proteasomal system and autophagy-lysosome pathway in Alzheimer's disease. Rev Neurosci 28(8):861–868
114. Rogaev EI et al (1995) Familial Alzheimer's disease in kindreds with missense mutations in a gene on chromosome 1 related to the Alzheimer's disease type 3 gene. Nature 376(6543):775–778
115. Sherrington R et al (1995) Cloning of a gene bearing missense mutations in early-onset familial Alzheimer's disease. Nature 375(6534):754–760
116. Goate A et al (1991) Segregation of a missense mutation in the amyloid precursor protein gene with familial Alzheimer's disease. Nature 349(6311):704–706
117. Bentahir M et al (2006) Presenilin clinical mutations can affect gamma-secretase activity by different mechanisms. J Neurochem 96(3):732–742
118. Citron M et al (1992) Mutation of the beta-amyloid precursor protein in familial Alzheimer's disease increases beta-protein production. Nature 360(6405):672–674
119. Kumar-Singh S et al (2006) Mean age-of-onset of familial alzheimer disease caused by presenilin mutations correlates with both increased Abeta42 and decreased Abeta40. Hum Mutat 27(7):686–695
120. Ling D et al (2009) Abeta42-induced neurodegeneration via an age-dependent autophagic-lysosomal injury in Drosophila. PLoS One 4(1):e4201
121. Esselens C et al (2004) Presenilin 1 mediates the turnover of telencephalin in hippocampal neurons via an autophagic degradative pathway. J Cell Biol 166(7):1041–1054
122. Lee JH et al (2010) Lysosomal proteolysis and autophagy require presenilin 1 and are disrupted by Alzheimer-related PS1 mutations. Cell 141(7):1146–1158
123. Beecham GW et al (2014) Genome-wide association meta-analysis of neuropathologic features of Alzheimer's disease and related dementias. PLoS Genet 10(9): e1004606
124. Naj AC et al (2014) Effects of multiple genetic loci on age at onset in late-onset Alzheimer disease: a genome-wide association study. JAMA Neurol 71(11):1394–1404
125. Pickford F et al (2008) The autophagy-related protein beclin 1 shows reduced expression in early Alzheimer disease and regulates amyloid beta accumulation in mice. J Clin Invest 118(6):2190–2199
126. Tebar F, Bohlander SK, Sorkin A (1999) Clathrin assembly lymphoid myeloid leukemia (CALM) protein: localization in endocytic-coated pits, interactions with clathrin, and the impact of overexpression on clathrin-mediated traffic. Mol Biol Cell 10(8):2687–2702

127. Moreau K et al (2014) PICALM modulates autophagy activity and tau accumulation. Nat Commun 5:4998
128. Salminen A et al (2013) Impaired autophagy and APP processing in Alzheimer's disease: the potential role of Beclin 1 interactome. Prog Neurobiol 106–107:33–54
129. Park JM et al (2018) ULK1 phosphorylates Ser30 of BECN1 in association with ATG14 to stimulate autophagy induction. Autophagy 1–28
130. Rocchi A et al (2017) A Becn1 mutation mediates hyperactive autophagic sequestration of amyloid oligomers and improved cognition in Alzheimer's disease. PLoS Genet 13(8):e1006962
131. Wang BJ et al (2017) ErbB2 regulates autophagic flux to modulate the proteostasis of APP-CTFs in Alzheimer's disease. Proc Natl Acad Sci U S A 114(15):E3129–E3138
132. Chang D et al (2017) A meta-analysis of genome-wide association studies identifies 17 new Parkinson's disease risk loci. Nat Genet 49(10):1511–1516
133. Kitada T et al (1998) Mutations in the parkin gene cause autosomal recessive juvenile parkinsonism. Nature 392(6676):605–608
134. Valente EM et al (2004) Hereditary early-onset Parkinson's disease caused by mutations in PINK1. Science 304(5674):1158–1160
135. Di Fonzo A et al (2009) FBXO7 mutations cause autosomal recessive, early-onset parkinsonian-pyramidal syndrome. Neurology 72(3):240–245
136. Cunningham CN et al (2015) USP30 and parkin homeostatically regulate atypical ubiquitin chains on mitochondria. Nat Cell Biol 17(2):160–169
137. Stolz A, Ernst A, Dikic I (2014) Cargo recognition and trafficking in selective autophagy. Nat Cell Biol 16(6):495–501
138. Narendra DP et al (2010) PINK1 is selectively stabilized on impaired mitochondria to activate Parkin. PLoS Biol 8(1):e1000298
139. Kane LA et al (2014) PINK1 phosphorylates ubiquitin to activate Parkin E3 ubiquitin ligase activity. J Cell Biol 205(2):143–153
140. Chang YF et al (2006) The F-box protein Fbxo7 interacts with human inhibitor of apoptosis protein cIAP1 and promotes cIAP1 ubiquitination. Biochem Biophys Res Commun 342(4):1022–1026
141. Burchell VS et al (2013) The Parkinson's disease-linked proteins Fbxo7 and Parkin interact to mediate mitophagy. Nat Neurosci 16(9):1257–1265
142. Zimprich A et al (2004) Mutations in LRRK2 cause autosomal-dominant parkinsonism with pleomorphic pathology. Neuron 44(4):601–607
143. Beilina A et al (2014) Unbiased screen for interactors of leucine-rich repeat kinase 2 supports a common pathway for sporadic and familial Parkinson disease. Proc Natl Acad Sci U S A 111(7):2626–2631
144. Angeles DC et al (2011) Mutations in LRRK2 increase phosphorylation of peroxiredoxin 3 exacerbating oxidative stress-induced neuronal death. Hum Mutat 32(12):1390–1397
145. MacLeod D et al (2006) The familial Parkinsonism gene LRRK2 regulates neurite process morphology. Neuron 52(4):587–593
146. Ramonet D et al (2012) PARK9-associated ATP13A2 localizes to intracellular acidic vesicles and regulates cation homeostasis and neuronal integrity. Hum Mol Genet 21(8):1725–1743
147. Ramirez A et al (2006) Hereditary parkinsonism with dementia is caused by mutations in ATP13A2, encoding a lysosomal type 5 P-type ATPase. Nat Genet 38(10):1184–1191
148. Dehay B et al (2012) Loss of P-type ATPase ATP13A2/PARK9 function induces general lysosomal deficiency and leads to Parkinson disease neurodegeneration. Proc Natl Acad Sci U S A 109(24):9611–9616
149. Dandana A et al (2016) Gaucher disease: clinical, biological and therapeutic aspects. Pathobiology 83(1):13–23
150. Mazzulli JR et al (2011) Gaucher disease glucocerebrosidase and alpha-synuclein form a bidirectional pathogenic loop in synucleinopathies. Cell 146(1):37–52
151. Sardi SP et al (2017) Glucosylceramide synthase inhibition alleviates aberrations in synucleinopathy models. Proc Natl Acad Sci U S A 114(10):2699–2704

152. Liao R et al (2001) Infusion of light chains from patients with cardiac amyloidosis causes diastolic dysfunction in isolated mouse hearts. Circulation 104(14):1594–1597
153. Comenzo RL et al (1996) Dose-intensive melphalan with blood stem cell support for the treatment of AL amyloidosis: one-year follow-up in five patients. Blood 88(7):2801–2806
154. Dember LM et al (2001) Effect of dose-intensive intravenous melphalan and autologous blood stem-cell transplantation on al amyloidosis-associated renal disease. Ann Intern Med 134(9 Pt 1):746–753
155. Sanchorawala V et al (2005) Serum free light-chain responses after high-dose intravenous melphalan and autologous stem cell transplantation for AL (primary) amyloidosis. Bone Marrow Transplant 36(7):597–600
156. Migrino RQ et al (2010) Systemic and microvascular oxidative stress induced by light chain amyloidosis. Int J Cardiol 145(1):67–68
157. Migrino RQ et al (2011) Human microvascular dysfunction and apoptotic injury induced by AL amyloidosis light chain proteins. Am J Physiol Heart Circ Physiol 301(6):H2305–H2312
158. Dorbala S et al (2014) Coronary microvascular dysfunction is related to abnormalities in myocardial structure and function in cardiac amyloidosis. JACC Heart Fail 2(4):358–367
159. Shin JT et al (2012) Overexpression of human amyloidogenic light chains causes heart failure in embryonic zebrafish: a preliminary report. Amyloid 19(4):191–196
160. Guan J et al (2013) Stanniocalcin1 is a key mediator of amyloidogenic light chain induced cardiotoxicity. Basic Res Cardiol 108(5):378
161. Lavatelli F et al (2015) Novel mitochondrial protein interactors of immunoglobulin light chains causing heart amyloidosis. FASEB J 29(11):4614–4628
162. Levinson RT et al (2013) Role of mutations in the cellular internalization of amyloidogenic light chains into cardiomyocytes. Sci Rep 3:1278
163. Scherz-Shouval R et al (2007) Reactive oxygen species are essential for autophagy and specifically regulate the activity of Atg4. EMBO J 26(7):1749–1760
164. Sciarretta S et al (2013) Activation of NADPH oxidase 4 in the endoplasmic reticulum promotes cardiomyocyte autophagy and survival during energy stress through the protein kinase RNA-activated-like endoplasmic reticulum kinase/eukaryotic initiation factor 2alpha/activating transcription factor 4 pathway. Circ Res 113(11):1253–1264
165. Song M et al (2014) Super-suppression of mitochondrial reactive oxygen species signaling impairs compensatory autophagy in primary mitophagic cardiomyopathy. Circ Res 115(3):348–353
166. Guan J et al (2014) Lysosomal dysfunction and impaired autophagy underlie the pathogenesis of amyloidogenic light chain-mediated cardiotoxicity. EMBO Mol Med 6(11):1493–1507
167. Arbustini E et al (1998) Restrictive cardiomyopathy, atrioventricular block and mild to subclinical myopathy in patients with desmin-immunoreactive material deposits. J Am Coll Cardiol 31(3):645–653
168. Li D et al (1999) Desmin mutation responsible for idiopathic dilated cardiomyopathy. Circulation 100(5):461–464
169. Sanbe A et al (2009) Protective effect of geranylgeranylacetone via enhancement of HSPB8 induction in desmin-related cardiomyopathy. PLoS One 4(4):e5351
170. Bhuiyan MS et al (2013) Enhanced autophagy ameliorates cardiac proteinopathy. J Clin Invest 123(12):5284–5297

Chapter 13
Regulation of Cell Death Signaling Pathways in Cardiac Myocytes by Mitochondrial Bnip3

Inna Rabinovich-Nikitin, Jonathon Gerstein, Rimpy Dhingra, Matthew Guberman, and Lorrie A. Kirshenbaum

Abstract The 19kd Bcl-2 interacting protein 3 (Bnip3) belongs to a subclass of the Bcl-2 gene family that regulates a wide range of cellular processes including apoptosis, necrosis and autophagy. However, while Bnip3 localizes to the mitochondria in different cell types including cancer cells, its biological function particularly in cardiac myocytes appears to be context specific. This varied response of Bnip3 may be due to differences in Bnip3 abundance, temporal or spatial activation or association with mitochondrial proteins in response to different cell stress conditions. Understanding the biological significance of Bnip3 and the signaling pathways it impinges upon to regulate apoptosis, necrosis and autophagy under disease states is of paramount importance toward developing new therapies to modulate cardiac cell death during cell stress conditions. Herein, we highlight the role of Bnip3 in cardiac cell death under two physiologically important and clinically relevant conditions where Bnip3 is known to be activated, namely ischemic stress and doxorubicin cardiotoxicity.

Keywords Bnip3 · Mitochondria · Apoptosis · Necrosis · Doxorubicin

Introduction

Historically, the adult myocardium has been viewed as a non-proliferative post-mitotic organ with a limited capacity for myocyte regeneration after injury. Hence, the loss of functional cardiac myocytes by programmed apoptosis or necrosis is postulated as a central underlying mechanism for ventricular remodeling and heart

I. Rabinovich-Nikitin · J. Gerstein · R. Dhingra · M. Guberman · L. A. Kirshenbaum
Department of Physiology and Pathophysiology, The Institute of Cardiovascular Sciences, St. Boniface Hospital Albrechtsen Research Centre Rm. 3016, 351 TachéAvenue, Winnipeg, MB R2H 2A6, Canada

L. A. Kirshenbaum (✉)
Max Rady Faculty of Health Sciences, University of Manitoba, Rady College of Medicine, Winnipeg, MB R226, Canada
e-mail: Lorrie@sbrc.ca

© Springer Nature Switzerland AG 2022

L. A. Kirshenbaum (ed.), *Biochemistry of Apoptosis and Autophagy*, Advances in Biochemistry in Health and Disease 18,
https://doi.org/10.1007/978-3-030-78799-8_13

failure. Based on morphological criteria alone, apoptosis was initially identified as a programmed death pathway typified by DNA fragmentation and cell shrinkage without loss of membrane integrity [1]. The fact that apoptosis is a regulated event and potentially amenable to therapeutic interventions, my laboratory and others investigated the signaling pathways and molecular factors that govern apoptotic cell death in the heart during disease. Because necrosis was traditionally viewed as an accidental or an unregulated passive response to injury [2], it was largely overlooked or even ignored as a potential form of cell death that could be manipulated by therapeutic or genetic interventions. Consequently, over the past several years a greater appreciation and detailed understanding of the molecular signaling pathways involved in apoptosis and necrosis, coupled with advanced biochemical criteria and cellular markers to discriminate between apoptotic and necrotic cell death, has resulted in an insurgence of recent studies exploring necrosis as a regulated form of cell death during disease.

Apoptosis and Necrosis Cell Death Pathways

Arguably, one of the most intriguing and compelling issues to impact contemporary biology to date is the concept that programmed cell death is a genetically regulated process. Observations made on the basis of distinct morphological criteria alone, more than 30 years ago by Kerr et al. [1], markedly distinguished apoptosis from classical cell death by necrosis. Apoptosis is a highly regulated evolutionary conserved genetic program of cell death essential for normal development and tissue homeostasis [3]. In contrast, necrosis has been traditionally considered as an unregulated passive response to injury [4]. While there is little doubt that necrosis induced by massive cellular trauma is likely an unregulated event, several lines of investigation including new exciting data from our laboratory have challenged this dogma and classical textbook definition that necrotic cell death is merely accidental and unregulated [5–7]. This emerging and contemporary view is a paradigm shift in our thinking about how cell fate is regulated. In fact, the concept that necrotic death is regulated has tremendous implications for understanding the pathogenesis of diseases that were previously unexplored as well as developing novel therapies for conditions where necrosis is known to play a significant role.

Bcl-2 Family and Molecular Regulation of Cell Death

The Bcl-2 gene family is highly conserved group of proteins that are found throughout evolution and associated with a number of cellular processes that include cell-cycle, metabolism and cell fate [8]. The archetypic member of this group of diverse proteins is Bcl-2 (B cell lymphoma 2). Bcl-2 was discovered as a translocation break-point mutation between chromosome (t14:18) in B-cell lymphomas and believed to confer resistance of cancer cells to anti-neoplastic agents [9–11]. Since its initial discovery, several other proteins with similar domain structures to Bcl-2 have been identified including the *c. elegans* ortholog *ced 9* [12]. The interesting feature among these proteins is their ability to influence cell fate through regulation of cellular processes involved in cell survival and programmed cell death pathways. Interestingly, these are relate through the presence of distinct structures known as Bcl-2 homology (BH) domains that are conserved among these proteins. Ostensibly, the presence or absence of a given domain confers the ability of these proteins to promote cell death of cell survival which is achieved through the formation of homotypic and heterotypic interactions which influences their cellular distribution and influence on cell fate in response to different physiological/pathophysiological conditions [13]. In this regard, in the most generalized state Bcl-2 and other family members such Bcl-xL or MCl-1 that promote cell survival contain an α-aliphatic N-terminal BH4 domains, BH3, BH2, and BH1 domain, and highly conserved carboxyl terminus domain required for insertion into membranes such as mitochondria, nuclear and endoplasmic reticulum [13]. Other family members such as Bax and Bak which induce cell death retain the N-and C- terminal as well as the conserved BH1, BH2 and BH3 domain but lack the α-aliphatic BH4 domain [13]. Notably, a sub-class of Bcl-2 family proteins known to as BH3 only exist which only contain the N- and -C-terminus and single BH3 like domain. While these proteins, like the other BH3 death proteins (Bax, Bak) can promote cell death in response to pathophysiological conditions, less is known of their ability to influence organelles such as mitochondria and endoplasmic reticulum to promote apoptosis, necrosis or autophagy, Fig. 13.1.

Mitochondrion Regulated Cell Death Pathways

In addition to energy production, the mitochondrion has been identified as major signaling platform for apoptosis and necrosis, respectively, Fig. 13.2. The mitochondrial events that discriminate apoptosis versus necrosis signaling events remain poorly understood. However, a growing body of evidence supports a paradigm in which permeability changes to the outer mitochondrial membrane (OMM) or inner mitochondrial membrane (IMM) distinguish apoptosis and necrosis respectively [14]. Hence, in response to cell stress such as genotoxic DNA damage, hypoxia, reactive oxygen species or nutrient deprivation many of the Bcl-2 proteins particularly Bax and Bak engage the intrinsic mitochondrial death pathway and activate apoptosis [15]. The activation and oligomerization of Bax and Bak proteins, results in permeabilization of the mitochondrial outer membrane (MOMP) through the formation of ion conductance Bax/Bak channels [15]. MOMP results in the dissipation of mitochondrial membrane potential ($\Delta\Psi m$) on the IMM where electron transport and respiration occurs and release of apoptogens from mitochondria including cytochrome c, through an undefined mechanism that associates with the mitochondrial protein SMAC, caspase 9 and dATP form the apoptosome in the cytosol [16]. Once formed the apoptosomes triggers the activation of death effector caspases 3, 6,

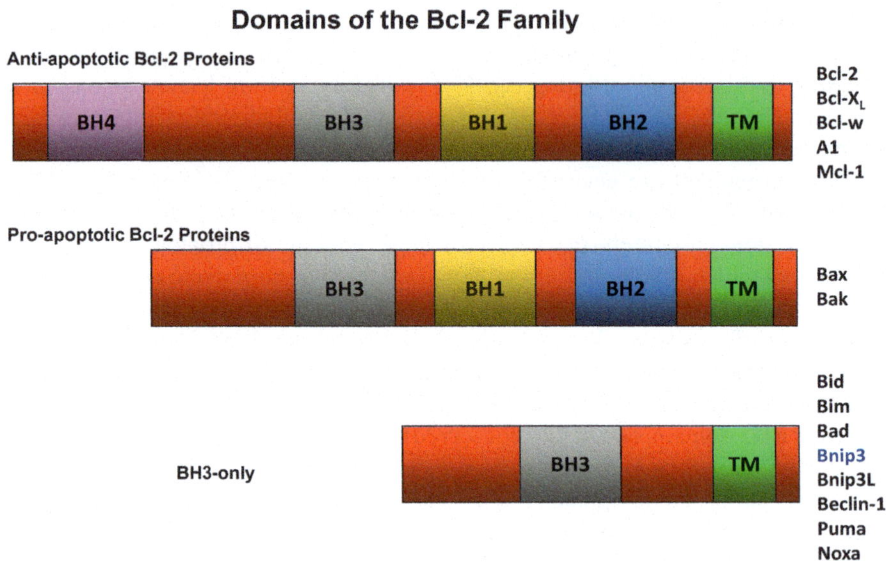

Fig. 13.1 Schematic Representation of Bcl-2 Protein Family. Diagram depicts Bcl-2 proteins and domain structure; N-terminus, Bcl-2 homology (BH), Carboxyl terminus trans membrane domain (TM). **Upper Panel** cell survival proteins; **Middle Panels** cell death proteins; **Lower Panel** BH3 only proteins

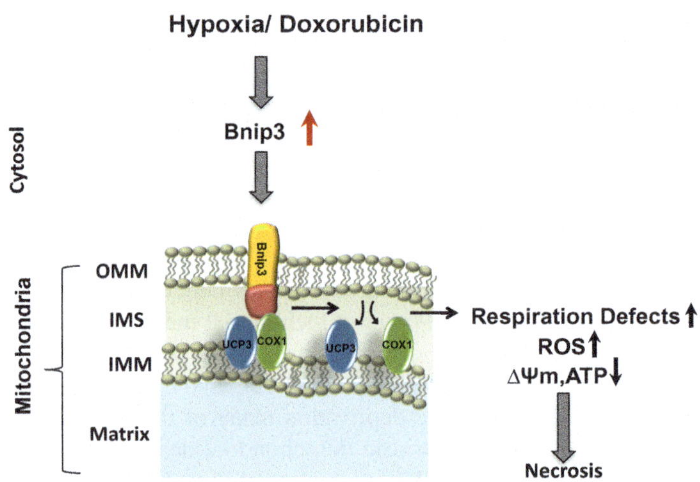

Fig. 13.2 Model of Bnip3 mediated cardiac cell death. Transcriptional activation of Bnip3 during hypoxia or in response to doxorubicin leads to the mitochondrial integration of Bnip3 where it disrupts interaction between UCP3 and COXI on the mitochondrial inner membrane (IMM). This leads to impaired mitochondrial respiration, loss of mitochondrial membrane potential $\Delta\Psi m$, ATP synthesis, and cell death

and 8 which through the proteolytic cleavage of intracellular substrates leads to the demise of the cell through apoptosis [16], Fig. 13.3. Notably, the loss of $\Delta\Psi m$ during apoptosis is a relatively late event compared to necrosis which is a relatively early event [17]. Conversely, early permeability changes resulting from the formation of a permeability transition pore on the IMM (mPTP) is considered an integral feature of necrotic cell death [17]. mPTP causes structural and functional derangements to the mitochondrion that include impaired respiration, loss of $\Delta\Psi m$, ATP synthesis, redistribution of ions, osmotic imbalance, mitochondrial matrix swelling, and increased Ca^{2+} that eventually leads to OMM rupture [17]. Linkage of the intrinsic death pathway to the extrinsic death receptor pathway is achieved through the processing and activation of initiation caspases such as 8 and 10 result in the proteolytic cleavage of Bid to truncated Bid (tBid) which upon translocation to the mitochondria triggers MOMP and apoptosis [16]. Further involvement of death receptors to necroptosis involves the intracellular adapter kinases Receptor Interacting Protein 1 Kinase,

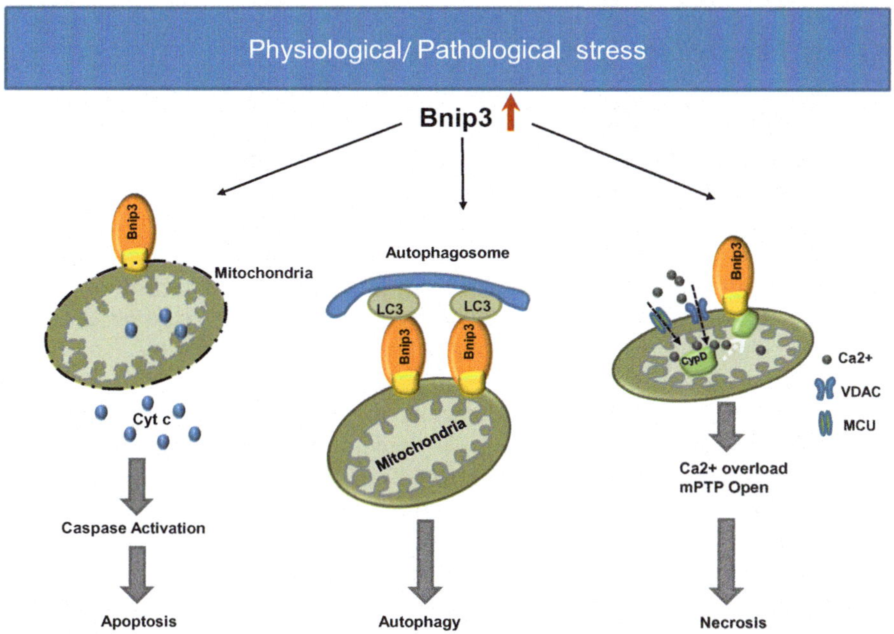

Fig. 13.3 Intersection of Bnip3 with Apoptosis, Necrosis and Autophagy. Left Panel: Bnip3 activates the intrinsic mitochondrial death pathway. Bnip3 triggers mitochondrial perturbations leading to cytochrome release, caspase activation and apoptosis, **Middle Panel**: Bnip3 promotes mitophagy and clearance of damaged mitochondria. Mitochondrial perturbations induced by Bnip3 serves as a docking site for the recruitment of LC3II to depolarized mitochondria along with other autophagy related proteins which target damaged mitochondria for removal by mitochondrial autophagy, **Right Panel**: Bnip3 promotes necrotic cell death of cardiac myocytes. The activation and integration of Bnip3 into mitochondrial membranes disrupts calcium regulation resulting in mitochondrial calcium loading possibly through the voltage dependent ion channel VDAC or mitochondrial calcium uniporter (MCU) leading to mPTP opening and necrotic cell death

(RIP1K RIP3) and mix linage kinase (MLKL) which presumably disrupt mitochondrial metabolism leading to necrotic cell death through a poorly understood mechanism involving RIP3 [18]. Importantly, many of the processes that lead to MOMP or IMM defects leading to apoptosis or necrosis are not well understood but can be suppressed by over-expression of Bcl-2 or B-cell lymphoma—extra larger (Bcl-XL) proteins. Together, these observations highlight the importance of Bcl-2 family proteins in regulating apoptotic and necrotic cell death, respectively. Given the overlapping nature of some Bcl-2 proteins to promote cell survival or cell death, it remains to be elucidated how the cell discriminates between apoptosis and necrosis pathways in response cell stress. Recent data suggests that selective post-translational modifications such as phosphorylation may alter the putative function and cell fate and context specific manner. For example, the site specific phosphorylation of Beclin-1 can switch its ability to interact with Bcl-2 and promote cell death through displacement of Bax from Bcl-2 or promote autophagy and cell survival [19].

Hypoxia-Inducible Expression of Bnip3

The adenovirus Bcl-2 19Kd interacting protein 3, (Bnip3) is a member of BH3 domain like members of the Bcl-2 gene family. Bnip3 was initially identified as cellular protein that formed protein–protein interactions with the 19kD anti-death gene of adenovirus in cells infected with adenovirus [20]. Indeed, to circumvent premature lysis of infected cells, adenovirus and several other human viruses produce anti-death genes that suppress cell death of the infected host cell to ensure sufficient viral progeny has replicated during lytic infection. Since these initial observations, several laboratories including our own, identified Bnip3 as critical regulator of cell death in a number of different cell types including cardiac myocytes [6]. To this end, we identified Bnip3 as a Bcl-2 interacting protein in cardiac myocytes. Notably, the 19kD anti-death gene of adenovirus is the viral homologue to Bcl-2 [21]. Our excitement surrounding the identification of Bnip3 is predicated on the fact that in contrast to other Bcl-2 family members known to promote cell death by apoptosis, Bnip3 is the only member of this family that is highly expressed in the heart and specifically cardiac myocytes under hypoxic conditions [6]. Indeed, in contrast to other death promoting proteins such as Bax, or Bak which are known to promote MOMP and cell death in response to a variety of cell stress signaling, Bnip3 was found to be the only protein activated in cardiac myocytes during hypoxia. The unique feature and hypoxia-inducible nature of Bnip3 identifies it a potentially important cellular target that could be manipulated therapeutically to prevent cell death and cardiac dysfunction during ischemic or hypoxic stress. During the course of our previous studies we established that Bnip3 resides under basal conditions loosely associate with the mitochondrial outer member, in contrast to Bax or Bak which reside in the cytoplasm and translocate to the mitochondrial outer membrane in response to apoptosis signaling [6, 13]. The hypoxia induced activation of Bnip3 involves not only induction of its promoter activity resulting in increased Bnip3 cellular content but also in its integration to mitochondrial inner membranes [6]. We have demonstrated that the integration of Bnip3 into the mitochondrial IMM is contingent upon its carboxy-terminal transmembrane domain [6]. Earlier studies from our group and others revealed that the BH3 like domain and the N-terminal domains of Bnip3 are dispensable for cell killing, again in contrast to Bax and Bak which require their respective BH3 domains for inducing apoptosis, however, the carboxyl terminal transmembrane (TM) was identified to be crucial for triggering mitochondrial and cell death of cardiac myocytes during hypoxic stress [6]. These findings are substantiated by studies in which engineered mutations or complete deletion of the TM domain completely abrogated mitochondrial dysfunction and cell death of cardiac myocytes. Further genetic ablation of Bnip3 through transgenesis resulted in mice with greater functional recovery and smaller infarct sizes following myocardial infarction than corresponding wild type animals [7]. Collectively, our data identify Bnip3 as an inducible death factor and critical regulator of mitochondrial dysfunction and cell death of cardiac myocytes during ischemic injury, Figs. 13.2 and 13.3.

Another interesting feature of Bnip3 that readily distinguishes it from the other Bcl-2 family members which predominately trigger apoptosis, is its ability to promote necrotic cell death by a mechanism that impinges on the IMM [5]. As stated above, the mitochondrion plays a critical role in oxidative metabolism and cellular respiration for ATP synthesis which is crucial for maintaining cellular homeostasis and day to day cardiac contractile function [22]. This is achieved through the stepwise univalent electron transport from the entry of NADH and FADH generated by cellular metabolism from glucose and fatty acid oxidation, which enter the respiratory chain complexes I to IV which produce an electrochemical proton gradient across the IMM. The flow of protons across the IMM through the mitochondrial F0/F1ATPase is responsible for ATP generation 23. Disruption of the electron transport chain proteins through oxidation or carbonylation of the mitochondrial complex proteins on the IMM triggers calcium mediated mPTP opening and necrotic cell death [23]. Importantly, our work has demonstrated that the mitochondrial integration of Bnip3 in response to cell stress disrupts the IMM resulting in cell death with features of necrosis [5], Fig. 13.3.

Another mode by which Bnip3 can provoke mPTP opening, involves Cyclophilin D (CypD), the known mPTP regulator. CypD is a peptidyl prolyl cis–trans-isomerase and normally resides in the mitochondrial matrix. Under mitochondrial stress conditions CypD translocates to the intermembrane space (IMS) of the mitochondria. How CypD translocates from matrix to intermembrane space under stress conditions is largely unknown, but post translational modifications of CypD, such as phosphorylation and acetylation, have been proposed as an underlying mechanisms. Calcium overload through the mitochondrial calcium uniporter (MCU) has also been suggested to activate CypD, resulting in its recruitment to IMS [24–27]. Increased interaction between CypD and F1F0 ATPase has been implicated in mitochondrial perturbations. We recently found that Bnip3 protein interacts with CypD in doxorubicin treated cardiomyocytes, Fig. 13.3. The significance of interaction between

Bnip3 and CypD is not clear at the moment, but the fact that Bnip3 induced mitochondrial defects and cell death could be supressed by genetic knock down of CypD or by use of chemical inhibitor Cyclosporin A (CSA), suggest that CypD might be the effector downstream from Bnip3. Taken together our studies unravel a new mechanism of Bnip3 mediated mitochondrial defects which are obligatorily linked to CypD [28].

Bnip3 and Autophagy

Autophagy is a cellular quality control process essential for maintaining tissue homeostasis and organ function [29]. Defects in autophagy have been linked to a variety of diseases entities including Danon Disease as well as heart failure [30]. Generally speaking, autophagy plays a key role in the removal of damaged organelles, proteins and pathogens such as viral particles and bacteria from the cell through an elaborate lysosome system [29, 31]. Autophagy can be classified as macro- micro and chaperone mediated autophagy. In most cases, macro-autophagy referred to as within is critical for the selective removal of oxidized proteins and lipids as well as damaged organelles such as mitochondria, Golgi and Endoplasmic reticulum [31]. In this regard, autophagy or mitophagy of damaged or irreparable mitochondria plays as an essential cellular quality control mechanism for ensuring a healthy pool of mitochondrial for oxidative metabolism, respiration and ATP production [32, 33]. The inability to effectively clear damaged mitochondria through defective quality control mechanisms leads to the accumulation of mitochondria producing reactive oxygen species which would be toxic to cells [32, 34]. Hence, clearing mitochondria is of paramount importance in maintaining cell viability and tissue quality control. In this regard, a less well defined feature of Bnip3 and related protein Bnip3L, is to serve as a docking site for the autophagy regulator proteins such as microtubule associated light chain 3 II (LC3II) and p62 for the autophagic clearance of damaged mitochondria through mitophagy [35, 36]. Indeed, the N-terminus of Bnip3 and Bnip3L contain LC3II-Interacting Regions which allow these proteins to interact with LC3II [35], Fig. 13.3.

In this way, Bnip3 may contribute to mitochondrial quality control mechanisms through association with LC3II or other autophagy related proteins (Atgs) to ensure the adequate removal of ROS producing mitochondria during stress conditions such as nutrient deprivation or hypoxia. However, the role of Bnip3 as a major participant in mitochondrial quality control is not well defined since germ-line deletion of Bnip3 does not appear to influence mitochondrial turnover under basal or stress conditions, suggesting that Bnip3 is either dispensable for mitochondrial autophagy or other proteins such as Bnip3L may functionally substitute for the absence of Bnip3 in controlling mitochondrial turn-over. Further, the relationship between Bnip3 and necrotic cell death in relation to mitophagy activation or inhibition requires further examination since the interplay between these two cellular processes remains poorly understood.

Bnip3 and Doxorubicin Cardiomyopathy

Doxorubicin (DOX) is a widely used chemotherapeutic used to treat cancers in both children and adults [5]. It has been proven to play a major role in the effective treatment of a number of cancers, including leukemia, breast cancer and non-Hodgkin's lymphoma [37]. Though its clinical efficacy is well documented, patients treated with DOX are observed to have a higher risk of aberrant arrhythmias, myocardial infarction and cardiomyopathy [38]. Furthermore, over a third of patients receiving a dosage greater than 601 mg/m^2 of body surface area suffered congestive heart failure [38]. In attempt to maximize the clinical reliability of DOX, the cardiotoxic effects of the chemotherapeutic should be minimized, while keeping its antitumor properties intact. The mechanisms that underlie DOX toxicity in cardiac myocytes remain poorly understood. There are reports demonstrating that DOX acts upon topoisomerase II activity to disrupt DNA synthesis and inhibit cell proliferation in cancer cells [39], while other reports demonstrate increased oxidative stress and reactive oxygen species (ROS) form iron overload [40]. Indeed, a recent report the Ardehali laboratory demonstrated that DOX treatment resulted in mitochondrial iron overloaded from defects in the ABC cassette iron transporter protein [41]. Restoration or normalization of iron homeostasis suppressed mitochondrial damage induced by DOX.

At the cellular level, close inspection by electron microscopy revealed severe ultrastructural defects that include disruption of the sarcomeres, mitochondrial swelling with rarefaction of mitochondrial cristae [5]. Interestingly, the cellular and mitochondrial defects induced by DOX were reminiscent of the cellular injury induced by Bnip3, raising the intriguing possibility that Bnip3 may underlie the cardiotoxic effects of DOX. In fact, because the heart is abundantly rich in mitochondria, predisposes it to potential damaging effects of drugs such as DOX. To explore this possibility as a step toward understanding the mechanisms that underlie DOX cardiotoxicity, we tested whether Bnip3 is involved in the mitochondrial defects and cardiac dysfunction in cardiac myocytes treated with DOX. For these studies we treated mice with saline or DOX 5 mg/kg for 4 days for a cumulative DOX dose of 20 mg [5]. One week following treatment, mice were assessed for cardiac function, parallel studies were conducted in isolated primary cultured cardiac myocytes to assess cell viability, mitochondrial respiration, mitochondrial membrane potential $\Delta\Psi$m and permeability transition pore opening mPTP following DOX (5ug/ml) treatment [5]. Interestingly, western blot analysis revealed a marked increase in Bnip3 expression in cardiac myocytes in vivo and in vitro which corresponded with a reduction in cell viability [5]. The increased cell death in vitro, was concordant with ultrastructural injury and impaired contractile function in vivo [5].

Further, mitochondria isolated from mice hearts following DOX treatment exhibited diminished oxygen consumption rates and respiratory reserve capacity, consistent with impaired mitochondrial respiration and ATP production [5]. Impaired respiration also coincided with a loss of mitochondrial membrane potential, mitochondrial calcium, and mPTP opening. A dramatic increase in LDH and cTnT release was observed in mice treated with DOX, a finding consistent with increased mitochondrial damage impaired contractile function [5]. These findings substantiate that defects to the mitochondrial IMM, trigger mPTP opening and necrotic cell death of cardiac myocytes. Given that Bnip3 was highly expressed and integrated into mitochondrial membranes, we were curious if Bnip3 was responsible for the cytotoxic effects induced by DOX in vitro and in vivo. Therefore, to test this possibility, we tested whether genetic knock-down or ablation of Bnip3 would influence the mitochondrial injury and cardiac dysfunction induced by DOX. To this end, we observed that in contrast to wild type hearts treated with DOX which displayed impaired mitochondrial morphology and contractile function, Bnip3−/− mice treated with DOX were relatively resistant to DOX treatment and were indistinguishable from vehicle treated saline controls with respect to cell morphology and cardiac contractility. Further, knock-down of Bnip3 or mutations of Bnip3 defective for integrating into the IMM, were sufficient to suppress DOX-induced mitochondrial injury. In fact, loss or inactivation of Bnip3 in the presence of DOX, normalized mitochondrial respiration, and mitochondrial $\Delta\Psi m$. Further, mitochondrial calcium overload which is critical for mPTP was suppressed in cells deficient for Bnip3 [5]. This suggests that Bnip3 may trigger intracellular calcium release from the endoplasmic reticulum to promote mPTP and necrotic cell death. Importantly, DOX induced mPTP opening and necrotic cell death in vitro and in vivo was abrogated following Bnip3 inactivation. Taken together, these data strongly suggest that Bnip3 is a critical effector of DOX induced mitochondrial injury and cell death of cardiac myocytes. To address the mode by which Bnip3 perturbs mitochondrial respiration on the IMM, we reasoned that Bnip3 may disrupt one or more of the electron chain transport complexes which would impair respiration and trigger mPTP. In this regards, we identified that intra-mitochondrial complexes between uncoupling protein 3 (UCP3) and cytochrome c oxidase subunit I (COXI) of complex IV were disrupted in cardiac myocytes following DOX treatment. Notably, genetic ablation of Bnip3 prevented DOX induced disruption of UCP3-COXI and impaired respiration in vivo and in vitro. These findings fully support a model in which the activation of Bnip3 in cardiac myocytes treated with DOX promotes mitochondrial perturbations resulting in mPTP opening and necrotic cell death [5], Fig. 13.2.

Concluding Comments

The Bcl-2 protein Bnip3 is a highly conserved evolutionary protein that is vastly transcriptionally induced in cardiac myocytes subjected to hypoxia. Bnip3 is also induced in cardiac myocytes treated with DOX. The activation and integration of Bnip3 into mitochondrial IMM promotes perturbations resulting in loss of mitochondrial $\Delta\Psi m$, impaired respiration, and mPTP. The ability of Bnip3 to provoke mitochondrial mPTP on the IMM is an underlying feature of necrotic cell death. The ability of Bnip3 to serve as a mitochondrial quality control mechanism by promoting mitophagy through the adapter protein for LC3II is another feature of Bnip3 that requires further investigation. How Bnip3 intersects with key cellular process that underlie apoptosis, necrosis and autophagy remains to be elucidated, Fig. 13.3.

Acknowledgements This work was supported by a Foundation grant to L.A.K from the Canadian Institute for Health Research (CIHR) L.A.K. holds a Canada Research Chair in Molecular Cardiology. IRN holds a post-doctoral fellowship from the CIHR.

Disclousures None.

References

1. Kerr JFR, Wyllie AH, Currie AR (1972) Apoptosis: a basic biological phenomenon with wide-ranging implications in tissue kinetics. Br J Cancer 26:239–257
2. Kung G, Konstantinidis K, Kitsis RN (2011) Programmed necrosis, not apoptosis, in the heart. Circ Res 108:1017–36
3. Elmore S (2007) Apoptosis: a review of programmed cell death. Toxicol Pathol 35:495–516
4. Danial NN, Korsmeyer SJ (2004) Cell death: critical control points. Cell 116:205–219
5. Dhingra R, Margulets V, Chowdhury SR, Thliveris J, Jassal D, Fernyhough P, Dorn GW, Kirshenbaum LA (2014) Bnip3 mediates doxorubicin-induced cardiac myocyte necrosis and mortality through changes in mitochondrial signaling. Proc Natl Acad Sci 111:E5537–E5544
6. Regula KM, Ens K, Kirshenbaum LA (2002) Inducible expression of BNIP3 provokes mitochondrial defects and hypoxia-mediated cell death of ventricular myocytes. Circ Res 91:226–31
7. Diwan A, Krenz M, Syed FM, Wansapura J, Ren X, Koesters AG, Li H, Kirshenbaum LA, Hahn HS, Robbins J, Jones WK, Dorn GW (2007) Inhibition of ischemic cardiomyocyte apoptosis through targeted ablation of Bnip3 restrains postinfarction remodeling in mice. J Clin Invest 117:2825–2833.
8. Marie Hardwick J, Soane L (2013) Multiple functions of BCL-2 family proteins. Cold Spring Harb Perspect Biol 5
9. Tsujimoto Y, Finger LR, Yunis J, Nowell PC, Croce CM (1984) Cloning of the chromosome breakpoint of neoplastic B cells with the t(14;18) chromosome translocation. Science 226:1097–1099
10. Tsujimoto Y, Cossman J, Jaffe E, Croce CM (1985) Involvement of the bcl-2 gene in human follicular lymphoma. Science 228:1440–1443
11. Nunez G, Seto M, Seremetis S, Ferrero D, Grignani F, Korsmeyer SJ, Dalla-Favera R (1989) Growth- and tumor-promoting effects of deregulated BCL2 in human B-lymphoblastoid cells. Proc Natl Acad Sci U S A 86:4589–4593
12. Hengartner MO, Horvitz HRC (1994) Elegans cell survival gene ced-9 encodes a functional homolog of the mammalian proto-oncogene bcl-2. Cell 76:665–676
13. Kvansakul M, Hinds MG (2013) Structural biology of the Bcl-2 family and its mimicry by viral proteins. Cell Death Dis 4:e909
14. Sesso A, Belizário JE, Marques MM, Higuchi ML, Schumacher RI, Colquhoun A, Ito E, Kawakami J (2012) Mitochondrial swelling and incipient outer membrane rupture in preapoptotic and apoptotic cells. Anat Rec 295:1647–1659
15. Fulda S, Gorman AM, Hori O, Samali A (2010) Cellular stress responses: cell survival and cell death. Int J Cell Biol 214074
16. Xiong S, Mu T, Wang G, Jiang X (2014) Mitochondria-mediated apoptosis in mammals. Protein Cell 5:737–749
17. Hurst S, Hoek J, Sheu SS (2017) Mitochondrial Ca^{2+} and regulation of the permeability transition pore. J Bioenerg Biomembr 49:27–47

18. Dhuriya YK, Necroptosis SD (2018) A regulated inflammatory mode of cell death. J Neuroinflammation 15:199
19. Marquez RT, Xu L (2012) Bcl-2: Beclin 1 complex: multiple, mechanisms regulating autophagy/apoptosis toggle switch. Am J Cancer Res 2:214–21
20. Chen G, Ray R, Dubik D, Shi L, Cizeau J, Bleackley RC, Saxena S, Gietz RD, Greenberg AH (1997) The E1B 19K/Bcl-2-binding protein Nip3 is a dimeric mitochondrial protein that activates apoptosis. J Exp Med 186:1975–1983
21. White E (1996) Life, death, and the pursuit of apoptosis. Genes Dev 10:1–15
22. Wang K, Xu Y, Sun Q, Long J, Liu J, Ding J (2018) Mitochondria regulate cardiac contraction through ATP-dependent and independent mechanisms. Free Radic Res 52:1256–1265
23. Osellame LD, Blacker TS, Duchen MR (2012) Cellular and molecular mechanisms of mitochondrial function. Best Pract Res Clin Endocrinol Metab 26:711–723
24. Bauer TM, Murphy E (2020) Role of mitochondrial calcium and the permeability transition pore in regulating cell death. Circ Res 126(2):280–293
25. Dhingra R, Lieberman B, Kirshenbaum LA. Cyclophilin D (2018) phosphorylation is critical for mitochondrial calcium uniporter regulated permeability transition pore sensitivity. Cardiovasc Res 115(2):261–263
26. Kwong JQ, Lu X, Correll RN, Schwanekamp JA, Vagnozzi RJ, Sargent MA, York AJ, Zhang J, Bers DM, Molkentin JD (2015) The mitochondrial calcium uniporter selectively matches metabolic output to acute contractile stress in the heart. Cell Rep 12:15–22
27. Lambert JP, Luongo TS, Tomar D, Jadiya P, Gao E, Zhang X, Lucchese AM, Kolmetzky DW, Shah NS, Elrod JW (2019) MCUB regulates the molecular composition of the mitochondrial calcium uniporter channel to limit mitochondrial calcium overload during stress. Circulation 140:1720–1733
28. Dhingra R, Guberman M, Rabinovich-Nikitin I, Gerstein J, Margulets V, Gang H, Madden N, Thliveris J, Kirshenbaum LA (2019) Impaired NF-κB signalling underlies cyclophilin D-mediated mitochondrial permeability transition pore opening in doxorubicin cardiomyopathy. Cardiovasc Res 116(6):1161–1174
29. Mughal W, Dhingra R, Kirshenbaum LA (2012) Striking a balance: autophagy, apoptosis, and necrosis in a normal and failing heart. Curr Hypertens Rep 14:540–547
30. Kirshenbaum LA (2012) Regulation of autophagy in the heart in health and disease. J Cardiovasc Pharmacol 60:109
31. Parzych KR, Klionsky DJ (2014) An overview of autophagy: morphology, mechanism, and regulation. Antioxidants Redox Signal. 20:460–473
32. Saito T, Sadoshima J (2015) Molecular mechanisms of mitochondrial autophagy/mitophagy in the heart. Circ Res 116:1477– 1490
33. Shirakabe A, Fritzky L, Saito T, Zhai P, Miyamoto S, Gustafsson ÅB, Kitsis RN, Sadoshima J (2016) Evaluating mitochondrial autophagy in the mouse heart. J Mol Cell Cardiol 92:134–139
34. Dorn GW, Kitsis RN (2015) The mitochondrial dynamism-mitophagy-cell death interactome: multiple roles performed by members of a mitochondrial molecular ensemble. Circ Res 116:167–182
35. Hanna RA, Quinsay MN, Orogo AM, Giang K, Rikka S, Gustafsson ÅB (2012) Microtubule-associated protein 1 light chain 3 (LC3) interacts with Bnip3 protein to selectively remove endoplasmic reticulum and mitochondria via autophagy. J Biol Chem 287:19094–19104
36. Ma X, Godar RJ, Liu H, Diwan A (2012) Enhancing lysosome biogenesis attenuates BNIP3-induced cardiomyocyte death. Autophagy 8:297–309
37. Young RC, Ozols RF, Myers CE (1981) The anthracycline antineoplastic drugs. N Engl J Med 305:139–153
38. Singal PK, Iliskovic N (1998) Doxorubicin-induced cardiomyopathy. N Engl J Med 339:900–905
39. Nitiss JL (2009) Targeting DNA topoisomerase II in cancer chemotherapy. Nat Rev Cancer 9:338–350

40. Gammella E, Recalcati S, Cairo G (2016) Dual role of ROS as signal and stress agents: iron tips the balance in favor of toxic effects. Oxid Med Cell Longev 8629024
41. Ichikawa Y, Ghanefar M, Bayeva M, Wu R, Khechaduri A, Naga Prasad SV, Mutharasan RK, Jairaj Naik T, Ardehali H (2014) Cardiotoxicity of doxorubicin is mediated through mitochondrial iron accumulation. J Clin Invest 124:617–630

Chapter 14
Role of Cardiomyocyte Apoptosis in Heart Failure

Sukhwinder K. Bhullar, Anureet K. Shah, and Naranjan S. Dhalla

Abstract Cardiomyocyte apoptosis has now been identified in a wide variety of patients with heart failure as well as in failing hearts due to experimentally induced pathophysiological situations such as myocardial infarction, hemodynamic alterations, and different types of cardiomyopathies. Several mechanisms including oxidative stress, inflammation and intracellular Ca^{2+}-overload have been suggested to induce the release of cytotoxic proteins such as cytochrome C for the activation of caspases -3/-9 and subsequent occurrence of apoptotic cell death. Extensive research has revealed both the up-regulation of pro-apoptotic pathway involving PKC isoforms -α/-ε, NFκB, p38-MAPK and BAX, and down-regulation of anti-apoptotic pathway involving Akt, BAD, Bcl-2 and Erk-1/2 in cardiomyocytes from failing hearts. In view of the critical role of inflammation in heart failure, some evidence has been presented to show the role of TNF-α in the activation and deactivation of cell survival and cell death signal transduction pathways for the occurrence of apoptosis in non-ischemic failing hearts due to volume overload and dilated cardiomyopathy. Although the occurrence of apoptosis in failing hearts can be seen to produce loss of some cardiomyocytes leading to heart dysfunction, no meaningful conclusion can be made regarding the exact role of apoptosis in contractile defects during the development of heart failure.

Keywords Pro-apoptotic pathway · Anti-apoptotic pathway · TNF-α signal transduction · Heart failure · Cardiomyopathy

S. K. Bhullar · N. S. Dhalla (✉)
Department of Physiology and Pathophysiology, Institute of Cardiovascular Sciences, Max Rady College of Medicine, St. Boniface Hospital Albrechtsen Research Centre, University of Manitoba, Winnipeg R2H 2A6, Canada
e-mail: nsdhalla@sbrc.ca

A. K. Shah
Department of Kinesiology, Nutrition and Food Science, California State University, Los Angeles, CA 90032, USA

© Springer Nature Switzerland AG 2022
L. A. Kirshenbaum (ed.), *Biochemistry of Apoptosis and Autophagy*, Advances in Biochemistry in Health and Disease 18,
https://doi.org/10.1007/978-3-030-78799-8_14

Introduction

Apoptosis, an energy-dependent programmed cell death process, was first reported in patients with heart failure by Narula et al. [1] in 1996. Later, the occurrence of this active process was demonstrated in different types of failing hearts by several other investigators [2–8]. The efficacy of apoptosis in the progression of heart failure was shown to be associated with apoptotic signaling activation and even the low rate of apoptosis was considered to cause cardiac dysfunction [9–17]. Although some studies have casted doubt on the involvement of apoptosis in heat failure [18–24], various other investigations have provided evidence that apoptosis in cardiomyocytes is related to the development of cardiac hypertrophy and is responsible for worsening of heart failure [25–30]. Despite this controversy regarding the role of apoptosis in the development of heart failure, there is ample evidence to indicate that in initial phases of apoptotic process, cells undergo the structural changes including shrinkage, plasma membrane blebbing and DNA fragmentation followed by disintegration into apoptotic bodies and their extraction by phagocytes. Since apoptotic cardiomyocytes neither push out their contents into the environment nor cause inflammation, major morphological deformity, as well as damage to the surrounding cells interstitium [31–37], it has been suggested that apoptosis can be easily manipulated [38–40]. In this regard, the understanding of precise measurement, detection, and monitoring of apoptosis could lead to its reversal or modulation with a therapeutic approach that could prove as an effective intervention to prevent heart failure [25–27, 41–43]. Thus targeting apoptosis was considered helpful to reduce the incidence, mortality, and economic burden associated with heart failure, which is a life-threatening among most cardiovascular diseases and is continuously growing around the globe [44–46]. It may be noted that the existence of cardiomyocyte apoptosis in the heart has been reported in various cardiac diseases, including ischemic heart failure, myocardial infarction, cardiac arrhythmias, pressure overload and volume overload. Furthermore, the most effective stimuli that provoke cardiomyocyte apoptosis include oxygen radicals, cytokines, growth factors and energy deprivation.

A variety of complex signal transduction pathways, which have been implicated in the execution of cardiomyocyte apoptosis, include Fas/TNF-α receptors signaling, G-protein coupled receptor signaling, mechanical stretch or mitogen-activated protein kinases, caspases, as well as phospholipase-C type biochemical reactions. In addition, the regulators of the apoptotic processes involve the activation and deactivation of pro-apoptotic and anti- apoptotic proteins such as the Bcl-2 family of proteins and the Bax associated proteins [2, 5, 47–58]. It is noteworthy that different cytokines have been shown to produce cardiomyocyte apoptosis in heart failure and the presence of very high levels of circulating TNF-α has been indicated [59]. TNF-α binds to its membrane receptors to activate downstream caspases and initiate the receptor-dependent cell death pathway [60], and provoke inhibition of the mitochondrial function [61] to induce apoptosis [62]. A member of the apoptotic signal protein family BAX enters the outer mitochondrial membrane to form a large conductance channel and allows the release of cytochrome C [63–65], while, Bcl-2 and other proteins of the anti-apoptotic family protect the release of cytochrome C [66] and target the protein kinase for cell survival [67]. Since TNF- α signaling pathway is now well known as a significant contributing factor in the progressive cardiomyocyte apoptosis for the development of heart failure [12, 17, 68–71], the present article is intended to focus discussion on the elevated levels of TNF-α—mediated signal transduction pathway and alterations in the activation and/or deactivation of anti- and pro-apoptotic proteins for the occurrence of cardiomyocyte apoptosis in the development of heart failure.

Development of Cardiomyocytes Apoptosis in Heart Failure

With progressive loss of cardiomyocytes and limited capacity of their regeneration, the heart is unable to sustain efficient contractile function as a consequence of fewer cardiomyocytes [4, 70, 72]. Thus, cardiomyocyte apoptosis is considered to play a significant role in the development of heart failure and this view is supported by numerous studies in humans and animals. Cardiomyocytes apoptosis in patients undergoing transplantation in heart failure due to dilated and ischemic cardiomyopathy is an excellent example in this regard [1, 2]. Activation of the mitochondrial apoptotic signal transduction pathway has been demonstrated to support the phenomenon of cardiomyocytes apoptosis in patients with congestive heart failure [4]. It should be recognized that cardiomyocyte apoptosis is only one of the mechanisms leading to cell death and cardiac dysfunction [1, 22, 47, 73, 74] and thus it is difficult to define its exact contribution in the development of heart failure. However, a link of cardiomyocytes apoptosis with transition from the left ventricular compensated to decompensated hypertrophy has been observed in response to chronic pressure overload [5, 75]. A progressive loss of cardiomyocytes in the failing heart due to hypertension is considered to contribute towards the development of heart failure [76]. Furthermore, increased serum Fas Ligands in patients with heart failure [77] and altered Bcl-2 activity, as well as activation of mitochondrial apoptotic signaling, leading to cardiomyocyte apoptosis seem to show association with the development of heart failure [78, 79]. It has also been pointed out that a persistent β-adrenergic receptor activation in failing heart may promote cardiomyocyte death by protein kinase A-dependent and protein kinase A-independent activation of calcium/calmodulin-dependent kinase II (CaMKII) [80] and can be seen to play a role in the progression of heart failure. These observations support the view that cardiomyocyte apoptosis may be of critical importance for the manifestation of contractile dysfunction during the development and progression of heart failure. In the following sections of this article, it is planned to provide some details regarding the signal transduction mechanisms of cardiomyocyte apoptosis in two experimental models of non-ischemic heart failure namely heart failure due to volume overload and heart failure in dilated cardiomyopathy.

Cardiomyocyte Apoptosis in Volume Overload Induced Heart Failure

The occurrence of cardiomyocytes apoptosis in the development of heart failure induced by volume overload in both the left ventricle and right ventricle under different conditions has now been well established [16, 29, 30, 54, 81–87]. In patients with left ventricular volume overload, a progressive increase in the expression of pro-apoptotic factors (Bax, p53, TNFR1), anti-apoptotic mitochondrial factor (Bcl-xL), and caspases -3, and -9 leading to cardiomyocyte apoptosis has been found at different stages of the heart failure [54]. Morphological evidence of myocardial apoptosis associated with activation of the mitochondrial apoptotic pathway has also been reported in an experimental heart failure model [81]. A reduced level of Akt and increased CaMKII signaling pathways induced cardiomyocyte apoptosis has been shown to occur during the transition to heart failure due to volume overload [83, 84]. Also, an increase in cardiomyocyte apoptosis in a murine model of right ventricular volume overload-induced heart failure has been demonstrated [85]. Since TNF-α family of cytokine receptors is considered to play a major role in cardiomyocyte apoptosis as a link to several signaling pathways [16], the following discussion will be focussed on changes in the levels of TNF-α signal transduction and associated alterations in both death and survival signals in heart failure induced by volume overload.

In our laboratory heart failure due to volume overload was induced by creating arteriovenous (AV) shunt in rats for a period of 16 weeks [15, 16] and some of the results are shown in Figs. 14.1 and 14.2. A marked increase in the cardiac TNF-α concentration was observed to be associated with increased occurrence of cardiomyocyte apoptosis in the failing heart (Fig. 14.1). Likewise, the content of pro-apoptotic proteins such as BAX, caspase -3 and caspase -9 content were increased in failing hearts of these animals (Fig. 14.1). On the other hand, phosphorylated Akt anti-apoptotic protein content was decreased significantly without any changes in non-phosphorylated Akt protein content in the failing heart (Fig. 14.2). Furthermore, phosphorylated form of Bcl-2, unlike the unphosphorylated form was reduced in the failing hearts due to volume overload (Fig. 14.2). It should be noted that several other studies have also demonstrated that TNF-α produces cardiomyocyte apoptosis by promoting activation of the BAX signaling pathway, reduction in phosphorylated Bcl-2 and Akt protein content, as well as increase in protein content for caspases in different types of failing hearts [60, 61, 66, 67]. These results indicate that elevated levels of TNF-α may serve as a trigger for both the activation of the cell death (involving BAX and caspase proteins) pathway and down-regulation of the cell survival (involving Akt and Bcl-2 proteins) pathway, and thus play a significant role in the occurrence of cardiomyocyte apoptosis in heart failure due to volume overload.

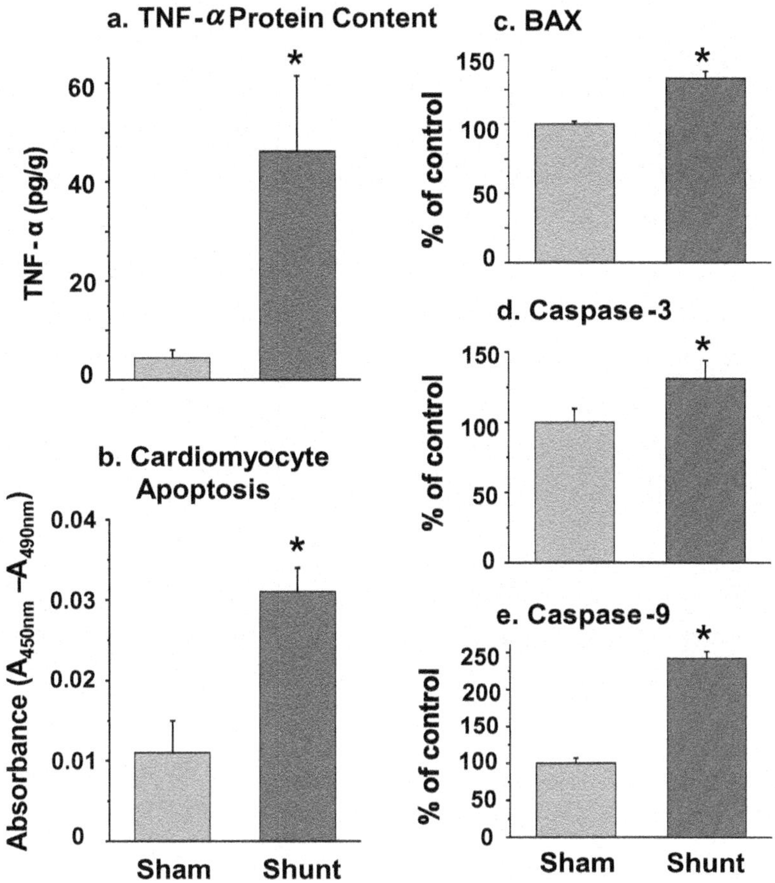

Fig. 14.1 TNF-α, BAX, caspase -3 and caspase -9 protein content as well as cardiomyocyte apoptosis in control (Sham) and failing hearts due to volume overload (Shunt) in rats. Data are taken from our paper: Dent MR, Das S, Dhalla NS. J Mol Cell Cardiol 43:726–732, 2007 [16]. *$P < 0.05$ versus sham control

Cardiomyocyte Apoptosis in Heart Failure Due to Dilated Cardiomyopathy

Dilated cardiomyopathy is a common clinical outcome of many prolonged cardiac insults including valvular defects. This form of cardiomyopathy is characterized by enlargement of the ventricular chamber and thinning of the ventricular walls, which make the heart to work inefficiently and reduce the amount of blood supply to the body [88, 89]. Triggered by functional abnormalities in the cardiomyocytes as a consequence of various etiologies such as gene mutations, mitochondrial irregularities, metabolic derangements, viral infections, and toxins, dilated cardiomyopathy is

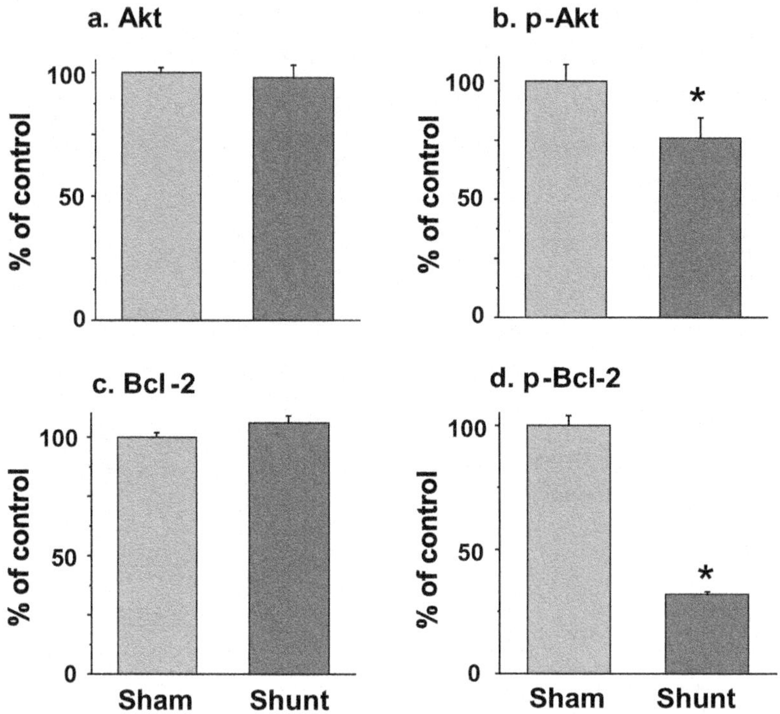

Fig. 14.2 Akt, p-Akt, Bcl-2, and p-Bcl-2 protein content in control (Sham) and failing hearts due to volume overload (Shunt) in rats. Data are taken from our paper: Dent MR, Das S, Dhalla NS. J Mol Cell Cardiol 43:726–732, 2007 [16]. p-phosphorylated; $*P < 0.05$ versus sham control

considered to be as the end-stage of severe cardiac disorder [90–92] and it is responsible for 40–50% of cases of heart failure [93]. The occurrence of cardiomyocytes apoptosis in heart failure due to dilated cardiomyopathy has been reported in different experimental models [55, 94–100]; a significantly higher apoptotic index has been found in patients with dilated cardiomyopathy [74].

It is well known that the extent of activation of mitochondrial caspases for the development of apoptosis is determined by the activation of some proteins such as p38-MAPK and NFκB as well as deactivation of Erk1/2, BAD and Bcl-2 due to phosphorylation/dephosphorylation processes [101–110]. In the process of cardiomyocyte apoptosis leading to the development of heart failure, TNF-α signal transduction [16, 17, 111–114] mediates, the up-regulation or down-regulation of pro- and anti-apoptotic proteins including BAD, Bcl-2 and caspases -3 and -9. Some other proteins such as PKC isozymes, p38-MAPK), nuclear factor (NFκB), extracellular protein kinases (Erk1/2) are also considered to be intimately involved in the genesis of cell death or cell survival [101–106, 114]. However, it is not clear whether TNF-α and PKC isozymes promote apoptosis due to the activation of p38-MAPK and deactivation of Erk1/2 pathways or do these changes further contribute towards the development of cardiac dysfunction in failing hearts. Accordingly, in order to understand the mechanisms, the occurrence of apoptosis and the status of TNF-α were assessed in the hearts of genetically modified 31-week old J2N-k strain of cardiomyopathic hamsters, which exhibited all the signs and symptoms of dilated cardiomyopathy and heart failure [17].

The dilated cardiomyopathic hamster hearts showed marked increases in TNF-α in the myocardium in association with increased caspase -9 and caspase -3 activities as well as the incidence of cardiomyocytes apoptosis (Fig. 14.3). An increase in phosphorylated NFκB proteins and the ratio of phosphorylated NFκB to total NFκB were increased whereas phosphorylated forms of both BAD and Bcl-2 protein content were decreased in the cardiomyopathic heart (Fig. 14.4). No change in the unphosphorylated forms of NFκB, BAD and Bcl-2 proteins were seen in the cardiomyopathic hearts [17]. It may also be noted that the protein content of PKC-α and PKC-ε and phosphorylated form of p38-MAPK were increased whereas the content for phosphorylated forms of both Erk1 and Erk2 were depressed in heart failure associated with dilated cardiomyopathy (Fig. 14.5). The unphophorylated form of p38-MAPK, Erk1

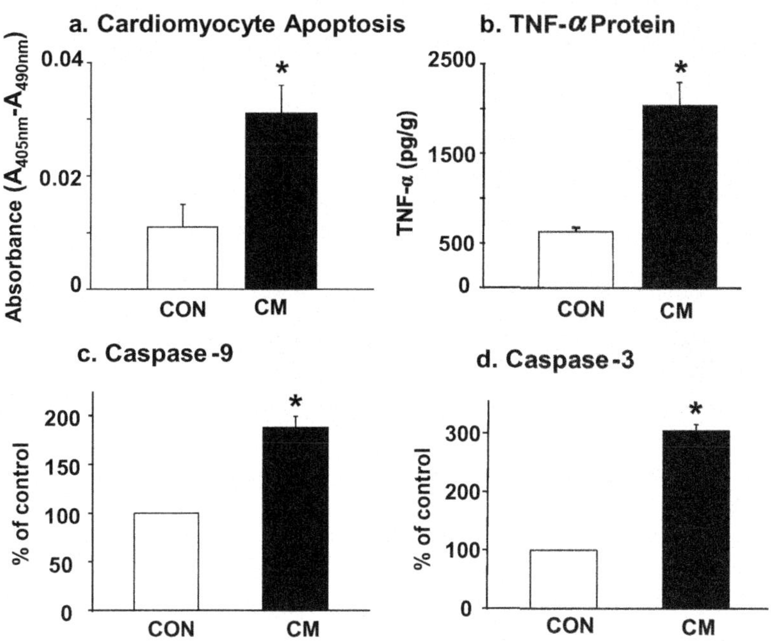

Fig. 14.3 Cardiomyocyte apoptosis and protein content of TNF-α, caspase -9 and caspase -3 in control (CON) and cardiomyopathic (CM) hamster hearts. Data are taken from our paper: Das S, Babick AP, Xu YJ, et al. J Cell Mol Med 14:1988–1997, 2010 [17]. *$P < 0.05$ versus control

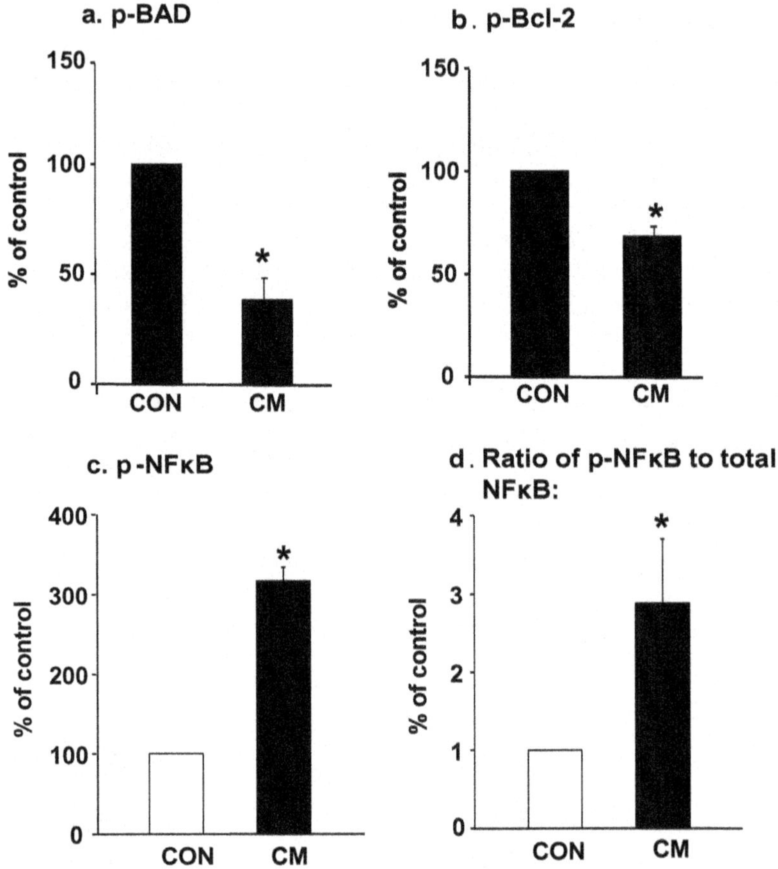

Fig. 14.4 Protein content of p-BAD, p-Bcl-2 and p-NFκB as well as the ratio of p- NFκB to total NFκB in control (CON) and cardiomyopathic (CM) hamster hearts. Data are taken from our paper: Das S, Babick AP, Xu YJ, et al. J Cell Mol Med 14:1988–1997, 2010 [17]. p-phosphorylated; *$P < 0.05$ versus control

and Erk 2 were unaltered in cardiomyopathic heart [17]. These observations indicate that TNF-α associated pro-apoptotic signal transduction pathway is up-regulated whereas the anti-apoptotic pathway is down-regulated in heart failure due to dilated cardiomyopathy.

Conclusions

The existence of cardiomyocytes apoptosis in different types of heart failure is now well established. Although apoptosis can be seen to result in the loss of some

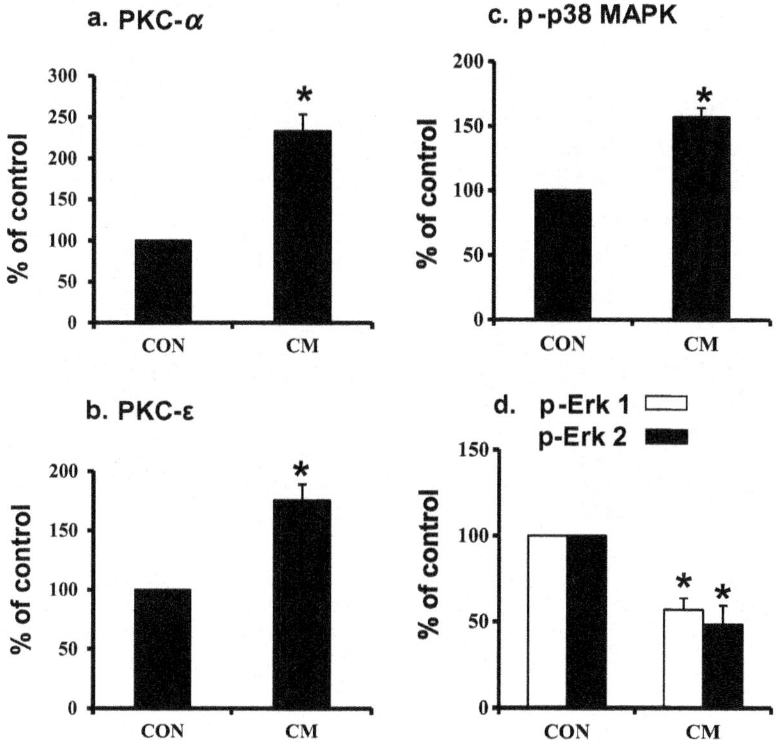

Fig. 14.5 Protein content of various PKC isoforms as well as in control (CON) and cardiomyopathic (CM) hamster hearts. Data are taken from our paper: Das S, Babick AP, Xu YJ, et al. J Cell Mol Med 14:1988–1997, 2010 [17]. $^{*}P < 0.05$ versus control

cardiomyocytes in the failing heart, the role of apoptosis in the development of contractile function in heart failure is not fully understood. Several mechanisms involving oxidative stress, intracellular Ca^{2+}-overload and inflammation have been suggested to participate in the genesis of cardiomyocyte apoptosis; however, the interactions of these signal transduction pathways are far from clear. In view of the critical role of TNF-α in the pathogenesis of heart failure, elevated level of this cytokine has been observed to be associated with changes in the activities of various pro-apoptotic and anti-apoptotic proteins for the development of cardiomyocyte apoptosis in the failing heart. A schematic representation of TNF-α associated cell death and cell survival pathways for the activation of caspases -3, -9 and subsequent cardiomyocyte apoptosis is depicted in Fig. 14.6. It appears that both the up-regulation of pro-apoptotic pathways and down-regulation of anti-apoptotic pathways by TNF-α lead to the activation of different caspases for the occurrence of apoptosis in the failing heart.

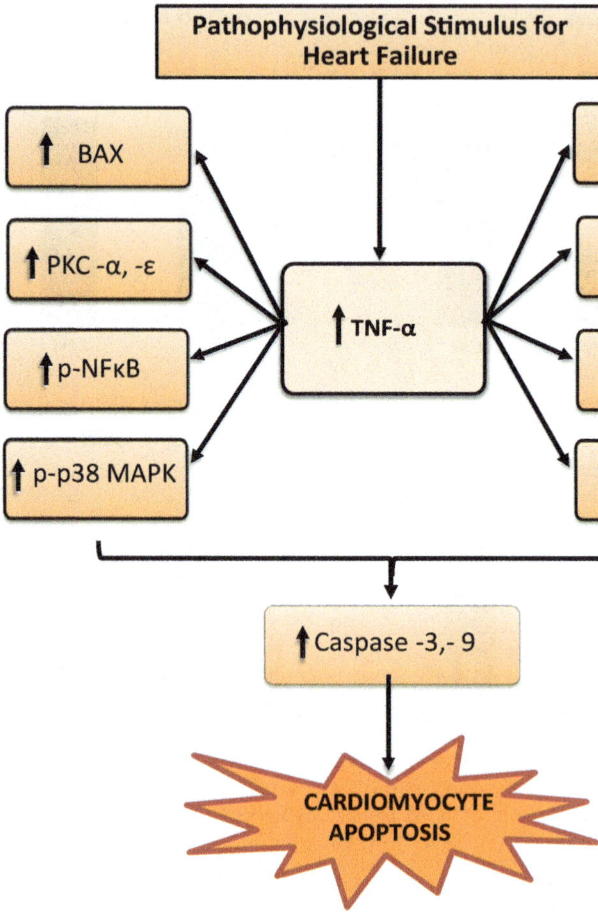

Fig. 14.6 Schematic representation of elevated levels of TNF-α and associated activation of some pro-apoptotic and deactivation of anti-apoptotic proteins in cardiomyocyte apoptosis in heart failure

Acknowledgements The infrastructure support for this project was provided by the St. Boniface Hospital Research Foundation, Winnipeg, Canada.

References

1. Narula J, Haider N, Virmani R et al (1996) Apoptosis in myocytes in end-stage heart failure. N Eng J Med 335:1182–1189
2. Olivetti G, Abbi R, Quaini F et al (1997) Apoptosis in the failing human heart. N Engl J Med 336:1131–1141
3. Anversa PL, Kajstura J, Guerra S, Beltrami CA (1998) Myocyte death and growth in the failing heart. Lab Invest 78:767–786

4. Narula J, Pandey P, Arbustini E et al (1999) Apoptosis in heart failure: release of cytochrome c from mitochondria and activation of caspase-3 in human cardiomyopathy. Proc Natl Acad Sci USA 96:8144–8149
5. Condorelli G, Morisco C, Stassi G et al (1999) Increased cardiomyocyte apoptosis and changes in proapoptotic and antiapoptotic genes Bax and Bcl-2 during left ventricular adaptation to chronic pressure overload in the rat. Circulation 99:3071–3078
6. Saraste A, Pulkki K, Kallajoki M et al (1999) Cardiomyocyte apoptosis and progression of heart failure to transplantation. Eur J Clin Invest 29:380–386
7. Frustaci A, Chimenti C, Setoguchi M et al (1999) Cell death in acromegalic cardiomyopathy. Circulation 99:1426–1434
8. Heling A, Zimmermann R, Kostin S et al (2000) Increased expression of cytoskeletal, linkage, and extracellular proteins in failing human myocardium. Circ Res 86:846–853
9. Hojo Y, Saito T, Kondo H (2012) Role of apoptosis in left ventricular remodeling after acute myocardial infarction. J Cardiol 60:91–92
10. Arbustini E, Brega A, Narula J (2008) Ultrastructural definition of apoptosis in heart failure. Heart Fail Rev 13:121–135
11. Dorn GW II (2009) Apoptotic and non-apoptotic programmed cardiomyocyte death in ventricular remodelling. Cardiovasc Res 81:465–473
12. Lee Y, Gustafsson AB (2009) Role of apoptosis in cardiovascular disease. Apoptosis 14:536–548
13. Philipp S, Pagel I, Hohnel K et al (2004) Regulation of caspase 3 and Fas in pressure overload-induced left ventricular dysfunction. Eur J Heart Fail 6:845–851
14. Regula KM, Kirshenbaum LA (2005) Apoptosis of ventricular myocytes: a means to an end. J Mol Cell Cardiol 38:3–13
15. Dent MR, Tappia PS, Dhalla NS (2010) Gender differences in apoptotic signaling in heart failure due to volume overload. Apoptosis 15:499–510
16. Dent MR, Das S, Dhalla NS (2007) Alterations in both death and survival signals for apoptosis in heart failure due to volume overload. J Mol Cell Cardiol 43:726–732
17. Das S, Babick AP, Xu YJ et al (2010) TNF-alpha-mediated signal transduction pathway is a major determinant of apoptosis in dilated cardiomyopathy. J Cell Mol Med 14:1988–1997
18. Kanoh M, Takemura G, Misao J et al (1999) Significance of myocytes with positive DNA end-labeling (TUNEL) in hearts with dilated cardiomyopathy. Circulation 99:2757–2764
19. Schaper J, Elsasser A, Kostin S (1999) The role of cell death in heart failure. Circ Res 85:867–869
20. Kang PM, Izumo S (2000) Apoptosis and heart failure: a critical review of the literature. Circ Res 86:1107–1113
21. Elsasser A, Suzuki K, Schaper J (2000) Unresolved issues regarding the role of apoptosis in the pathogenesis of ischemic injury and heart failure. J Mol Cell Cardiol 32:711–724
22. Knaapen MW, Davies MJ, De Bie M et al (2001) Apoptotic versus autophagic cell death in heart failure. Cardiovasc Res 51:304–312
23. Takemura G, Fujiwara H (2006) Morphological aspects of apoptosis in heart diseases. J Cell Mol Med 10:56–75
24. Sanchis D, Llovera M, Ballester M, Comella JX (2008) An alternative view of apoptosis in heart development and disease. Cardiovasc Res 77:448–451
25. Cristobal C, Segovia J, Alonso-Pulpon LA et al (2010) Apoptosis and acute cellular rejection in human heart transplants. Rev Esp Cardiol 63:1061–1069
26. Vahasilta T, Malmberg M, Saraste A et al (2011) Cardiomyocyte apoptosis after antegrade and retrograde cardioplegia during aortic valve surgery. J Thorac Cardiovasc Surg 92:1351–1357
27. Khoynezhad A, Jalali Z, Tortolani AJ (2007) A synopsis of research in cardiac apoptosis and its application to congestive heart failure. Tex Heart Inst J 34:352–359
28. Gelpi RJ, Park M, Gao S et al (2011) Apoptosis in severe, compensated pressure overload predominates in nonmyocytes and is related to the hypertrophy but not function. Am J Physiol Heart Circ Physiol 300: H1062–H1068

29. Kolpakov MA, Seqqat R, Rafiq K et al (2009) Pleiotropic effects of neutrophils on myocyte apoptosis and left ventricular remodeling during early volume overload. J Mol Cell Cardiol 47:634–645
30. Fiorillo C, Nediani C, Ponziani V et al (2005) Cardiac volume overload rapidly induces oxidative stress-mediated myocyte apoptosis and hypertrophy. Biochim Biophys Acta 1741:173–182
31. Kerr JFR, Wyllie AH, Currie AR (1972) Apoptosis: a basic biological action and prevented cell death. Br J Cancer 26:239–257
32. Wyllie AH, Kerr JF, Currie AR (1980) Cell death: the significance of apoptosis. Int Rev Cytol 68:251–306
33. Searle J, Kerr JFR, Bishop CJ (1982) Necrosis and apoptosis: distinct modes of cell death with fundamentally different significance. Pathol Annu 17:229–259
34. Arends MJ, Wyllie AH (1991) Apoptosis: mechanisms and roles in pathology. Int Rev Exp Pathol 32:223–254
35. Alter P, Jobmann M, Meyer E et al (2001) Apoptosis in myocarditis and dilated cardiomyopathy: does enterovirus genome persistence protect from apoptosis? An endomyocardial biopsy study. Cardiovasc Pathol 10:229–234
36. Krijnen PAJ, Nijmeijer R, Meijer CJL et al (2002) Apoptosis in myocardial ischaemia and infarction. J Clin Pathol 55:801–811
37. Hughes SE (2003) Detection of apoptosis using in situ markers for DNA strand breaks in the failing human heart. Fact or epiphenomenon? J Pathol 201:181–186
38. Nagata S (1997) Apoptosis by death factor. Cell 88:355–365
39. Green DR, Llambi F (2015) Cell death signaling. Cold Spring Harb Perspect Biol 7:1–24
40. Takemura G, Kanamori H, Okada H et al (2018) Anti-apoptosis in nonmyocytes and pro-autophagy in cardiomyocytes: two strategies against postinfarction heart failure through regulation of cell death/degeneration. Heart Fail Rev 23:759–772
41. Shekhar A, Heeger P, Reutelingsperger C et al (2018) Targeted imaging for cell death in cardiovascular disorders. JACC Cardiovasc Imaging 11:476–493
42. Jiang X, Guo CX, Zeng XJ et al (2015) A soluble receptor for advanced glycation end-products inhibits myocardial apoptosis induced by ischemia/reperfusion via the JAK2/STAT3 pathway. Apoptosis 20:1033–1047
43. Gogiraju R, Xu X, Bochenek ML et al (2015) Endothelial p53 deletion improves angiogenesis and prevents cardiac fibrosis and heart failure induced by pressure overload in mice. J Am Heart Assoc 4:1–21
44. Benjamin EJ, Blaha MJ, Chiuve SE et al (2017) Heart disease and stroke statistics-2017 update: a report from the American heart association. Circulation 135:146–603
45. Foex P (2017) Innovations in management of cardiac disease: drugs, treatment strategies and technology. Br J Anaesth 119:23–33
46. Taylor CJ, Ryan R, Nichols L et al (2017) Survival following a diagnosis of heart failure in primary care. Fam Pract 34:161–168
47. Bennett MR (2002) Apoptosis in the cardiovascular system. Heart 87:480–487
48. Nishida K, Otsu K (2008) Cell death in heart failure. Circ J 72:17–21
49. Haudek SB, Taffet GE, Schneider MD et al (2007) TNF provokes cardiomyocyte apoptosis and cardiac remodeling through activation of multiple cell death pathways. J Clin Invest 117:2692–2701
50. Haunstetter A, Izumo S (1998) Apoptosis basic mechanisms and implications for cardiovascular disease. Circ Res 82:1111–1128
51. Torre-Amione G, Kapadia S, Lee J et al (1996) Tumor necrosis factor-alpha and tumor necrosis factor receptors in the failing human heart. Circulation 93:704–711
52. Baldi A, Abbate A, Bussani R et al (2002) Apoptosis and post-infarction left ventricular remodeling. J Mol Cell Cardiol 34:165–174
53. Chang W, Kajstura J, Nitahara JA et al (1996) Programmed cell death affects the viable myocardium after infarction in rats. Exp Cell Res 226:316–327

54. Moorjani N, Westaby S, Narula J et al (2009) Effects of left ventricular volume overload on mitochondrial and death-receptor-mediated apoptotic pathways in the transition to heart failure. Am J Cardiol 103:1261–1268
55. Cesselli D, Jakoniuk I, Barlucchi L et al (2001) Oxidative stress-mediated cardiac cell death is a major determinant of ventricular dysfunction and failure in dog dilated cardiomyopathy. Circ Res 89:279–286
56. Salvesen GS, Dixit VM (1997) Caspases: intracellular signaling by proteolysis. Cell 91:443–446
57. Black SC, Huang JQ, Rezaiefar P et al (1998) Co-Localization of the cysteine protease caspase-3 with apoptotic myocytes after in vivo myocardial ischemia and reperfusion in rats. J Mol Cell Cardiol 30:733–742
58. Kirshenbaum LA, Meissac D (1997) The bcl-2 gene product prevents programmed cell death of ventricular myocytes. Circulation 96:1580–1585
59. Feuerstein GZ (1999) Apoptosis in cardiac diseases-new opportunities for novel therapeutics for heart diseases. Cardiovasc Drugs Ther 13:289–294
60. Carlson DL, Willis MS, White J et al (2005) Tumor necrosis factor-α-induced caspase activation mediates endotoxin-related cardiac dysfunction. Crit Care Med 33:1021–1028
61. Zell R, Geck P, Werdan K, Boekstegers P (1997) TNF-alpha and IL-1 alpha inhibit both pyruvate dehydrogenase activity and mitochondrial function in cardiomyocytes: evidence for primary impairment of mitochondrial function. Mol Cell Biochem 177:61–67
62. Comstock KL, Krown KA, Page MT et al (1998) LPS-induced TNF-alpha release from and apoptosis in rat cardiomyocytes: obligatory role for CD14 in mediating the LPS response. J Mol Cell Cardiol 30:2761–2775
63. Kluck RM, Bossy-Wetzel E, Green DR, Newmeyer DD (1997) The release of cytochrome c from mitochondria: a primary site for Bcl-2 regulation of apoptosis. Science 275:1132–1136
64. Kroemer G, Reed JC (2000) Mitochondrial control of cell death. Nat Med 6:513–519
65. Murphy E, Imahashi K, Steenbergen C (2005) Bcl-2 regulation of mitochondrial energetics. Trends Cardiovasc Med 15:283–290
66. Das S, Tosaki A, Bagchi D et al (2005) Resveratrol-mediated activation of cAMP response element-binding protein through adenosine A3 receptor by Akt-dependent and—independent pathways. J Pharmacol Exp Ther 314:762–769
67. Das S, Cordis GA, Maulik N, Das DK (2005) Pharmacological preconditioning with resveratrol: role of CREB-dependent Bcl-2 signaling via adenosine A3 receptor activation. Am J Physiol Heart Circ Physiol 288:H328–H335
68. Krown KA, Page MT, Nguyen C et al (1996) Tumor necrosis factor alpha-induced apoptosis in cardiac myocytes. Involvement of the sphingolipid signaling cascade in cardiac cell death. J Clin Invest 98:2854–2865
69. Engel D, Peshock R, Armstong RC et al (2004) Cardiac myocyte apoptosis provokes adverse cardiac remodeling in transgenic mice with targeted TNF overexpression. Am J Physiol Heart Circ Physiol 287:H1303–H1311
70. Orogo AM, Gustafsson AB (2013) Cell death in the myocardium: my heart won't go on. IUBMB Life 65:651–656
71. Mughal W, Dhingra R, Kirshenbaum LA (2012) Striking a balance: autophagy, apoptosis, and necrosis in a normal and failing heart. Curr Hypertens Rep 14:540–547
72. Hofstra L, Liem IH, Dumont EA et al (2000) Visualisation of cell death in vivo in patients with acute myocardial infarction. Lancet 356:209–212
73. Kostin S, Pool L, Elsasser A et al (2003) Myocytes die by multiple mechanisms in failing human hearts. Circ Res 92:715–724
74. Glumac S, Pejic S, Kostadinovic S et al (2016) Apoptosis in endomyocardial biopsies from patients with dilated cardiomyopathy. Folia Biol (Praha) 62:207–211
75. Chen C, Zou LX, Lin QY et al (2019) Resveratrol as a new inhibitor of immunoproteasome prevents PTEN degradation and attenuates cardiac hypertrophy after pressure overload. Redox Biol 20:390–401

76. Liu JJ, Peng L, Bradley CJ et al (2000) Increased apoptosis in the heart of genetic hypertension, associated with increased fibroblasts. Cardiovasc Res 45:729–735
77. Nishigaki K, Minatoguchi S, Seishima M et al (1997) Plasma Fas ligand, an inducer of apoptosis, and plasma soluble Fas, an inhibitor of apoptosis, in patients with chronic congestive heart failure. J Am Coll Cardiol 29:1214–1220
78. Chao W, Shen Y, Li L et al (2002) Importance of FADD signaling in serum-deprivation- and hypoxia-induced cardiomyocyte apoptosis. J Biol Chem 277:31639–31645
79. Grazette LP, Rosenzweig A (2005) Role of apoptosis in heart failure. Heart Fail Clin 1:251–261
80. Zhang X, Szeto C, Gao E et al (2013) Cardiotoxic and cardioprotective features of chronic β-adrenergic signaling. Circ Res 112:498–509
81. Chen YW, Pat B, Gladden JD et al (2011) Dynamic molecular and histopathological changes in the extracellular matrix and inflammation in the transition to heart failure in isolated volume overload. Am J Physiol Heart Circ Physiol 300:H2251–H2260
82. Treskatsch S, Shakibaei M, Feldheiser A et al (2015) Ultrastructural changes associated with myocardial apoptosis, in failing rat hearts induced by volume overload. Int J Cardiol 197:327–332
83. Mohamed BA, Elkenani M, Jakubiczka-Smorag J et al (2019) Genetic deletion of calcium/calmodulin-dependent protein kinase type II delta does not mitigate adverse myocardial remodeling in volume-overloaded hearts. Sci Rep 9:9889
84. Mohamed BA, Schnelle M, Khadjeh S et al (2016) Molecular and structural transition mechanisms in long-term volume overload. Eur J Heart Fail 18:362–371
85. Reddy S, Zhao M, Hu DQ et al (2013) Physiologic and molecular characterization of a murine model of right ventricular volume overload. Am J Physiol Heart Circ Physiol 304:H1314–H1327
86. Aboryag NB, Mohamed DM, Dehe L et al (2017) Histopathological changes in the kidney following congestive heart failure by volume overload in rats. Oxid Med Cell Longev 1–10:2017
87. Shaqura M, Mohamed DM, Aboryag NB et al (2017) Pathological alterations in liver injury following congestive heart failure induced by volume overload in rats. PLoS One 12:e0184161
88. Wencker D, Chandra M, Nguyen K et al (2003) A mechanistic role for cardiac myocyte apoptosis in heart failure. J Clin Invest 111:1497–1504
89. Harvey PA, Leinwand LA (2011) The cell biology of disease: cellular mechanisms of cardiomyopathy. J Cell Biol 194:355–365
90. Gu G, Na Y, Chung H et al (2017) Zebrafish larvae model of dilated cardiomyopathy induced by terfenadine. Korean Circ J 47:960–969
91. Weintraub RG, Semsarian C, Macdonald P (2017) Dilated cardiomyopathy. The Lancet 390:400–414
92. Maron BJ, Towbin JA, Thiene G et al (2006) Contemporary definitions and classification of thecardiomyopathies. Circulation 113:1807–1816
93. Hantson P (2019) Mechanisms of toxic cardiomyopathy. Clin Toxicol (Phila) 57:1–9
94. Sakamoto A, Ono K, Abe M et al (1997) Both hypertrophic and dilated cardiomyopathies are caused by mutation of the same gene, delta-sarcoglycan, in hamster: an animal model of disrupted dystrophin-associated glycoprotein complex. Proc Natl Acad Sci USA 94:13873–13878
95. Mitsuhashi S, Saito N, Watano K et al (2003) Defect of delta-sarcoglycan gene is responsible for development of dilated cardiomyopathy of a novel hamster strain, J2N-k: calcineurin/PP2B activity in the heart of J2N-k hamster. J Biochem 134:269–276
96. Takagi C, Urasawa K, Yoshida I et al (1999) Enhanced GRK5 expression in the hearts of cardiomyopathic hamsters, J2N-k. Biochem Biophys Res Commun 262:206–210
97. Zorc M, Vraspir-Porenta O, Zorc-Pleskovic R et al (2003) Apoptosis of myocytes and proliferation markers as prognostic factors in end-stage dilated cardiomyopathy. Cardiovasc Pathol 12:36–39
98. Jain M, Jakubowski A, Cui L et al (2009) A novel role for tumor necrosis factor-like weak inducer of apoptosis (TWEAK) in the development of cardiac dysfunction and failure. Circulation 119:2058–2068

99. Wu B, Li J, Ni H et al (2018) TLR4 Activation promotes the progression of experimental autoimmune myocarditis to dilated cardiomyopathy by inducing mitochondrial dynamic imbalance. Oxid Med Cell Longev 3181278
100. Chen SN, Lombardi R, Karmouch J et al (2019) DNA damage response/TP53 pathway is activated and contributes to the pathogenesis of dilated cardiomyopathy associated with LMNA (Lamin A/C) mutations. Circ Res 124:856–873
101. Das DK (2003) Protein kinase C isozymes signaling in the heart. J Mol Cell Cardiol 35:887–889
102. Wang J, Liu X, Arneja AS et al (1999) Alterations in protein kinase A and protein kinase C levels in heart failure due to genetic cardiomyopathy. Can J Cardiol 15:683–690
103. Valen G, Yan ZQ, Hansson GK (2001) Nuclear factor kappa-B and the heart. J Am Coll Cardiol 38:307–314
104. Badrichani AZ, Stroka DM, Bilbao G et al (1999) Bcl-2 and Bcl-XL serve an anti-inflammatory function in endothelial cells through inhibition of NF-kappaB. J Clin Invest 103:543–553
105. Cook SA, Sugden PH, Clerk A (1999) Activation of c-Jun N-terminal kinases and p38-mitogen-activated protein kinases in human heart failure secondary to ischaemic heart disease. J Mol Cell Cardiol 31:1429–1434
106. Bueno OF, Molkentin JD (2002) Involvement of extracellular signal-regulated kinases 1/2 in cardiac hypertrophy and cell death. Circ Res 91:776–781
107. Yang E, Zha J, Jockel J et al (1995) Bad, a heterodimeric partner for Bcl-XL and Bcl2, displaces Bax and promotes cell death. Cell 80:285–291
108. Scheid MP, Duronio V (1998) Dissociation of cytokine-induced phosphorylation of BAD and activation of PKB/akt: involvement of MEK upstream of Bad phosphorylation. Proc Natl Acad Sci USA 95:7439–7444
109. Cardone MH, Roy N, Stennicke HR et al (1998) Regulation of cell death protease caspase-9 by phosphorylation. Science 282:1318–1321
110. Nicolson WD, Thornberry NA (1997) Caspases: killer proteases. Trends Biochem Sci 257:299–306
111. Ruvolo PP, Deng X, May WS (2001) Phosphorylation of Bcl2 and regulation of apoptosis. Leukemia 15:515–522
112. Saini HK, Xu Y-J, Zhang M et al (2005) Role of tumour necrosis factor-alpha and other cytokines in ischemia-reperfusion-induced injury in the heart. Exp Clin Cardiol 10:213–222
113. Zhang M, Xu Y-J, Saini HK et al (2005) TNF alpha as a potential mediator of cardiac dysfunction due to intracellular Ca^{2+}-overload. Biochem Biophys Res Commun 327:57–63
114. Sack M (2002) Tumor necrosis factor-alpha in cardiovascular biology and the potential role for anti-tumor necrosis factor-alpha therapy in heart disease. Pharmacol Ther 94:123–135

Chapter 15
The Role of FGF2 isoforms in Cell Survival in the Heart

Elissavet Kardami and Navid Koleini

Abstract Multiple cardiac pathologies culminate in heart failure which is a major cause of morbidity and mortality world-wide. Although great advances have been made, there remain a need to identify additional strategies to manage heart disease and improve outcomes. Identifying the various signals and pathways that contribute to cardiac cell pathology and death, as well as those that promote cell survival and overall protection can lead to new ways to intervene and ameliorate the adverse consequences of various pathological stimuli. This chapter will focus on fibroblast growth factor 2 (FGF2) isoforms as potential agents and/or targets of cardioprotective therapies. In the first part we will provide an overview of the biological properties of FGF2 isoforms, including regulation of expression and secretion, localization, and signalling pathways triggered by FGF2, based on literature from diverse cell types but especially cardiac cells. Our focus will be on molecular signals associated with the causation of, or prevention from, various forms of cells death, by apoptosis, necrosis and dysregulated autophagy, and their relationship to different FGF2 isoforms. In the second part we will present an overview of major experimental models of cardiac pathology that have addressed the effects of endogenous and/or administered FGF2 isoforms. We anticipate that a picture will emerge highlighting both the complexity as well as the potential of FGF2 isoforms in the context of heart disease prevention and management.

E. Kardami (✉) · N. Koleini
Institute of Cardiovascular Sciences, St. Boniface Albrechtsen Research Centre, 351 Tache Avenue, Winnipeg, MB R22A, Canada
e-mail: ekardami@sbrc.ca

E. Kardami
Department of Human Anatomy & Cell Sciences, University of Manitoba, Winnipeg, Canada

E. Kardami · N. Koleini
Department of Physiology & Pathophysiology, University of Manitoba, Winnipeg, Canada

N. Koleini
School of Medicine, Division of Cardiology, Johns Hopkins University, Baltimore, USA

© Springer Nature Switzerland AG 2022
L. A. Kirshenbaum (ed.), *Biochemistry of Apoptosis and Autophagy*,
Advances in Biochemistry in Health and Disease 18,
https://doi.org/10.1007/978-3-030-78799-8_15

Keywords FGF2 isoforms · Cardioprotection · Cell death · Signal transduction · Dysregulated autophagy · Ischemia–reperfusion · Doxorubicin cardiotoxicity

General Introduction

Fibroblast growth factor 2 (FGF2), or, as was originally described, basic FGF, is a multifunctional protein which is expressed in variable amounts by many cell types and is implicated in multiple cellular pathways that affect normal physiology as well as pathophysiology by regulating wound healing, cell proliferation, migration, differentiation and cell survival [1, 2]. FGF2 belongs to the larger family of heparin-binding growth factors (FGF1 to 23) [2]. One single FGF2 mRNA produces, through alternate translation, CUG-initiated high molecular weight (>20 kDa) isoforms (Hi-FGF2), and a AUG-initiated low molecular weight (18 kDa, Lo-FGF2) isoform. In the heart, several studies have pointed to FGF2 as an important factor modulating cardiac response to injury and cardiomyocyte cell death; however, the majority of the literature represents findings regarding Lo-FGF2, while the role of Hi-FGF2, which represents the bulk of tissue FGF2, has not received much attention [1]. Lo-FGF2 has been extensively studied as a cytoprotective agent in several systems including brain injury [3] and as promoting dermal wound healing as reviewed in [4].

FGF2 expression and regulation. The human *FGF2* gene is located on chromosome 4 and consists of two introns and three exons [5]. The human and rat *FGF2* promoters lack the common TATA or CCAAT binding sites that are present in most promoters; instead the *FGF2* promoter contains multiple rich G/C sequences which could act as a binding site for the specificity protein 1 (Sp1) transcription factor. The promoter structure resembles that of constitutively active genes, the so-called housekeeping genes [6]. In addition, the *FGF2* promoter contains early growth response protein 1 (Egr-1) binding elements and increases in Egr-1 activity stimulates *FGF2* mRNA transcription in rat cardiomyocytes. The overlap between the consensus sequences of Sp1 and Egr-1 results in a complex regulation of the promoter activity allowing the gene to be regulated in early and intermediate development as well as in response to stress. Various stimuli and transcription factors have been reported to increase *FGF2* transcription including β-adrenergic agonists, endothelin-1, TGF-β, angiotensin II, and transcription factor NF-κB [7].

Human, rodent and avian FGF2 mRNAs contain multiple translation initiation sites: CUG-initiated sites are located upstream of the more commonly used AUG-initiation consensus site, and produce high molecular weight (Hi, >20 kDa) FGF2 isoforms. Translation from the one AUG start site produces the low molecular weight (Lo, 17–18 kDa) FGF2 isoform. Human FGF2 mRNA contains four CUG start sites, corresponding to 22, 22.5, 24 and 34 kDa isoforms, while rodent FGF2 mRNA has two CUG start sites, corresponding to 21.2–23 kDa isoforms [1, 2]. The 22–24 kDa human Hi-FGF2 22–24 kDa isoforms represent the bulk of human Hi-FGF2 [8].

Internal ribosome entry sequence (IRES) -mediated translation initiation occurs under conditions of cellular stress (heat shock, hypoxia, apoptosis) providing a strategy for producing the 18–24 kDa isoforms of FGF2 when most other protein synthesis is shut down [9].Only the 34 kDa human FGF2 isoform is translated through the more common cap- dependent mechanism of translation initiation [10]. At least two IRES trans-acting factors have been shown to regulate FGF2 IRES: p53 can bind to the FGF2 mRNA leader region and block IRES as well as the cap dependent translation; heterogeneous nuclear ribonucleoprotein A1 can directly bind to the FGF2 IRES facilitating 40S ribosomal binding and protein translation [11].

Scant information exists as to whether there is selective translational regulation of the AUG-initiated versus the CUG-initiated FGF2 isoforms. An older study posited that stressed or transformed cells preferentially produce the CUG-initiated Hi-FGF2 isoforms, while 'normal cells' produce mainly the AUG-initiated Lo-FGF2 [12]; we have not found this to be the case since adult organs such as the brain [13] and the heart [8, 14], accumulate predominantly Hi-FGF2.

Factors interacting preferentially with FGF-2 mRNA may promote translation initiation via CUG rather than AUG; a 'translational enhancer' sequence in the 3' untranslated sequence of human FGF2 mRNA was identified [15]. These studies have not however been confirmed by other laboratories. One might consider the possibility that the CUG-initiated Hi-FGF2 isoforms are the predominant products of translation, and that post-translational processing converts Hi-FGF2, partially or in total, to a Lo-FGF2 protein. We and others have found that tissues contain serinelike protease(s) that exert limited degradation of the CUG-initiated isoforms at their unique N-terminal region producing smaller 17–18 kDa Lo-FGF-2 [8, 16]. Inadequate prevention of endogenous proteolysis, which persists even in heparin-bound tissue extracts, is likely responsible for the underestimation of the actual levels of Hi- FGF2 present in various tissues and cells [13, 17]. The limited proteolysis of the N-terminal extension of Hi-FGF2 by thrombin is likely of physiological significance as it would be expected to convert an anti-angiogenic FGF2 isoform (Hi-FGF2) to a pro-angiogenic Lo-FGF2 [16]. Santiago and colleagues have demonstrated that thrombin can cleave fibroblast-secreted and pro- hypertrophic Hi-FGF2 to an 18 kDa, non-hypertrophic, Lo-FGF2-like protein [8].

FGF2 localization and secretion/export. FGF2 accumulates in smooth muscles, adipose tissue, kidneys and heart, especially atrial tissue [18]. The large amounts of FGF2 found in the atria implicates this growth factor in the normal physiology of tissue and may be related to the higher proliferative capacity of atrial fibroblasts [19]. Cells of mesenchymal origin are rich in FGF2; fibroblasts are the main producers of FGF2 [8, 14].

FGF2 isoforms are present in extracellular and intracellular spaces, including the nucleus and cytosol [8, 14]. Heparan sulfate proteoglycans (HSPGs) in the extracellular matrix, the basal lamina, and even the plasma membrane can act as sinks of secreted FGF2, concentrating it in the vicinity of various cells, and releasing it for interaction and activation of the plasma membrane tyrosine kinase receptors, as needed [20].

The FGF2 isoforms share a common nuclear localization-like sequence, while Hi-FGF2 contains an additional NLS in its N-terminus. The traditional belief is that Hi-FGF2 isoforms localize only to the nucleus, and thus they have been referred to as "nuclear" FGF2, while Lo-FGF2 could accumulate in the nucleus and cytosol as well as be secreted to the extracellular space [21]. This view has been challenged; a number of studies using mostly rat or human cells and tissues demonstrated that Hi-FGF2 is also exported to the extracellular space, and can be localized in the cytosol, as well as the nucleus of cells [8]. Rat cardiac ventricular, as well as human atrial and ventricular fibroblasts export Hi-FGF2, in addition to, and in excess of, Lo-FGF2, to the extracellular milieu; importantly, extracellular Hi-FGF2 can exist in the soluble phase, but also in association with the cell surface, due to its affinity to HSPGs [14]. Retention of FGF2 isoforms by the extracellular face of the plasma membrane and adjacent matrix may explain the failure to detect secreted Hi-FGF2 in a mouse model [22]. Importantly, fibroblast-secreted, extracellular Hi-FGF2, is biologically active, by promoting cardiomyocyte hypertrophy in vitro; or by promoting fibroblast activation to a pro-fibrotic phenotype [8].

The effects of fibroblast-secreted Hi-FGF2 on cardiomyocytes and fibroblasts can be blocked by neutralizing antibodies selective for Hi-FGF2 [14, 23]; or by converting Hi-FGF2 to a Lo-FGF2-like version through limited proteolysis. Importantly Hi-FGF2, as well as Lo-FGF2 are present in human biological fluids, such as pericardial fluid [8]. Therefore, in contrast to a prevailing dogma [21], both Hi and Lo-FGF2 are capable of paracrine and autocrine activities, and can be targeted/neutralized by reagents, such as monoclonal antibodies, that can reach the extracellular space.

Unlike proteins secreted via conventional pathways, FGF2 does not possess a signal peptide for secretion. A variety of mechanisms could control FGF2 release from cells. Minor and reversible plasma membrane injury, as might occur during cellular contraction, results in FGF2 release to the extracellular space [24]. Involvement of the plasma membrane Na^+/K^+ ATPase was inferred in earlier studies, where it was shown that FGF2 export was inhibited by ouabain [25]. Importantly, FGF2 secretion was also shown to occur by a mechanism similar to that of interleukin-1 and other proteins that lack a conventional signal peptide for secretion; the mechanism requires the activity of caspase-1, a major enzyme involved in innate inflammation [14, 26, 27]. Release of FGF2 by stromal bone marrow cells via exosomes has also been reported [28].

FGF2 signal transduction. The biological effects of extracellular FGF2 are initiated by binding and activating plasma membrane tyrosine kinase FGF receptors, FGFR1-4 [2]. A fifth FGFR isoform, lacking the cytosolic domain, FGFR5, can act as a co-receptor [29]. In addition to FGFR, HSPGs, at the plasma membrane are an integral part of FGF signalling. HSPGs can independently bind both FGFs and FGFR1 and are thought to stabilize the FGF/FGFR1 complexes [2]. FGFR1 represents the most abundant FGF2 receptor in the immature and adult heart and cardiomyocytes [1]. FGF2 isoforms are also capable of "intracrine", cytosolic/nuclear, signalling in conjunction (or not) with intracellular FGFR1 [30].

FGF2 isoforms and FGFR1. Hi- and Lo-FGF2 share a core sequence which contains the FGFR binding site. This would suggest that both types of isoforms are capable of activating FGFR1 at the plasma membrane. Early studies showed that that Hi- and Lo-FGF2 were both capable of stimulating proliferative growth of endothelial cells [31] and cardiomyocytes [32] presumably via activation of FGFR; however only Hi-FGF2 inhibited cell migration, by a process requiring its unique N-terminal region, as it was blocked by antibodies raised against a Hi-FGF2-specific sequence; a peptide containing the N-terminal region of human Hi-FGF2 was reported to be anti-angiogenic in vivo, signalling via the co-receptor Neurophilin-1 [33].

There is evidence that FGFR1 activation (phosphorylation) is downregulated in mice chronically overexpressing human Hi-FGF2 [22]; our own work has shown that endogenous Hi-FGF2 expression decreases relative FGFR1 activity [34]. Therefore it would appear that prolonged or constitutive exposure to Hi-FGF2, but not Lo-FGF2, downregulates FGFR1 activation.

Extracellular-acting Hi-FGF2 and Lo-FGF2 have some effects in common and some that are isoform-specific on cardiac myocytes and fibroblasts . Engagement of co-receptors by the unique N-terminal region of Hi-FGF2 could provide a way to modify downstream signal transduction and specific endpoints. It is also important to consider that, separately from plasma membrane FGFR1, intracellular/nuclear FGFR1 is found in association with Hi-FGF2, and can form nuclear complexes regulating gene expression [30]; cardiac mitochondrial FGFR1, regulating mitochondrial integrity in direct interaction with FGF2 isoforms has also been detected [35]. Overall it is possible that intracellular FGF2 isoforms can interact with intracellular FGFR1 at various intracellular locations, producing signals that may reinforce or even oppose the plasma membrane originating signal transduction. The mode of differential FGF2 isoform signalling via FGFR1 presents a complexity that is not as yet understood or in fact explored in sufficient detail.

In the remaining of this section we will outline the major signalling pathways activated downstream of FGF2, based mainly in studies on Lo-FGF2. FGFR belongs to the larger family of receptors containing immunoglobulin-like domain in their ligand binding region; cardiac FGFR1 has 2 or 3 immunoglobulin-like domains [36] results of alternative splicing; it also has a transmembrane domain and a cytosolic (catalytic) domain at the C-terminal of the molecule. Following FGF2 ligand binding FGFR dimerizes, an event that facilitates trans-phosphorylation of several tyrosine (Y) residues at the FGFR1 cytosolic domain, including residues 653/54, 583, 463, 585, and 766. Phosphorylation at Y766 is essential for binding and activation of PKC gamma and downstream pathways [37].

ERK. Activation of FGFR1 results in sequential recruitment of substrates, adaptors and kinases and the recruitment of the RAS-GTPase; activated RAS uses its effector RAF to then activate mitogen-activated protein kinases (MAPKs) in different downstream branches of the pathway, including the ERK1/ERK2 pathway, the p38 pathway, and the JNK pathway. The ERK pathway has been studied extensively regarding FGF2/FGFR signaling, and is considered to mediate FGF2-induced cell proliferation, cytoprotection, and hypertrophy [38–40]. It should be noted that different FGF2 isoforms may engage the ERK pathway to achieve different endpoints; it has been shown that both Hi- and Lo-FGF2 can activate the ERK pathway with equal potency but to contrasting ends. In the case of Hi-FGF2, the ERK pathway was required to induce apoptosis caused by increased nuclear accumulation of Hi-FGF2 [41].

AKT. Phosphoinositide 3-kinase (PI3K)-AKT is another major signaling pathway activated downstream of FGFR1. In the heart, AKT is reported to be essential for preconditioning and cardioprotection [42]. PI3K catalyzes formation of phosphatidylinositol (3,4,5)-trisphosphate (PIP3) from PI 4,5-bisphosphate (PIP2). PIP3 acts as a docking site for AKT (also known as protein kinase B) and phosphoinositide-dependent protein kinase (PDK). PDK1 and PDK2 mediate activation of AKT via phosphorylation at threonine 308 and serine 473, respectively, [43]. Dephosphorylation of PIP3 to PIP2 by phosphatases and tensin homolog (PTEN) blocks AKT phosphorylation resulting in its inactivation. Phosphorylated AKT has a wide range of effects in cells which include stimulation of cell proliferation, cell survival, changes in gene transcription, inhibition of glycogen synthase kinase (GSK) which in turn increases glucose metabolism [44]. The promotion of a pro-survival/anti-apoptotic pathway can occur through the upregulation of Bcl-2 (B-cell lymphoma 2) [45] and inhibition of Bcl-2-associated death promoter (Bad) and bcl-2-like protein 4 (Bax) [44]. In muscles, FGF2 secreted from vascular smooth muscle cells in response to hypoxia/reoxygenation protected co-cultured H9C2 cells via the AKT pathway [46]. The FGF2 stimulated migration of Sca1+ cardiac stem cells were attenuated by a PI3K-AKT inhibitor [47]. The FGF2- protection from hydrogen peroxide-induced necrosis and mitochondrial dysfunction in H9C2 cells required the AKT pathway [48]. The AKT/ERK pathway was shown to protect from myocardial damage due to ischemia-reperfusion [49–51].

mTOR. Growth factors such as FGF2 are known to activate (via phosphorylation on serine 2448) the mammalian target of Rapamycin (mTOR) serine/threonine kinase; AKT is implicated in this event. mTOR signaling is a central regulator of various anabolic processes, and an inhibitor of autophagy initiation [52]. mTOR forms two complexes, mTORC1 and mTORC2, through assembly of different adaptors. mTOR complexes are master regulators of protein synthesis, autophagy, as well as mitochondrial and lysosomal biogenesis. mTORC1 is inactivated by rapamycin through its Raptor subunit; in contrast mTORC2 contains Rictor which makes this complex insensitive to rapamycin. Well known substrates of mTORC1 are S6 kinase 1 (S6K1) and 4E (eIF4E)-binding protein 1 (4E-BP1). mTORC1 phosphorylates S6K1 which in turn stimulates mRNA biogenesis and protein translation, while it inhibits 4E-BP1 allowing formation of eIF4F complex and, therefore, cap-dependent protein translation. mTORC1 also promotes ribosomal biogenesis as well as lipid membrane formation through sterol regulatory element-binding protein dependent lipid synthesis [53]. Active mTOR signaling is important for lysosomal biogenesis and completion of autophagy through the lysosomal degradation of autophagic cargo [54].

PLC/PKC. Phospholipase C/protein kinase C activation mediates several of the effects of FGF2 on cardiomyocytes and the heart, including proliferation of immature cardiomyocytes and increased cardiac resistance against multiple types of injury. The FGF2-stimulated transient negative inotropism in perfused adult rat and protection from ischemia-reperfusion injury were linked to the activation of PKC [55]. The FGF2-triggered and PKCepsilon-mediated phosphorylation of the channel and hemichannel protein Connexin-43 (Cx43) at serine 262 was essential for development of increased resistance to injury, and could be observed at both intercalated disk sites and also in association with subsarcolemmal mitochondria [56, 57]. Perfusion of rat hearts with Lo-FGF2 exerts a preconditioning-like cardiorotective effect against ischemia-reperfusion (I-R) injury; it also results in the translocation of PKCepsilon to cardiac mitochondria, which renders them resistant to calcium-overload -induced mPTP [35]. The ability of FGF2 to protect cardiomyocytes against genotoxic stress by Doxorubicin was also found to require PKC activity [58].

CK2. Protein kinase 2 (CK2) is a serine/threonine kinase which can regulate cell proliferation, cell survival, and chromatin remodelling [59]. It has been established that CK2, which interacts with both FGF2 isoforms, is essential for cell, including cardiomyocyte, proliferation stimulated by Lo-FGF2. A single mutation, a serine 117 to alanine substitution in the core sequence of FGF2, prevents interaction of FGF2 isoforms with CK2, and at the same time blocks the mitogenic ability of Lo-FGF2 [60, 61] Using thus mutated Lo-FGF2 is was shown that CK2, while not essential for acute cardiomyocyte protection in vitro or in vivo [61, 62], it was essential for long term cardioprotection and angiogenesis by Lo-FGF2 [61]. Furthermore, the ability of overexpressed Hi-FGF2 to promote nuclear condensation and apoptosis were also dependent on CK2, as they were prevented by emodin, a CK2 inhibitor, and in the presence of a mutant Hi-FGF2 incapable of interacting with CK2 [63].

FGF2 Isoforms in Heart Pathophysiology

Early studies aimed at determining the role of FGF2 in cardiac pathology made use of mouse models where both Hi- and Lo-FGF2 isoforms were knocked out [64–66]. As a consequence, it was not possible to assign isoform-specific or non-specific functions to endogenous FGF2 expression. Thus the remaining of this section will deal mainly with models that allow a degree of distinction between FGF2 isoforms. There is a large body of evidence showing that treatment with Lo-FGF2 protects cardiac cells against various types of injurious conditions including ischemia-reperfusion [55, 67–70], oxidative stress [48, 49], Doxorubicin toxicity, pressure-overload remodelling, both acutely and long-term [58, 71]. In comparison, a series of studies have identified certain differences between Lo- and Hi-FGF2. Ectopic expression of Hi- or Lo-FGF2 in immature chicken or rat cardiomyocytes in culture showed that while increases in extracellular FGF2 (Lo-as well as Hi-) increase myocyte DNA synthesis and proliferation, increases in intracellular, nuclear Hi-FGF2 (but not Lo-FGF2) decreased DNA synthesis, increased bi-nucleation, and promoted chromatin condensation typical of apoptosis [32, 41, 72, 73]. The pro-apoptotic effects of nuclear Hi-FGF2 on cardiomyocytes were prevented by the overexpression of Bcl-2, indicative of mitochondrial involvement [41]. Overexpression of a dominant-negative FGFR1 was also able to attenuate the pro-apoptotic effects of intracellular Hi-FGF2 in cardiomyocytes [74], supporting the notion that signalling by intracellular Hi-FGF2/FGFR1/ERK can promote cell death. As mentioned earlier, CK2, normally a pro-survival kinase, mediated the pro-apoptotic function of intracellular Hi-FGF2 [63].

Myocardial infarction, together with ischemia-reperfusion (I-R)-caused myocardial damage and ensuing remodeling represent major causes of heart failure. Several experimental models in vitro and in vivo have been used to address the mechanism, prevention, and management of ischemic injury. Administration of Lo-FGF2 to the ex vivo heart by perfusion either prior to global ischemia, or during reperfusion following global ischemia protects from I-R associated loss of contractile function, cell damage and cell death [55, 68, 75]. In vivo, direct injection of Lo-FGF2 into rat myocardium rendered ischemic due to irreversible coronary occlusion ameliorated, both acutely and in the long term, scar size (and therefore myocyte loss), decline in contractile function, and hypertrophy development; it also increased small vessel formation at the infarct border, increasing circulation and perfusion in the area [55, 68, 75]. Similar results testifying to the protective effect of administered or overexpressed Lo-FGF2 have been obtained using different animal models and different modes of administration [1, 76]. Central cardioprotective signals such as PKC and ERK have been documented by several reports as mediators of the protective effect of Lo-FGF2 [77]. In addition, administration of Lo-FGF2 in mice prevented loss of contractile function, maladaptive hypertrophy, fibrosis and cell death caused by M-I; the mechanism included downregulation of the anti-oncogene p53 and upregulation of endothelial angiogenesis and the AKT/HIF-1a pathway [78].

Administration of Hi-FGF2 to the ischemic myocardium had distinct long term effects compared to Lo-FGF2. Exogenous Hi-FGF2 exerted acute protection, similar to that of Lo-FGF2, but not sustained protection: at 8 weeks after myocardial infarction, Hi-FGF2-injected hearts displayed deteriorating function, increased hypertrophy and scar size, and reduced formation of small vessels at the infarct border, reducing heart perfusion in the area [75]. Furthermore, only Hi-FGF2 (but not Lo-FGF2) administration caused the accumulation of cardiotrophin-1, a pro-inflammatory cytokine associated with heart failure [79] in the hearts post-infarction. The ability of administered Hi-FGF2 to directly promote myocyte hypertrophy has been further documented in several studies in vitro [14, 75]. A detrimental effect for endogenous or overexpressed Hi-FGF2 in the heart subjected to I-R has been also documented in ex vivo working heart model where lack of endogenous Hi-FGF2 expression protected from I-R induced loss of contractile function, while hearts from a transgenic mouse model overexpressing the 24 kDa human Hi-FGF2 were more vulnerable to I-R induced dysfunction [22].

Anti-cancer genotoxic drugs are another cause of myocardia damage leading to heart failure. Doxorubicin is representative of a group of drugs used against various types of cancer. Unfortunately, the effectiveness of Doxorubicin is compromised by its cardiotoxic effects, especially when higher doses of the drug are needed. Adverse effects can manifest acutely as well as years after treatment and include ventricular dysfunction, dilated cardiomyopathy and heart failure [23]. At the cellular level, Doxorubicin causes cardiomyocyte damage, and death by apoptosis, necrosis and dysregulated autophagy [54, 80]. The ability of Hi-FGF2 to exert short term, but not long term, protection has been documented using in vitro and in vivo models of Doxorubicin induced cardiotoxicity. In acute studies (24 hour), in vitro, Lo- and Hi-FGF2 isoforms, administered to cardiomyocytes prior to Doxorubicin insult, were equally capable of attenuating cardiomyocyte injury and cell death, indicated by their ability to decrease levels of: released cytochrome-c, active caspase 3, the tumour suppressor p53, oxidative stress, loss of ATP and mitochondrial damage. Both isoforms prevented the Doxorubicin induced downregulation of the transcription factor TFEB, a master regulator of lysosomal biogenesis [54], and a regulator of endoplasmic reticulum (ER) autophagy [81]. It has been proposed that downregulation of TFEB creates a lysosomal deficit in the Doxorubicin-treated cells, by blocking autophagy flux and successful elimination of defective proteins and organelles [54]. By preventing TFEB downregulation, added FGF2 would be expected to contribute to prevention of cell death caused by dysregulated autophagy.

Bnip3 (BH3-only protein Bcl-2-like 19 kDa-interacting protein-3), a member of the Bcl-2 family of proteins is an important effector of Doxorubicin-induced cell death [82]. Bnip3 causes depolarization of mitochondria by promoting mPTP formation which leads to cell death [82]. Both p53 and Bnip3 can also regulate autophagy and mitophagy [82]. Pre-treatment with FGF2 isoforms prevented the Doxorubicin-induced p53 and Bnip3 upregulation [54]. The protective effects of FGF2 isoforms , at least in the acute setting, required the activity of mTORC1 (which promotes anabolic processes and prevents autophafy) and its downstream targets NRF2 and HO-1 acting as detoxification agents [54]. Prevention of dysregulated autophagy, and apoptosis, through the activation of the PI3K/Akt/mTOR pathway by Lo-FGF2 has also been observed in a model of I-R myocardial injury [67]. The ability of Lo-FGF2 to prevent excessive autophagy and apoptosis has also been reported in brain spinal cord injury and the neuronal cell line PC12 [3, 83].

The effects of endogenous FGF2 on cardiac damage and dysfunction observed at 10 days after Doxorubicin injection in vivo were isoform-specific: A bolus dose of Doxorubicin administered to wild type mice, (expressing Hi- and Lo-FGF2 isoforms in a 7:3 ratio), elicited decrease in contractile function, and increased accumulation of the pro-cell death protein Bnip3. In contrast, Hi-FGF-2 knockout mice, constitutively expressing only endogenous Lo-FGF2, were protected from the adverse effects of Dox [84]. These results reinforce the notion that chronic exposure to Hi-FGF2 has adverse effects during prolonged stress in the heart.

Non-myocytes, being the major producers of FGF2 in the heart, may be responsible, at least in part for the effects of endogenous FGF2 isoform expression observed in the in vivo model of Dox cardiotoxicity: wild type rodent cardiomyocytes co-cultured with fibroblasts from wild type mice showed increased vulnerability to Dox cellular and mitochondrial damage compared to those co-cultured with Hi-FGF2-knockout mice [84]. Further in vitro studies suggested that neutralization of fibroblast-produced Hi-FGF2 with specific antibodies protected cardiomyocytes from Doxorubicin-induced injury, in vitro [84]. The latter finding is significant as it opens the possibility to attenuate Doxorubicin-induced cardiotoxicity by using neutralizing anti-Hi-FGF2 antibodies.

Pressure overload hemodynamic stress, as occurs in hypertension, is a frequent cause of cardiac functional and structural deterioration and heart failure. Our recent studies indicated that elimination of endogenous Hi-FGF2, allowing unopposed action for endogenous Lo-FGF2, is protective against pressure overload-induced decrease in systolic function, and myocardial damage, observed at 4–8 weeks from trans-aortic constriction surgery in mice [34]. Microarray analysis of the whole transcriptome of hearts from wild type mice and mice expressing only endogenous Lo-FGF2, showed that absence of Hi-FGF2 promoted a significant increase in the expression of heat shock protein 70 (hsp70), in non-stressed as well as in heart subjected to pressure overload [85]. Hsp70 is reported to prevent caspase-mediated apoptosis, as well as programmed necrosis (necroptosis) by modulating autophagy in cardiomyocytes subjected to I-R injury [86], and can therefore be added to the signals used by Lo-FGF2 to protect hearts from chronic stress induced pathologies [85].

Conclusion

After many years of research on FGF2, several issues have been established, while several more remain to be understood. Thus, it is clear that FGF2 isoforms are capable of affecting cardiac response to injurious stimuli either in a beneficial manner, typical of Lo-FGF2, or adversely, as is the case for Hi-FGF2 especially in the context of constitutive exposure and chronic stress. The ability of administered Lo-FGF2 to protect from cardiac cell death (by apoptosis, necrosis, and dysregulated autophagy) provides an important avenue to effect mitochondrial and cardiac protection during cardiac stresses. There is a need to expand studies on the potential of Lo-FGF2 to protect from I-R or genotoxic cardiac injury in higher mammals, including humans. Targeting/neutralizing endogenous Hi-FGF2 expression and production, could be considered as another strategy for augmenting cardiac resistance to injury.

The role of context (age, sex, species, cell type) needs to be taken into consideration in understanding the role and potential of FGF2. Along these lines, the cross talk between various FGFR1-dependent or independent signalling pathways activated by extracellular, and/or intracellular FGF2 isoforms, are fundamental aspects of FGF2 biology in urgent need of untangling.

Acknowledgements This work was funded by a grant from the Canadian Institutes for Health Research (FRN-74733) and the Molson Women's Heart Health Foundation.

References

1. Kardami E, Detillieux K, Ma X, Jiang Z, Santiago JJ, Jimenez SK, Cattini PA (2007) Fibroblast growth factor-2 and cardioprotection. Heart Fail Rev 12(3–4):267–277
2. Ornitz DM, Itoh N (2015) The fibroblast growth factor signaling pathway. Wiley Interdiscip Rev Dev Biol 4(3):215–266
3. Sun D, Wang W, Wang X, Wang Y, Xu X, Ping F, Du Y, Jiang W, Cui D (2018) bFGF plays a neuroprotective role by suppressing excessive autophagy and apoptosis after transient global cerebral ischemia in rats. Cell Death Dis 9(2):172

4. Abdelhakim M, Lin X, Ogawa R (2020) The Japanese experience with basic fibroblast growth factor in cutaneous wound management and scar prevention: a systematic review of clinical and biological aspects. Dermatol Ther (Heidelb) 10(4):569–587
5. Shibata F, Baird A, Florkiewicz RZ (1991) Functional characterization of the human basic fibroblast growth factor gene promoter. Growth factors (Chur, Switzerland) 4(4):277–287
6. Jimenez SK, Sheikh F, Jin Y, Detillieux KA, Dhaliwal J, Kardami E, Cattini PA (2004) Transcriptional regulation of FGF-2 gene expression in cardiac myocytes. Cardiovasc Res 62(3):548–557
7. Lee JG, Kay EP (2012) NF-kappaB is the transcription factor for FGF-2 that causes endothelial mesenchymal transformation in cornea. Invest Ophthalmol Vis Sci 53(3):1530–1538
8. Santiago JJ, McNaughton LJ, Koleini N, Ma X, Bestvater B, Nickel BE, Fandrich RR, Wigle JT, Freed DH, Arora RC, Kardami E (2014) High molecular weight fibroblast growth factor-2 in the human heart is a potential target for prevention of cardiac remodeling, PLoS One 9(5):e97281
9. Holcik M, Sonenberg N (2005) Translational control in stress and apoptosis. Nat Rev Mol Cell Biol 6(4):318–327
10. Arnaud E, Touriol C, Boutonnet C, Gensac MC, Vagner S, Prats H, Prats AC (1999) A new 34-kilodalton isoform of human fibroblast growth factor 2 is cap dependently synthesized by using a non-AUG start codon and behaves as a survival factor. Mol Cell Biol 19(1):505–514
11. Bonnal S, Pileur F, Orsini C, Parker F, Pujol F, Prats AC, Vagner S (2005) Heterogeneous nuclear ribonucleoprotein A1 is a novel internal ribosome entry site trans-acting factor that modulates alternative initiation of translation of the fibroblast growth factor 2 mRNA. J Biol Chem 280(6):4144–4153
12. Vagner S, Touriol C, Galy B, Audigier S, Gensac MC, Amalric F, Bayard F, Prats H, Prats AC (1996) Translation of CUG- but not AUG-initiated forms of human fibroblast growth factor 2 is activated in transformed and stressed cells. J Cell Biol 135(5):1391–1402
13. Liu L, Doble BW, Kardami E (1993) Perinatal phenotype and hypothyroidism are associated with elevated levels of 21.5–22-kDa basic fibroblast growth factor in cardiac ventricles. Dev Biol 157(2):507–516
14. Santiago JJ, Ma X, McNaughton LJ, Nickel BE, Bestvater BP, Yu L, Fandrich RR, Netticadan T, Kardami E (2011) Preferential accumulation and export of high molecular weight FGF-2 by rat cardiac non-myocytes. Cardiovasc Res 89(1):139–147
15. Touriol C, Roussigne M, Gensac MC, Prats H, Prats AC (2000) Alternative translation initiation of human fibroblast growth factor 2 mRNA controlled by its 3'-untranslated region involves a Poly(A) switch and a translational enhancer. J Biol Chem 275(25):19361–19367
16. Yu PJ, Ferrari G, Pirelli L, Galloway AC, Mignatti P, Pintucci G (2008) Thrombin cleaves the high molecular weight forms of basic fibroblast growth factor (FGF-2): a novel mechanism for the control of FGF-2 and thrombin activity. Oncogene 27(18):2594–2601
17. Doble BW, Fandrich RR, Liu L, Padua RR, Kardami E (1990) Calcium protects pituitary basic fibroblast growth factors from limited proteolysis by co-purifying proteases. Biochem Biophys Res Commun 173(3):1116–1122
18. Kardami E, Fandrich RR (1989) Basic fibroblast growth factor in atria and ventricles of the vertebrate heart. J Cell Biol 109(4 Pt 1):1865–1875
19. Rizvi F, DeFranco A, Siddiqui R, Negmadjanov U, Emelyanova L, Holmuhamedov A, Ross G, Shi Y, Holmuhamedov E, Kress D, Tajik AJ, Jahangir A (2016) Chamber-specific differences in human cardiac fibroblast proliferation and responsiveness toward simvastatin. Am J Physiol Cell Physiol 311(2):C330–C339
20. Bashkin P, Doctrow S, Klagsbrun M, Svahn CM, Folkman J, Vlodavsky I (1989) Basic fibroblast growth factor binds to subendothelial extracellular matrix and is released by heparitinase and heparin-like molecules. Biochemistry 28(4):1737–1743
21. Okada-Ban M, Thiery JP, Jouanneau J (2000) Fibroblast growth factor-2. Int J Biochem Cell Biol 32(3):263–267
22. Liao S, Bodmer JR, Azhar M, Newman G, Coffin JD, Doetschman T, Schultz Jel J (2010) The influence of FGF2 high molecular weight (HMW) isoforms in the development of cardiac ischemia-reperfusion injury. J Mol Cell Cardiol 48(6):1245–1254

23. Koleini N, Kardami E (2017) Autophagy and mitophagy in the context of doxorubicin-induced cardiotoxicity. Oncotarget 8(28):46663–46680
24. Clarke MS, Caldwell RW, Chiao H, Miyake K, McNeil PL (1995) Contraction-induced cell wounding and release of fibroblast growth factor in heart. Circ Res 76(6):927–934
25. Dahl JP, Binda A, Canfield VA, Levenson R (2000) Participation of Na, K-ATPase in FGF-2 secretion: rescue of ouabain-inhibitable FGF-2 secretion by ouabain-resistant Na. K-ATPase alpha subunits, Biochemistry 39(48):14877–14883
26. Keller M, Ruegg A, Werner S, Beer HD (2008) Active caspase-1 is a regulator of unconventional protein secretion. Cell 132(5):818–831
27. Martin-Sanchez F, Diamond C, Zeitler M, Gomez AI, Baroja-Mazo A, Bagnall J, Spiller D, White M, Daniels MJ, Mortellaro A, Penalver M, Paszek P, Steringer JP, Nickel W, Brough D, Pelegrin P (2016) Inflammasome-dependent IL-1beta release depends upon membrane permeabilisation. Cell Death Differ 23(7):1219–1231
28. Javidi-Sharifi N, Martinez J, English I, Joshi SK, Scopim-Ribeiro R, Viola SK, Edwards DKT, Agarwal A, Lopez C, Jorgens D, Tyner JW, Druker BJ, Traer E (2019) FGF2-FGFR1 signaling regulates release of Leukemia-Protective exosomes from bone marrow stromal cells. eLife 8
29. Regeenes R, Silva PN, Chang HH, Arany EJ, Shukalyuk AI, Audet J, Kilkenny DM, Rocheleau JV (2018) Fibroblast growth factor receptor 5 (FGFR5) is a co-receptor for FGFR1 that is up-regulated in beta-cells by cytokine-induced inflammation. J Biol Chem 293(44):17218–17228
30. Stachowiak MK, Birkaya B, Aletta JM, Narla ST, Benson CA, Decker B, Stachowiak EK (2015) Nuclear FGF receptor-1 and CREB binding protein: an integrative signaling module. J Cell Physiol 230(5):989–1002
31. Piotrowicz RS, Maher PA, Levin EG (1999) Dual activities of 22–24 kDA basic fibroblast growth factor: inhibition of migration and stimulation of proliferation. J Cell Physiol 178(2):144–153
32. Pasumarthi KB, Kardami E, Cattini PA (1996) High and low molecular weight fibroblast growth factor-2 increase proliferation of neonatal rat cardiac myocytes but have differential effects on binucleation and nuclear morphology. Evidence for both paracrine and intracrine actions of fibroblast growth factor-2, Circ Res 78(1):126–136
33. Zhang L, Parry GC, Levin EG (2013) Inhibition of tumor cell migration by LD22-4, an N-terminal fragment of 24-kDa FGF2, is mediated by Neuropilin 1. Cancer Res 73(11):3316–3325
34. Koleini N, Nickel BE, Nagalingam RS, Landry NM, Fandrich RR, Cheung DYC, Dixon IM, Czubryt MP, Jassal DS, Cattini PA, Kardami E (2021) Elimination of endogenous high molecular weight FGF2 prevents pressure-overload-induced systolic dysfunction linked to increased FGFR1 activity and NR1D1 expression. Cell Tissue Res https://doi.org/10.1007/s00441-021-03465-0
35. Srisakuldee W (2014) Studies on the role of connexin 43 phosphorylation in the injury—resistant heart. University of Manitoba, Physiology
36. Jin Y, Pasumarthi KB, Bock ME, Lytras A, Kardami E, Cattini PA (1994) Cloning and expression of fibroblast growth factor receptor-1 isoforms in the mouse heart: evidence for isoform switching during heart development. J Mol Cell Cardiol 26(11):1449–1459
37. Peters KG, Marie J, Wilson E, Ives HE, Escobedo J, Del Rosario M, Mirda D, Williams LT (1992) Point mutation of an FGF receptor abolishes phosphatidylinositol turnover and Ca^{2+} flux but not mitogenesis. Nature 358(6388):678–681
38. Iwai-Kanai E, Hasegawa K, Fujita M, Araki M, Yanazume T, Adachi S, Sasayama S (2002) Basic fibroblast growth factor protects cardiac myocytes from iNOS-mediated apoptosis. J Cell Physiol 190(1):54–62
39. House SL, House BE, Glascock B, Kimball T, Nusayr E, Schultz JE, Doetschman T (2010) Fibroblast growth factor 2 mediates isoproterenol-induced cardiac hypertrophy through activation of the extracellular regulated Kinase. Mol Cell Pharmacol 2(4):143–154
40. House SL, Branch K, Newman G, Doetschman T, Schultz Jel J (2005) Cardioprotection induced by cardiac-specific overexpression of fibroblast growth factor-2 is mediated by the MAPK cascade. Am J Physiol Heart Circ Physiol 289(5):H2167–H2175

41. Ma X, Dang X, Claus P, Hirst C, Fandrich RR, Jin Y, Grothe C, Kirshenbaum LA, Cattini PA, Kardami E (2007) Chromatin compaction and cell death by high molecular weight FGF-2 depend on its nuclear localization, intracrine ERK activation, and engagement of mitochondria. J Cell Physiol 213(3):690–698
42. Hausenloy DJ, Tsang A, Yellon DM (2005) The reperfusion injury salvage kinase pathway: a common target for both ischemic preconditioning and postconditioning. Trends Cardiovasc Med 15(2):69–75
43. Ediriweera MK, Tennekoon KH, Samarakoon SR (2019) Role of the PI3K/AKT/mTOR signaling pathway in ovarian cancer: biological and therapeutic significance. Semin Cancer Biol
44. Hers I, Vincent EE, Tavare JM (2011) Akt signalling in health and disease. Cell Signal 23(10):1515–1527
45. Mortenson MM, Galante JG, Gilad O, Schlieman MG, Virudachalam S, Kung HJ, Bold RJ (2007) BCL-2 functions as an activator of the AKT signaling pathway in pancreatic cancer. J Cell Biochem 102(5):1171–1179
46. Ye G, Fu Q, Jiang L, Li Z (2018) Vascular smooth muscle cells activate PI3K/Akt pathway to attenuate myocardial ischemia/reperfusion-induced apoptosis and autophagy by secreting bFGF. Biomed Pharmacother Biomed Pharmacother 107:1779–1785
47. Ling L, Gu S, Cheng Y, Ding L (2018) bFGF promotes Sca1+ cardiac stem cell migration through activation of the PI3K/Akt pathway. Mol Med Rep 17(2):2349–2356
48. Chen Q, Chen X, Han C, Wang Y, Huang T, Du Y, Dong Z (2016) FGF-2 Transcriptionally down-regulates the expression of BNIP3L via PI3K/Akt/FoxO3a signaling and inhibits necrosis and mitochondrial dysfunction induced by high concentrations of hydrogen peroxide in H9c2 Cells. Cell Physiol Bioche: Int J Exp Cell Physiol Biochem Pharmacol 40(6):1678–1691
49. Liu MH, Li GH, Peng LJ, Qu SL, Zhang Y, Peng J, Luo XY, Hu HJ, Ren Z, Liu Y, Tang H, Liu LS, Tang ZH, Jiang ZS (2016) PI3K/Akt/FoxO3a signaling mediates cardioprotection of FGF-2 against hydrogen peroxide-induced apoptosis in H9c2 cells. Mol Cell Biochem 414(1–2):57–66
50. Cheng Y, Li Z, Kardami E, Loh YP (2016) Neuroprotective effects of LMW and HMW FGF2 against amyloid beta toxicity in primary cultured hippocampal neurons. Neurosci Lett 632:109–113
51. Wang Z, Wang Y, Ye J, Lu X, Cheng Y, Xiang L, Chen L, Feng W, Shi H, Yu X, Lin L, Zhang H, Xiao J, Li X (2015) bFGF attenuates endoplasmic reticulum stress and mitochondrial injury on myocardial ischaemia/reperfusion via activation of PI3K/Akt/ERK1/2 pathway. J Cell Mol Med 19(3):595–607
52. Saxton RA, Sabatini DM (2017) mTOR signaling in growth, metabolism, and disease. Cell 168(6):960–976
53. Sciarretta S, Volpe M, Sadoshima J (2014) Mammalian target of rapamycin signaling in cardiac physiology and disease. Circ Res 114(3):549–564
54. Koleini N, Nickel BE, Wang J, Roveimiab Z, Fandrich RR, Kirshenbaum LA, Cattini PA, Kardami E (2017) Fibroblast growth factor-2-mediated protection of cardiomyocytes from the toxic effects of doxorubicin requires the mTOR/Nrf-2/HO-1 pathway. Oncotarget 8(50):87415–87430
55. Padua RR, Merle PL, Doble BW, Yu CH, Zahradka P, Pierce GN, Panagia V, Kardami E (1998) FGF-2-induced negative inotropism and cardioprotection are inhibited by chelerythrine: involvement of sarcolemmal calcium-independent protein kinase C. J Mol Cell Cardiol 30(12):2695–2709
56. Doble BW, Ping P, Kardami E (2000) The epsilon subtype of protein kinase C is required for cardiomyocyte connexin-43 phosphorylation. Circ Res 86(3):293–301
57. Srisakuldee W, Makazan Z, Nickel BE, Zhang F, Thliveris JA, Pasumarthi KB, Kardami E (2014) The FGF-2-triggered protection of cardiac subsarcolemmal mitochondria from calcium overload is mitochondrial connexin 43-dependent. Cardiovasc Res 103(1):72–80
58. Wang J, Nachtigal MW, Kardami E, Cattini PA (2013) FGF-2 protects cardiomyocytes from doxorubicin damage via protein kinase C-dependent effects on efflux transporters. Cardiovasc Res 98(1):56–63

59. Litchfield DW (2003) Protein kinase CK2: structure, regulation and role in cellular decisions of life and death. Biochem J 369(Pt 1):1–15
60. Bailly K, Soulet F, Leroy D, Amalric F, Bouche G (2000) Uncoupling of cell proliferation and differentiation activities of basic fibroblast growth factor. FASEB J 14(2):333–344
61. Jiang ZS, Srisakuldee W, Soulet F, Bouche G, Kardami E (2004) Non-angiogenic FGF-2 protects the ischemic heart from injury, in the presence or absence of reperfusion. Cardiovasc Res 62(1):154–166
62. Koleini N, Nickel BE, Edel AL, Fandrich RR, Ravandi A, Kardami E (2018) Non-mitogenic FGF2 protects cardiomyocytes from acute doxorubicin-induced toxicity independently of the protein kinase CK2/heme oxygenase-1 pathway. Cell Tissue Res 374(3):607–617
63. Ma X (2004) Casein kinase 2 mediates nuclear disruption by high molecular weight fibroblast growth factor 2. University of Manitoba, Anatomy
64. Pellieux C, Foletti A, Peduto G, Aubert JF, Nussberger J, Beermann F, Brunner HR, Pedrazzini T (2001) Dilated cardiomyopathy and impaired cardiac hypertrophic response to angiotensin II in mice lacking FGF-2. J Clin Invest 108(12):1843–1851
65. Schultz JE, Witt SA, Nieman ML, Reiser PJ, Engle SJ, Zhou M, Pawlowski SA, Lorenz JN, Kimball TR, Doetschman T (1999) Fibroblast growth factor-2 mediates pressure-induced hypertrophic response. J Clin Invest 104(6):709–719
66. Zhou M, Sutliff RL, Paul RJ, Lorenz JN, Hoying JB, Haudenschild CC, Yin M, Coffin JD, Kong L, Kranias EG, Luo W, Boivin GP, Duffy JJ, Pawlowski SA, Doetschman T (1998) Fibroblast growth factor 2 control of vascular tone. Nat Med 4(2):201–207
67. Wang ZG, Wang Y, Huang Y, Lu Q, Zheng L, Hu D, Feng WK, Liu YL, Ji KT, Zhang HY, Fu XB, Li XK, Chu MP, Xiao J (2015) bFGF regulates autophagy and ubiquitinated protein accumulation induced by myocardial ischemia/reperfusion via the activation of the PI3K/Akt/mTOR pathway. Sci Rep 5:9287
68. Padua RR, Sethi R, Dhalla NS, Kardami E (1995) Basic fibroblast growth factor is cardioprotective in ischemia-reperfusion injury. Mol Cell Biochem 143(2):129–135
69. Cuevas P, Carceller F, Martinez-Coso V, Asin-Cardiel E, Gimenez-Gallego G (2000) Fibroblast growth factor cardioprotection against ischemia-reperfusion injury may involve K+ ATP channels. Eur J Med Res 5(4):145–149
70. Unger EF, Banai S, Shou M, Lazarous DF, Jaklitsch MT, Scheinowitz M, Correa R, Klingbeil C, Epstein SE (1994) Basic fibroblast growth factor enhances myocardial collateral flow in a canine model. Am J Physiol 266(4 Pt 2):H1588–H1595
71. Sontag DP, Wang J, Kardami E, Cattini PA (2013) FGF-2 and FGF-16 protect isolated perfused mouse hearts from acute doxorubicin-induced contractile dysfunction. Cardiovasc Toxicol 13(3):244–253
72. Hirst CJ, Herlyn M, Cattini PA, Kardami E (2003) High levels of CUG-initiated FGF-2 expression cause chromatin compaction, decreased cardiomyocyte mitosis, and cell death. Mol Cell Biochem 246(1–2):111–116
73. Pasumarthi KB, Doble BW, Kardami E, Cattini PA (1994) Over-expression of CUG- or AUG-initiated forms of basic fibroblast growth factor in cardiac myocytes results in similar effects on mitosis and protein synthesis but distinct nuclear morphologies. J Mol Cell Cardiol 26(8):1045–1060
74. Ma X, Hirst C, Santiago JJ, Fandrich RR, Sussman MA, Kardami E (2009) The high molecular weight FGF2-induced cardiomyocyte cell death is prevented by nuclear Akt and Pim-1 kinases, and requires the activity of intracellular FGF2 receptor. Circ Res 105(P13):E13
75. Jiang ZS, Jeyaraman M, Wen GB, Fandrich RR, Dixon IM, Cattini PA, Kardami E (2007) High- but not low-molecular weight FGF-2 causes cardiac hypertrophy in vivo; possible involvement of cardiotrophin-1. J Mol Cell Cardiol 42(1):222–233
76. Jimenez SK, Jassal DS, Kardami E, Cattini PA (2011) Protection by endogenous FGF-2 against isoproterenol-induced cardiac dysfunction is attenuated by cyclosporine A. Mol Cell Biochem 357(1–2):1–8
77. Liao S, Bodmer J, Pietras D, Azhar M, Doetschman T, Schultz Jel J (2009) Biological functions of the low and high molecular weight protein isoforms of fibroblast growth factor-2 in cardiovascular development and disease. Dev Dyn 238(2):249–264

78. Rao Z, Shen D, Chen J, Jin L, Wu X, Chen M, Li L, Chu M, Lin J (2020) Basic fibroblast growth factor attenuates injury in myocardial infarction by enhancing hypoxia-inducible factor-1 alpha accumulation. Front Pharmacol 11:1193
79. Martínez-Martínez E, Brugnolaro C, Ibarrola J, Ravassa S, Buonafine M, López B, Fernández-Celis A, Querejeta R, Santamaria E, Fernández-Irigoyen J, Rábago G, Moreno MU, Jaisser F, Díez J, González A, López-Andrés N (2019) CT-1 (Cardiotrophin-1)-Gal-3 (Galectin-3) axis in cardiac fibrosis and inflammation. Hypertension 73(3):602–611
80. Wang EY, Biala AK, Gordon JW, Kirshenbaum LA (2012) Autophagy in the heart: too much of a good thing? J Cardiovasc Pharmacol 60(2):110–117
81. Cinque L, De Leonibus C, Iavazzo M, Krahmer N, Intartaglia D, Salierno FG, De Cegli R, Di Malta C, Svelto M, Lanzara C, Maddaluno M, Wanderlingh LG, Huebner AK, Cesana M, Bonn F, Polishchuk E, Hubner CA, Conte I, Dikic I, Mann M, Ballabio A, Sacco F, Grumati P, Settembre C (2020) MiT/TFE factors control ER-phagy via transcriptional regulation of FAM134B. EMBO J 39(17):e105696
82. Dhingra R, Margulets V, Chowdhury SR, Thliveris J, Jassal D, Fernyhough P, Dorn GW, Kirshenbaum LA (2014) Bnip3 mediates doxorubicin-induced cardiac myocyte necrosis and mortality through changes in mitochondrial signaling, (2nd edn). Proc Natl Acad Sci U S A 111(51):E5537–E5544
83. Zhang HY, Wang ZG, Wu FZ, Kong XX, Yang J, Lin BB, Zhu SP, Lin L, Gan CS, Fu XB, Li XK, Xu HZ, Xiao J (2013) Regulation of autophagy and ubiquitinated protein accumulation by bFGF promotes functional recovery and neural protection in a rat model of spinal cord injury. Mol Neurobiol 48(3):452–464
84. Koleini N, Santiago JJ, Nickel BE, Sequiera GL, Wang J, Fandrich RR, Jassal DS, Dhingra S, Kirshenbaum LA, Cattini PA, Kardami E (2019) Elimination or neutralization of endogenous high-molecular-weight FGF2 mitigates doxorubicin-induced cardiotoxicity. Am J Physiol Heart Circu Physiol 316(2):H279-H288
85. Kardami E, Koleini N, Nickel B, Nagalingam R, Landry NM, Fandrich RR, Cheung DYC, Dixon IM, Czubryt MP, Jassal DS, Cattini PA (2020) Changes in gene expression caused by genetic elimination of high molecular weight FGF2 are associated with prevention of stress-induced cardiac systolic dysfunction. FASEB J 34:1–1
86. Liu X, Zhang C, Zhang C, Li J, Guo W, Yan D, Yang C, Zhao J, Xia T, Wang Y, Xu R, Wu X, Shi J (2016) Heat shock protein 70 inhibits cardiomyocyte necroptosis through repressing autophagy in myocardial ischemia/reperfusion injury. Vitro Cell Dev Biol Anim 52(6):690–698

The manufacturer's authorised representative in the EU is Springer Nature Customer Service Centre GmbH, Europaplatz 3, 69115 Heidelberg, Germany. If you have any concerns regarding our products, please contact ProductSafety@springernature.com

Printed and bound by CPI Group (UK) Ltd, Croydon, CR0 4YY
22/09/2026
02220946-0008